Preface

During the last eight years this textbook of endocrinology has been used by both graduate and medical students as they learn physiology, anatomy, biochemistry, and cell biology. The book was written with the intention of providing a solid, clear, and succinct account of basic endocrine function, emphasizing the impact that advances in the biological sciences have on our understanding of physiological processes. The text addresses clinical disorders only as a means of illustrating the divergence from normal physiology and normal regulation of endocrine function. We hope the basic insights into endocrinology that the book offers will pave the way to a better understanding of the complex array of endocrine disorders that medical students will encounter in later years. It should also provide a framework for those, who, planning to pursue a career in research, are seeking to identify critical subjects of investigation.

The contributors are well-recognized authorities in the fields of neuroendocrinology, reproductive endocrinology, endocrinology of growth and pregnancy, intermediary metabolism, diabetes and lipid physiology, calcium homeostasis, adrenal steroidogenesis, and mechanisms of both peptide and steroid hormone action. For over a decade members of these groups have given a course in endocrinology and reproduction to first-year medical students. We have now taken advantage of these resources by developing a textbook from this course.

The first five chapters of the book consider general aspects of endocrine physiology: the organization of the endocrine system, genetic control of hormone formation, mechanisms of hormone action, immune–endocrine interactions, and assessment of endocrine function. The next two chapters concern the neuroendocrinology of the hypothalamus and its relationship to the pituitary gland. Subsequent chapters are devoted to reproductive physiology (sexual differentiation, female reproductive function, male reproductive function, endocrinology of pregnancy), somatic growth and development, thyroid gland physiology, and the adrenal glands. The last two chapters deal with the multihormonal control of calcium homeostasis and of glucose, lipid, and protein metabolism. Selected clinical examples of deranged endocrine physiology are provided throughout the book, without clinical jargon. We have tried to keep the level of discussion in each chapter appropriate for students who are just beginning their medical and graduate education.

For this third edition, there are two new chapters, one titled ''Cytokines and

Immune–Endocrine Interactions'' and the other ''Growth Regulation.'' There are 25 new figures and 12 revised figures. We would like to express our special appreciation to the contributors. We also thank Christy K. Gonzales for secretarial assistance and Marty Burgin for preparation of the illustrations.

Dallas   J.E.G.
Beaverton, Ore.   S.R.O.
1995

Textbook of
Endocrine Physiology

Third Edition

Edited by

JAMES E. GRIFFIN, M.D.

Professor of Internal Medicine
University of Texas Southwestern Medical Center at Dallas

SERGIO R. OJEDA, D.V.M.

Head, Division of Neuroscience
Oregon Regional Primate Research Center

New York   Oxford   OXFORD UNIVERSITY PRESS   1996

Oxford University Press

Oxford   New York
Athens   Auckland   Bangkok   Bombay
Calcutta   Cape Town   Dar es Salaam   Delhi
Florence   Hong Kong   Istanbul   Karachi
Kuala Lumpur   Madras   Madrid   Melbourne
Mexico City   Nairobi   Paris   Singapore
Taipei   Tokyo   Toronto

and associated campanies in

Berlin   Ibadan

Library of Congress Cataloging-in-Publication Data
Textbook of endocrine physiology /
edited by James E. Griffin, Sergio R. Ojeda.—3rd ed.
p. cm.   Includes bibliographical references and index.
ISBN 0-19-510754-3.—ISBN 0-19-510755-1 (pbk.)
1. Endocrine glands—Physiology.
I. Griffin, James E.
II. Ojeda, Sergio R.
[DNLM: 1. Endocrine Glands—physiology.
WK 102 T355 1996]   QP187.T43   1996   612.4—dc20
DNLM/DLC for Library of Congress   95-43189

9 8 7 6 5 4 3 2 1

Printed in the United States of America
on acid-free paper

Contents

Contributors

Neil A. Breslau, M.D.
Professor of Internal Medicine
Univ. of Texas Southwestern Medical
  Center

Bruce R. Carr, M.D.
Professor of Obstetrics and
  Gynecology
Univ. of Texas Southwestern Medical
  Center

M. Linette Casey, M.D.
Associate Professor of Biochemistry
Univ. of Texas Southwestern Medical
  Center

Pinchas Cohen, M.D.
Assistant Professor of Pediatrics
Children's Hospital of Philadlephia

Daniel W. Foster, M.D.
Professor and Chairman of Internal
  Medicine
Univ. of Texas Southwestern Medical
  Center

Fredrick W. George, Ph.D.
Senior Research Scientist, Dept. of
  Internal Medicine
Univ. of Texas Southwestern Medical
  Center

James E. Griffin, M.D.
Professor of Internal Medicine
Univ. of Texas Southwestern Medical
  Center

Norman M. Kaplan, M.D.
Professor of Internal Medicine
Univ. of Texas Southwestern Medical
  Center

Samuel M. McCann, M.D.
United Companies Distinguished
  Professor Of Biomedical Science
Pennington Biomedical Research
  Center

J. Denis McGarry, M.D.
Professor of Internal Medicine and
  Biochemistry
Univ. of Texas Southwestern Medical
  Center

Carole R. Mendelson, Ph.D.
Professor of Biochemistry and
  Obstetrics and Gynecology
Univ. of Texas Southwestern Medical
  Center

Sergio R. Ojeda, D.V.M.
Head, Division of Neuroscience
Oregon Regional Primate Research
  Center

Ron G. Rosenfeld, M.D.
Professor and Chairman of Pediatrics
Oregon Health Sciences University

Willis K. Samson, Ph.D.
Professor and Chairman of
  Physiology
University of North Dakota Medical
  School

Michael R. Waterman, Ph.D.
Professor and Chairman of
  Biochemistry
Vanderbilt University School of
  Medicine

Textbook of
Endocrine Physiology

Organization of the Endocrine System

SERGIO R. OJEDA
JAMES E. GRIFFIN

The evolutionary appearance of multicellular organisms dictated the necessity of establishing coordinating systems to regulate and integrate the function of the different cells. Two basic regulatory mechanisms developed to perform this delicate function: the nervous system and the endocrine system. Whereas the former employs electrochemical signals to send commands to peripheral organs and to receive information from them, the latter performs its regulatory function by producing chemical agents that, in general, are transported by the bloodstream to the target organs.

The two systems are, nevertheless, closely linked. The best known connection is that of the hypothalamus and the pituitary gland. Hypothalamic neurosecretory cells produce substances that are delivered to the portal blood vessels (see Chapter 6) and transported to the anterior pituitary where they regulate the secretion of adenohypophyseal hormones. Other hypothalamic neurons send their axons to the posterior pituitary where they end in close proximity to the vascular bed of the gland and release their neurosecretory products directly into the bloodstream. The nervous and endocrine systems are further related by the innervation of endocrine glands. Most, if not all, endocrine glands, including the gonads, the thyroid, and the adrenals, receive nerves that appear to control both their blood flow and their secretory activity. In turn, the endocrine system regulates the function of the nervous system. For example, the gonadal and adrenocortical steroids act directly on the central nervous system to either inhibit or stimulate the secretory activity of those neurons that produce releasing hormones involved in the control of the pituitary–gonadal and pituitary–adrenal axis, respectively (i.e., luteinizing hormone-releasing hormone [LHRH] and corticotropin-releasing hormone [CRH]; see Chapter 6).

Although the conventional definitions of the nervous and endocrine systems clearly delineate their differences, in recent years it has become increasingly obvious that an absolute distinction between the two systems cannot be made. Thus, for instance, the nervous system produces substances that do not act across synapses, but rather are released into the bloodstream and travel to distant target cells. Among these substances of neural origin are the hypothalamic-releasing factors that control the secretion of anterior pituitary hormones; epinephrine of adrenomedullary origin, which upon release into the bloodstream can affect the function of various organs, such as the liver and muscle; and oxytocin and vasopressin, which are secreted by hypothalamic neurons and act on the mammary gland and uterus (oxytocin) to stimulate the activity of contractile tissue and on the kidney (vasopressin) to regulate extracellular fluid volume. Conversely, several peptides originally discovered in the gastrointestinal tract where they exert hormonal actions are now known to be produced in both the central and peripheral nervous systems where they appear to act as neurotransmitters or neuromodulators. Examples of these hormones are gastrin, secretin, and vasoactive intestinal peptide.

Because of their interrelationships, endocrine glands and the components of the nervous system that regulate endocrine function are referred to as the neuroendocrine system. The discipline devoted to the study of this system is called *neuroendocrinology* (see Chapter 6). Recent developments in this field have led to the realization that the neuroendocrine system does not operate alone; instead, it is subjected to regulatory input from an unexpected source, the immune system (see Chapter 4). The relationship between the neuroendocrine system and the immune system is reciprocal: hormones influence the cellular components of the immune system, and cytokines (the secretory products of the immune system) affect neuroendocrine functions. Most remarkably, immunologically reactive cells have been found to secrete certain hormones such as adrenocorticotropin (ACTH) and β-endorphin, which until recently were thought to be produced only in the brain and pituitary gland.

Perhaps the best characterized neuroendocrine–immune system relationship is that of the immune–hypothalamic–pituitary–adrenal axis. Cytokines, particularly interleukin-1 (produced by antigenically challenged macrophages), act on the hypothalamus to stimulate the secretion of CRH which, in turn, stimulates the pituitary gland to secrete ACTH. This hormone acts on the adrenal cortex to elicit glucocorticoid secretion. Glucocorticoids then promote the formation of neutrophils in the bone marrow, decrease the formation of monocytes/macrophages and lymphocytes, and inhibit cytokine production as well as antibody production.

The immune system also affects the secretion of other pituitary hormones via the hypothalamus. Thus, in addition to the stimulatory effect of interleukin-1 on CRH release, cytokines have been shown to depress thyrotropin-releasing hormone (TRH) and LHRH production. In turn, several pituitary hormones and neuroendocrine peptides, including somatostatin, thyroid-stimulating hormone (TSH), and growth hormone (GH), are able to affect specific immunological functions (such as immunoglobulin synthesis, which is enhanced by TSH, and lymphocyte production, which is increased by GH).

The ovary provides another example of the interrelationships between the endocrine and the immune systems. Cytokines directly regulate certain ovarian functions and, as in the case of the neuroendocrine system, ovarian hormones have the ability to affect the immune system. For instance, interleukin-1 synthesis increases in the ovary during the hours preceding ovulation, resulting in a greater production of progesterone and prostaglandins, which in turn have a stimulatory effect on ovulation. Conversely, low doses of progesterone enhance the expression of the interleukin-1 gene in macrophages.

In their definition of a hormone, Baylis and Starling specified that for a substance to be considered a hormone, it should meet the requirement of being produced by an organ in small amounts, released into the bloodstream, and transported to a distant organ to exert its specific actions. This definition indeed applies to most of the "classical hormones" produced by the endocrine glands. It is now apparent, however, that hormones can act on contiguous cells, performing a *paracrine* function, and that they can modify the secretory activity of the cells that produce them, performing an *autocrine* function. In acting locally they do not need to be transported in the bloodstream, but they still can exert their action in a manner commensurate with their hormonal nature, that is, through binding to specific receptors.

ENDOCRINE GLANDS AND HORMONES

The endocrine system is composed of several glands located in different areas of the body that produce hormones with different functions. The major morphological feature of endocrine glands is that they are ductless, that is, they release their secretory products directly into the bloodstream and not into a duct system. Since they are richly vascularized, each secreting cell can deliver its products efficiently into the circulation.

The classical endocrine glands and their known secretory products are listed in Table 1-1. In recent years, the traditional view of the endocrine system's glandular nature has been broadened to include production of recognized hormones in organs whose primary function is not endocrine. Perhaps the earliest example of this was

Table 1-1 Classical Endocrine Glands and Their Hormones

Gland		Hormone
Pituitary	Anterior lobe	Luteinizing hormone (LH), follicle-stimulating hormone (FSH), prolactin (PRL), growth hormone (GH), adrenocorticotropin (ACTH), β-lipotropin, β-endorphin, thyroid-stimulating hormone (TSH)
	Intermediate lobe	Melanocyte-stimulating hormone (MSH), β-endorphin
	Posterior lobe	Vasopressin (AVP) or antidiuretic hormone (ADH), oxytocin
Thyroid		Thyroxine (T_4), 3,5,3'-triiodothyronine (T_3), calcitonin
Parathyroid		Parathyroid hormone (PTH)
Adrenal	Cortex	Cortisol, aldosterone, dehydroepiandrosterone, androstenedione
	Medulla	Epinephrine, norepinephrine
Gonads	Testis	Testosterone, estradiol, androstenedione, inhibin, activin, müllerian-inhibiting substance
	Ovary	Estradiol, progesterone, testosterone, androstenedione, inhibin, activin, FSH-releasing peptide, relaxin
Placenta		Human chorionic gonadotropin (hCG), human placental lactogen (hPL), progesterone, estrogen
Pancreas		Insulin, glucagon, somatostatin, pancreatic polypeptide, gastrin, vasoactive intestinal peptide (VIP)
Pineal		Melatonin, biogenic amines, several peptides

Table 1-2 Nonclassical "Endocrine Organs" and Their Hormones

Organ	Hormone
Brain (especially hypothalamus)	Corticotropin-releasing hormone (CRH), thyrotropin-releasing hormone (TRH), luteinizing hormone-releasing hormone (LHRH), growth hormone-releasing hormone (GHRH), somatostatin, growth factors[a] (fibroblast growth factors, transforming growth factor-α (TGF-α), others
Heart	Atrial natriuretic peptides
Kidney	Erythropoietin, renin, 1,25-dihydroxyvitamin D
Liver, other organs, fibroblasts	Insulin-like growth factor I (IGF-I)
Gastrointestinal tract	Cholecystokinin (CCK), gastrin, secretin, vasoactive intestinal peptide (VIP), enteroglucagon, gastrin-releasing peptide
Platelets	Platelet-derived growth factor (PDGF), transforming growth factor-β (TGF-β)
Macrophages, lymphocytes	Cytokines, TGF-β, proopiomelanocortin (POMC) derived peptides
Various sites	Epidermal growth factor (EGF), TGF-α

[a]Not considered to be hormones, but they can act as such.

the recognition that hormones are produced by cells scattered along the mucosa of the stomach and small intestine. The brain, heart, kidney, liver, and certain blood elements also produce peptides or form active steroid metabolites from circulating precursors that deserve the designation of hormones (Table 1-2). Not all of these hormones will be considered in this volume.

CHEMICAL NATURE OF HORMONES

Chemically, the hormones fall into three general categories. The first comprises hormones derived from single amino acids. They are the amines, such as norepinephrine, epinephrine, and dopamine, which derive from the amino acid tyrosine, and the thyroid hormones, $3,5,3'$-triiodothyronine (T_3) and $3,5,3',5'$-tetraiodothyronine (thyroxine, T_4), which derive from the combination of two iodinated tyrosine amino acid residues.

The second category is composed of peptides and proteins. These can be as small as thyrotropin-releasing hormone (three amino acids) and as large and complex as growth hormone and follicle-stimulating hormone, which have about 200 amino acid residues and molecular weights in the range of 25,000–30,000.

The third category comprises the steroid hormones, which are derivatives of cholesterol and can be grouped into two types: (1) those with an intact steroid nucleus such as the gonadal and adrenal steroids and (2) those with a broken steroid nucleus (the B ring) such as vitamin D and its metabolites. Figure 1-1 provides examples of these three categories of hormones.

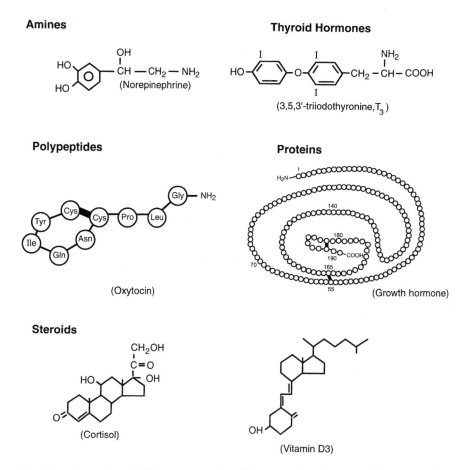

Fig. 1-1. Examples of different categories of hormones. In the case of the protein hormone, each circle represents an amino acid, as shown for the polypeptide hormone.

FUNCTION OF HORMONES

Regardless of their chemical nature, hormones are present in the bloodstream in very low concentrations (10^{-7}–10^{-12} M). Thus, it is not surprising that a prerequisite for any hormone to exert its actions is that it must first bind to specific, high-affinity cellular receptors. These receptors may be located at the cell membrane, as in the case of protein hormones and amines, or in the nucleus, as with thyroid and steroid hormones (see Chapter 3). Another characteristic feature of hormones is that a single hormone can exert various effects in different tissues; conversely, a single function can be regulated by several hormones. Estradiol exemplifies the versatility of some hormones. It is produced by the ovary and can act on the ovarian follicles themselves to promote granulosa cell differentiation, on the uterus to stimulate its growth and maintain the cyclic change of the uterine mucosa, on the mammary gland to simulate ductal growth, on bone to promote linear growth and closure of the epiphysial plates, on the hypothalamic–pituitary system to regulate the secretion of gonadotropins and prolactin, and on general metabolic processes to affect adipose tissue distribution, volume of extracellular fluid, etc. In each case estradiol acts through the same common mechanism, namely binding to high-affinity, specific nuclear receptors followed by attachment of the receptor–steroid complex to DNA regions of genes that are expressed in a tissue-specific manner.

An example of a single function regulated by more than one hormone is the release of fatty acids (lipolysis) from adipose tissue stores. A variety of hormones including catecholamines, glucagon, secretin, prolactin, and β-lipotropin stimulate lipolysis within minutes by activating, via cyclic AMP, a rate-limiting triglyceride hydrolase known as *hormone-sensitive lipase*. Growth hormone and glucocorticoids also stimulate lipolysis, but after a time lag of 2 hours because they induce the synthesis of hormone-sensitive lipase rather than activating this hydrolase. Other hormones such as insulin, insulin-like growth factors (IGFs) or somatomedins, oxytocin, and gastric-inhibitory polypeptide inhibit lipolysis.

An example of a complex function regulated by different hormones is the development of the mammary gland, which is under the primary influence of prolactin, estradiol, and progesterone and the permissive influence of glucocorticoids, thyroid hormones, and insulin. The effect of these latter hormones is considered permissive because, by themselves, they have little effect, but when they are present the actions of prolactin, estrogen, and progesterone become fully manifested.

Hormones exert their functions in four broad physiological areas. These are reproduction, growth and development, maintenance of the internal environment, and the regulation of energy availability.

Reproduction

Hormones produced by the gonads (androgens, estrogen, progestogens) and the anterior pituitary gland (luteinizing hormone [LH], follicle-stimulating hormone [FSH], growth hormone [GH], and prolactin) interact to regulate the growth and structural integrity of the reproductive organs, the production of gametes, the patterns of sexual behavior, the phenotypic difference between the sexes, and the continuation of the species (through their effects on ovulation, spermatogenesis, pregnancy, and lactation).

Several examples of these interactions can be provided. For instance, estradiol induces hypertrophy and hyperplasia of both the muscular and endothelial layers of the uterine wall; in turn the secretion of estradiol from the ovary is under the control of the pituitary gonadotropins. Testosterone, which in males is under the control of LH, exerts tropic effects on sex accessory glands such as the prostate and seminal vesicles. Both estradiol in females and testosterone in males play fundamental roles in determining the female and male external appearance at puberty. Testosterone promotes the growth of the testes, scrotum, and penis, and stimulates muscle development, particularly that of the pectoral region and shoulders. Estradiol promotes the development of the female external genitalia and the redistribution of adipose tissue, which localizes more noticeably in the thighs, hips, and breasts (see Chapter 9). Although in humans gonadal steroids do not induce a stereotyped pattern of sexual behavior as in animals, there is no doubt that they are important in maintaining the libido.

Both ovulation and spermatogenesis are processes tightly controlled by the pituitary gonadotropins LH and FSH, which act either directly on the gonads to promote follicular development (ovary) and formation of sperm (testis) or indirectly through their stimulatory effects on estrogen and testosterone secretion. During pregnancy a variety of hormones of placental origin including estradiol, progesterone, chorionic gonadotropin, placental lactogen, and several others (see Chapter 11) contribute to maintaining the normal progression of pregnancy. After delivery another group of hormones, most noticeably prolactin (see above), maintains the structure and function of the lactating breast.

Growth and Development

A variety of hormones play primary and permissive roles in the timing and progression of growth both for overall body size and for individual tissues. In many cases the local production of growth factors is a result of hormone action. In other instances the production of a growth factor may be hormone independent, but the growth factor interacts with hormones to promote or reduce growth.

The classical hormones involved in the process of growth are GH, thyroid hormones, insulin, glucocorticoids, androgens, and estrogens. Although the stimulatory effect of GH on bodily growth is mediated by a family of peptides known collectively as *insulin-like growth factors* (see Chapter 12), it is unclear to what extent these growth factors are involved in mediating the stimulatory effect of androgens and estrogens on linear growth and acquisition of muscle mass. An example of a hormone that plays both permissive and primary roles in growth is thyroxine. In its absence GH fails to stimulate skeletal growth, a phenomenon that appears to be related to a reduced ability of the tissue to respond to IGFs. In the central nervous system, however, thyroxine plays a primary role in inducing growth and cellular differentiation. Some of its actions in the brain appear to be mediated by the production of a tissue-specific growth factor, nerve growth factor (NGF). Experimental evidence indicates that the synthesis of some peptide growth factors such as the IGFs, NGF, and the epidermal growth factor (EGF) is induced by specific hormones (GH, thyroxine, and androgens, respectively). It is not known whether other peptide growth factors such as fibroblast growth factor (FGF) and platelet-derived growth factor (PDGF) are hormonally regulated. A more detailed discussion of the hormones and peptide growth factors involved in regulating growth is presented in Chapter 12.

Maintenance of Internal Environment

The maintenance of the internal environment involves the control of extracellular fluid volume and blood pressure, the electrolyte composition of bodily fluids, the regulation of plasma and tissue levels of calcium and phosphate ions, and the maintenance of bone, muscle, and body stores of fat.

A multitude of hormones participates in the regulation of these processes. For example, vasopressin or antidiuretic hormone, which is synthesized in the hypothalamus and released from the posterior pituitary, acts on the kidney to induce water reabsorption; aldosterone produced in the adrenal cortex stimulates sodium reabsorption and potassium excretion in the kidney. Thus, both hormones contribute to regulating blood pressure, extracellular fluid volume, and electrolyte composition of bodily fluids.

Plasma levels of calcium and phosphate ions are controlled by parathyroid hormone (PTH) from the parathyroid glands (see Chapter 15). PTH increases serum calcium concentration mainly through its stimulatory actions on calcium transport in bone and kidney, but also by enhancing intestinal calcium absorption through its stimulatory influence on the renal formation of 1,25-dihydroxyvitamin D. PTH also acts on the kidney to increase phosphate excretion.

The functions of the bone, muscle, and adipose tissue are regulated by hormones as diverse as PTH, estrogens, androgens, and GH (bone), and catecholamines, insulin, glucagon, and glucocorticoids (muscle and adipose tissue). Pertinent details are given in Chapters 14, 15, and 16.

Regulation of Energy Availability

For an organism to survive, it must be able to convert the calories contained in food into energy, to store part of that energy for subsequent use, and to mobilize it when necessary. These functions are regulated by several hormones, among which insulin and glucagon play prominent roles (see Chapter 16). In a more general sense energy metabolism as a whole is influenced by hormones through their effects on cellular processes such as transport across membranes, motility, and release of secretory products, all of which require energy. For example, thyroid hormones stimulate the synthesis of the cell membrane enzyme Na^+, K^+-ATPase, resulting in increased oxygen consumption (see Chapter 13). An example of a hormone influencing energy metabolism through the activation of cellular motility is estrogen, which greatly enhances the motility of cilia on epithelial cells lining the oviduct and, thus, facilitates the transport of the ovum to the site of fertilization.

SYNTHESIS AND RELEASE OF HORMONES

Peptide and protein hormones are synthesized in the rough endoplasmic reticulum. As with all proteins, the amino acid sequence of protein hormones is determined by specific messenger RNAs that are synthesized in the nucleus and have a nucleotide sequence dictated by a specific gene. We shall see in Chapter 2 that translation of the specific messenger RNA sequence results in the ribosomal synthesis of a protein larger than the mature hormone. This precursor form may be either a *prohormone* or a *preprohormone*. Prohormones are extended at their amino termini by a hydrophobic amino acid sequence called *leader* or *signal* peptide. Preprohormones are also extended by a signal peptide, but in addition contain internal cleavage sites that upon enzymatic action yield different bioactive peptides. In some instances, preprohormones contain peptide sequences that may not have a known biological activity. These sequences are called *cryptic* peptides. In other instances these peptides may act as "spacers" between two bioactive peptides.

Because of its hydrophobic nature, the *leader* peptide sequence of both preprohormones and prohormones permits the newly synthesized protein to move across the membrane of the endoplasmic reticulum in order to be transported to the Golgi apparatus. The leader peptide is removed from the hormone before the synthesis of the polypeptide chain is ended and this permits the protein to assume its secondary structure during its transport to the Golgi apparatus. Once it arrives at this organelle, the hormone may be further processed by proteolytic enzymes, which cleave the prohormone generating one or more mature hormones, and/or by other enzymes that add nonprotein residues such as carbohydrates. Whatever the case, the hormone is always stored in granules that fuse with the cell membrane during the release process and allow their contents to be extruded into the extracellular perivascular space (see Chapter 2). This process is called *exocytosis* and involves the participation of microtubules and the mobilization of calcium across the cell membrane. During release, not only the hormone but also the cleaving enzymes and peptides associated with the prohormone are discharged from the granule. Like proteins, amines are packed into granules and are released by a similar process of exocytosis.

Amine and steroid hormones are synthesized in a manner much different than proteins. They originate from a precursor molecule (tyrosine and cholesterol, respectively) that is either totally (tyrosine) or partially (cholesterol) transported to the cell of synthesis via the bloodstream. Once inside the cell, these precursor molecules are subjected to the sequential action of several enzymes resulting in the formation of various intermediate products that themselves may be hormones. In contrast to protein hormones, thyroid hormones and steroids, once produced, can freely cross the cell membrane without having to be packed in granules and actively exocytosed.

Although we have been discussing mechanisms involved in the formation of hormones at their site of origin, there are well-documented examples of hormones that are produced at sites other than those in which their precursor is formed using these same mechanisms (i.e., proteolytic cleavage of peptide hormone precursors and enzymatic modification of steroid hormones). In some instances, a hormone with little activity is converted into an active or more active form by the action of enzymes in the circulation or other tissues. For example, angiotensin II is formed in the circulation through the sequential action of the enzyme renin, produced in the kidney to convert angiotensinogen to angiotensin I and a converting enzyme produced in the lungs to cleave two amino acids from angiotensin I to form angiotensin II. The weak adrenal androgen androstenedione can be metabolized to the estrogen estrone in adipose tissue, and testosterone is converted into the more potent androgen dihydrotestosterone in androgen target tissues such as the prostate.

Although protein, amine, and steroid hormones differ in their mechanisms of synthesis and release, a feature they all share is the variability of their blood levels. As we shall see next, fluctuating plasma hormone levels are the consequence of the episodic nature of a hormone secretion.

PATTERNS OF HORMONE SECRETION

The concentration of hormones in the circulation is generally regulated by control loops or feedback mechanisms to allow response to physiologic needs (see below). In addition, it is now established that the basal secretion of most hormones is not a continuous process but rather has a pulsatile nature. When hormone release is induced by a secretagogue the pattern of response is episodic. Conversely, inhibition of hormone secretion by other hormones always results in suppression of the episodic increases in hormone levels. The pulsatile pattern of hormone secretion is characterized by episodes of release that can be as frequent as every 5–10 minutes; each episode is followed by a quiescent period during which plasma levels of the hormone fall toward basal values. Another discharge then occurs and the cycle repeats itself, often varying in both amplitude and frequency of the pulses. In the case of hormones subjected to negative feedback control (see below), removal of the inhibitory feedback signal results in a marked enhancement of the amplitude and frequency of the episodes of secretion.

Secretory episodes may occur with different periodicities. For example, the most prominent episodes of release may occur with a frequency of about an hour. This mode of release is called *circhoral.* If the episodes of release occur at intervals longer than an hour but less than 24 hours the rhythm is called *ultradian,* if the periodicity is of about a day the rhythm is called *circadian,* and if it recurs every day it is called *quotidian.* Release patterns of the latter type are usually referred to as *diurnal* because the increase in secretory activity becomes expressed at defined periods of the day. For example, ACTH has a characteristic diurnal pattern of release with plasma levels rising sharply during the early morning hours.

The release pattern of a hormone may have a much less frequent periodicity. For example, the monthly preovulatory discharge of gonadotropins recurs approximately every 30 days, a pattern of release that has been called *circatrigintan.* Other hormones such as thyroxine exhibit changes in plasma levels that occur over months. If the changes take place on a yearly basis the rhythm is called *circannual* or *seasonal,* as it occurs in relation to the seasonal phases of the year.

Although it is clear that these low periodicity rhythms of hormone release are dictated by either environmental or hormonal cues, the mechanisms underlying the pulsatile release of hormones are unknown. It does appear, however, that secretory cells, including neurons, have the intrinsic ability to secrete their products in a discontinuous, episodic manner. Superimposed on this basic mode of release the amount of hormone produced as well as the frequency of the secretory episodes are modulated by specific negative and positive feedback mechanisms (see below).

The physiological importance of pulsatile hormone release has been best demonstrated in humans and subhuman primates treated with LHRH, the hypothalamic peptide that stimulates the secretion of gonadotropins from the anterior pituitary (see Chapter 6). Only when LHRH is given in a pulsatile fashion at a frequency of about one pulse per hour is gonadotropin secretion and gonadal function maintained normally; a slower frequency fails to maintain gonadotropin secretion at a level sufficient to support normal gonadal production of steroids and gametes; a faster frequency or continuous administration of LHRH inhibits the secretion of gonadotropins and effectively prevents normal gonadal function. This latter phenomenon, which results from a reduced sensitivity of the pituitary to a frequent or continuous LHRH stimulation, has been applied to the treatment of precocious puberty and to attempts at fertility control using long-acting LHRH analogs (see Chapters 9 and 10).

TRANSPORT AND METABOLISM OF HORMONES

Once a hormone is released into the bloodstream it may circulate freely, if it is water soluble, or it may be bound to a carrier protein. In general, amines, peptides, and proteins circulate in free form whereas steroids and thyroid hormones are bound to transport proteins. A well-known exception to this rule is provided by the insulin-like growth factors, which, in spite of being polypeptides, circulate tightly attached to specific binding proteins. Some plasma proteins such as albumin and prealbumin have the capacity to transport nonselectively a variety of low-molecular-weight hormones. In contrast, specific transport proteins that are globulins have saturable, high-affinity binding sites for the hormones they carry. These proteins include thyroid hormone-binding globulin (TBG), testosterone-binding globulin (TeBG), and cortisol-binding globulin (CBG).

Binding of hormones to carrier proteins has a profound impact on the hormone clearance rate from the circulation. The greater the binding capacity of the specific protein carrier, the slower the clearance rate of the hormone. Thus, hormones that circulate mostly in bound form, such as thyroxine, have a much slower disappearance rate from the plasma than hormones that circulate either in free form or only weakly bound. In general, changes in the plasma levels of binding proteins are rapidly followed by adjustments in the secretion rate of the corresponding hormone, so that the fraction of hormone readily available for tissue delivery remains constant and endocrine function remains normal. One well-known example of this is the increase in CBG concentration that occurs during pregnancy as a consequence of estradiol stimulation. While the total plasma cortisol rises as a result of the increased CBG levels, the cortisol available to the tissues remains normal. This is because as the concentration of CBG increases there is a temporary decrease in the cortisol available to target tissues as more is bound to CBG. This results in a temporary increase in ACTH by activation of feedback mechanisms (see below) and increased cortisol secretion to bring the total plasma concentration of cortisol to a higher level and return tissue delivery of cortisol to normal. Thus, in the steady state with intact control mechanisms, alterations in hormone-binding proteins do not affect endocrine status.

The metabolic clearance rate (MCR) of a hormone defines quantitatively its removal from plasma. Under steady-state conditions, the MCR represents the volume of plasma cleared of the hormone per unit of time; usually the units employed are milliliters per minute. If a radioactive hormone is infused into the bloodstream until a constant level is reached and the infusion is then stopped, the disappearance rate of the labeled hormone from the plasma can be determined and the plasma half-life of the hormone calculated. The plasma half-life of a hormone is inversely related to its MCR. Table 1-3 lists the half-lives of several hormones.

Only a small portion of the circulating hormones is removed from the circulation by most target tissues. The bulk of hormone clearance is done by the liver and the kidneys. This process includes degradation by a variety of enzymatic mechanisms such as hydrolysis, oxidation, hydroxylation, methylation, decarboxylation, sulfation, and glucuronidation. In general, only a small fraction ($<1\%$) of any hormone is excreted intact in the urine or feces.

The interaction of hormones with their target tissues is apparently followed by

Table 1-3. Half-life of Protein, Amine and
Steroid Hormones in Plasma

Hormone	Half-life
Amines	2–3 min
Thyroid hormones T_4	6.7 days
T_3	0.75 days
Polypeptides	4–40 min
Proteins	15–170 min
Steroids	4–120 min

intracellular degradation of the hormone. In the case of protein hormones and amines, degradation occurs after their binding to membrane receptors, internalization of the hormone–receptor complex, and the dissociation of this complex into its two components (see Chapter 3). In the case of steroid or thyroid hormones, degradation may occur after binding of the hormone–receptor complex to nuclear chromatin.

FEEDBACK MECHANISMS

The secretion of most, if not all, hormones is regulated by closed-loop systems known as *feedback mechanisms*. Indeed, the endocrine system as a whole is organized in a hierarchy of closed-loop systems that not only operate between cells but also constitute an essential feature of intracellular regulation. Feedback mechanisms are of two types: *negative feedback* and *positive feedback*. Negative feedback is the prevailing control mechanism regulating endocrine function; in its simplest form it is a closed loop in which hormone A stimulates the production of hormone B, which in turn acts on the cells producing hormone A to decrease its rate of secretion as cortisol feeds back to inhibit the hypothalamic–pituitary system and decrease ACTH levels (Fig. 1-2). In the less common positive feedback mechanisms, hormone B further stimulates the production of hormone A instead of diminishing it (Fig. 1-2). A typical example of a positive feedback loop is that which exists between LH and estradiol. During the menstrual cycle a gradual increase in plasma LH levels stim-

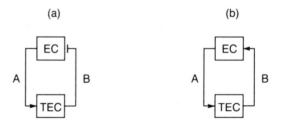

Fig. 1-2 Schematic representation of a negative (a) and a positive (b) feedback loop. The arrows represent stimulation. The blunt-ended line represent inhibition. EC, endocrine cells; TEC, target endocrine cell; A, hormone A; B, hormone B.

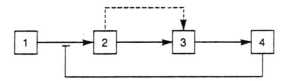

Fig. 1-3 Schematic representation of a feed-forward loop (broken line) that activates the progression of reaction in a hypothetical metabolic pathway. The loop, however, operates within a system tightly controlled by negative feedback systems (blunt-ended line). (Adapted from Rasmussen H: In: *Textbook of Endocrinology,* 5th ed., RH Williams, ed., Saunders, Philadelphia, p. 2, 1974, with permission.)

ulates the production of estradiol by the ovary; after reaching a certain level, estradiol induces an abrupt increase in LH secretion, known as the preovulatory surge of LH, because it induces ovulation. Upon reaching maximal levels plasma LH declines despite the presence of elevated estrogen concentrations. This latter phenomenon exemplifies the self-limiting nature of positive feedback systems. The secretion of hormone A declines rapidly after its initial activation because the secretory cell has a limited capacity to produce the hormone and because there are always additional negative feedback loops that limit the magnitude of the response.

Endocrine control systems also have *feed-forward loops,* which can be negative or positive, that direct the flow of hormonal information. These feed-forward loops are intrinsically unstable because they do not function as closed-loop systems. Since they always form part of a large, more complex feedback circuit their performance is closely regulated by subserving closed-loop mechanisms. This is diagrammed in Fig. 1-3, which shows a feed-forward loop that accelerates the conversion of product 2 into product 3. The latter results in a product 4 that then inhibits the formation of product 2.

A well-recognized example of a feed-forward loop is the release of insulin by the β-cells of the islets of Langerhans in response to an increase in plasma glucose concentration. Upon stimulation of its secretion by the elevated glucose levels, insulin acts on the liver to enhance the uptake of glucose. When plasma glucose levels fall to basal levels, insulin mobilization decreases, and glucose uptake is reduced.

As one would suspect, there are different levels of complexity in the organization of endocrine control systems. The simplest is that of a single negative feedback loop. Much more complex is the situation in which one hormone controls the production of another hormone from a different cell type, which, in turn, controls a third secretory cell that produces a hormone(s) that both feeds back upon the first and/or second component of the system and regulates the function of still another cell type. An example of such a system is the hypothalamic control of gonadotropin secretion from the anterior pituitary (Fig. 1-4). While the hypothalamus stimulates the secretion of gonadotropins through the delivery of LHRH to the pituitary gland, the gonadotropins control the secretion of ovarian steroids, which in turn act on the uterus and breast and also feed back on the hypothalamus and anterior pituitary to regulate the secretion of LHRH and gonadotropins, respectively.

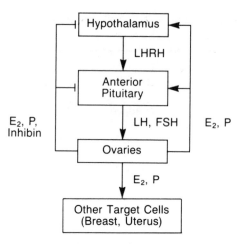

Fig. 1-4 The hypothalamic–pituitary–gonadal axis as an example of a complex endocrine control system. Arrows indicate stimulation and blunt-ended lines represent inhibition. E_2, estradiol; P, progesterone; LH, luteinizing hormone; FSH, follicle-stimulating hormone; LHRH, luteinizing hormone-releasing hormone.

MECHANISMS OF ENDOCRINE DISEASE

Endocrine disorders can result from hormone deficiency, hormone excess, or hormone resistance. With some notable exceptions (e.g., calcitonin), hormone deficiency always causes disease. Hormone deficiency is usually the result of a destructive process occurring in the gland in which the hormone is produced. Thus, infection by viruses or bacteria, infarction due to impaired blood supply, physical compression by tumor growth, or attack by cellular or humoral immune mechanisms all may lead to impaired hormone production in specific situations in most endocrine glands. Alternatively, hormone deficiency states can result from genetic defects in hormone formation such as gene deletion, failure to cleave a peptide hormone precursor to the active hormone, or a specific enzymatic defect in the formation of thyroid or steroid hormones.

Hormone excess usually results in disease. The hormone may be overproduced by the gland that normally secretes it or by a tissue that is not normally an endocrine organ. Malignancies are often involved in each of these types of hormone excess. Some tumors of endocrine glands (e.g., pituitary, adrenal) are functional and secrete the appropriate hormone for the gland but in an unregulated manner. In the malignant transformation occurring in nonendocrine tissues (e.g., lung cancer), dedifferentiation may lead to the production of certain peptide hormones. Other mechanisms of hormone excess include the effects of antireceptor antibodies stimulating a receptor instead of blocking its activation, as in the common form of hyperthyroidism, and the ingestion of exogenous hormones, as in the glucocorticoid excess resulting from its therapeutic use.

Hormone resistance as a mechanism of disease has now been described for almost all hormones. In these disorders the hormone is present in normal or increased amounts but the expected actions of the hormones are not present. In some cases a structurally abnormal peptide hormone is present so that the resistance is only to the endogenous hormone (e.g., insulin, PTH), and response to exogeneous hormone is normal. In other instances there are antibodies to the hormone or hormone receptor (e.g., insulin and its receptor). Finally, hormone resistance may also occur as the result of primary receptor defects (e.g., androgen and vitamin D receptors) or defects in the postreceptor mechanisms of hormone action (e.g., insulin, PTH).

SUGGESTED READING

Adashi EY: The potential relevance of cytokines to ovarian physiology: The emerging role of resident ovarian cells of the white blood cell series. Endocr Rev 11:454–464, 1990.

Blalock JE: A molecular basis for bidirectional communication between the immune and neuroendocrine systems. Physiol Rev 69:1–32, 1989.

Cahill GF: Origin, evolution, and role of hormones. In: *Endocrinology*, 2nd ed., LJ Degroot, ed., Saunders, Philadelphia, pp. 3–5, 1989.

Follett BK, and Follett DE: *Biological Clocks in Seasonal Reproductive Cycles,* Wright & Sons, Bristol, 1981.

Pardridge WM: Transport of protein bound hormones into tissues *in vivo.* Endocr Rev 2:102–123, 1981.

Rasmussen H: Organization and control of endocrine systems. In: *Textbook of Endocrinology,* 5th ed., RH Williams, ed., Saunders, Philadelphia, pp. 2–30, 1974.

Scarborough DE: Cytokine modulation of pituitary hormones secretion. Ann NY Acad Sci 594:169–187, 1990.

Veldhuis JD, Evans WS, Rogol AD, Drake CR, Thorner MO, Merrian GR, and Johnson ML: Performance of LH pulse-detection algorithms at rapid rates of venous sampling in humans. Am J Physiol: Endocrinol Metab 24:E554–E563, 1984.

Weitzman ED: Biologic rhythms and hormone secretion patterns. In: *Neuroendocrinology,* DT Krieger and JC Hughes, eds., Sinauer, Sunderland, MA, pp. 85–92, 1980.

Wilson JD, and Foster DW: *Williams Textbook of Endocrinology,* 8th ed., Saunders, Philadelphia, 1992.

2

Genes and Hormones

MICHAEL R. WATERMAN

The endocrine functions discussed in this textbook are regulated, in large part, by the expression of genes encoding a diverse set of hormones and enzymes. Major advances in biological and medical sciences have led to the elucidation of gene structure and to our understanding of the relationship between gene structure and function (i.e., transcription). Thus we have begun to appreciate the molecular bases of endocrine physiology as our biochemical understanding of these processes continues to develop at a rapid pace. This chapter briefly reviews gene structure and its relationship to transcription. Post-transcriptional aspects of gene expression will also be considered, with emphasis on their involvement in peptide hormone biosynthesis.

GENE STRUCTURE

A *gene* may be defined as a unit of DNA within a chromosome that can be transcribed to yield RNA that serves a particular function in a cell, such as ribosomal RNA or transfer RNA that function directly or messenger RNA that is translated into a protein. A surprising and important discovery in biology was the finding that in most eukaryotic genes, the DNA sequence encoding the primary sequence of a protein is interrupted by segments of noncoding DNA. This is in direct contrast to prokaryotic genes in which the complete protein-coding region is contained within an uninterrupted DNA sequence. The genomic segments whose RNA complements are found in mature messenger RNA and hence encode primary protein sequence are called *exons*, whereas those genomic segments whose RNA complements are removed during maturation of the RNA are called *introns*. Thus, as illustrated in Fig. 2-1, eukaryotic genes usually consist of alternating exons and introns. The numbers of these alternating segments, their lengths, and consequently, the overall length of eukaryotic genes vary considerably and are not necessarily related to the size of their protein product. A single intron, in fact, can be longer than the mature messenger RNA produced by the gene in which it resides. Prokaryotic genes contain no introns, while lower eukaryotic genes such as those in yeast contain few and frequently no introns, in contrast to genes from higher eukaryotes which usually contain several introns.

The transcriptional unit of a gene is defined by its primary transcript, which

A. Gene

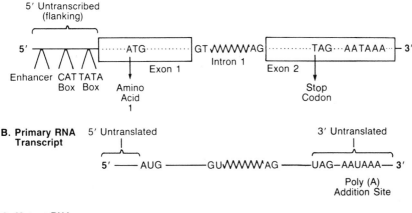

B. Primary RNA Transcript

C. Mature RNA

Fig. 2-1 Schematic representation of the structure of a eukaryotic gene (A), the primary product of this gene produced by transcription (B), and the resultant mature messenger RNA (mRNA) molecule produced by the post-transcriptional events of splicing, capping, and poly(A) addition (C). As seen in (A), this hypothetical gene contains two exons separated by an intron. At the 5'-end of exon 1 lies an untranslated DNA sequence. Translocation of the mature mRNA (C) begins at the initiator AUG and ends at the stop codon UAG. In addition, the gene contains in exon 2 a 3'-untranslated region that contains a poly(A) addition signal shown as AAUAAA in the mature mRNA (C). Also shown associated with the gene is a 5'-untranscribed region that contains a TATA box, a CAT box, and an enhancer sequence. The primary transcriptional product of this gene is shown in (B). This primary transcript is then converted into a mature mRNA molecule by splicing, which removes the DNA sequence contained in intron 1, plus addition of a cap structure at the 5'-end and a poly(A) tail at the 3'-end (C).

begins at base position 1 in the first exon and ends at the last base of the last exon. The signals that instruct RNA polymerases to bind to the DNA and initiate transcription lie beyond the limits of this transcription unit in the 5'-flanking (upstream) region of the gene. Usually, the DNA sequences that are important in regulating transcription are also located in this 5'-flanking region of a gene, although sequences that enhance transcription have also been found in intronic sequences of a few genes and even at the 3'-flanking region of genes. The ending of transcription, the 3'-end of the gene, cannot be identified as precisely as can the position at which transcription begins.

Although the primary transcript of a gene contains both exonic and intronic sequences and resides in the nucleus of the cell, the mature messenger RNA contains only the exonic sequences and resides in the cytoplasm where translation takes place. Maturation of the primary transcript occurs in the nucleus, leading to production of a mature RNA, which is then transported to the cytoplasm. A key aspect of the maturation of RNA is the removal of the intronic (intervening) sequences by a process known as *splicing*. The order in which the intronic sequences are removed varies from one RNA species to another, but the process is sequential. The biochemical mechanism by which splicing occurs continues to be an active area of investigation and involves small RNA molecules and proteins, the complexes being called *splicosomes*. These RNA molecules recognize specific sequences and/or structures at intron–exon boundaries, thereby signaling precisely where the RNA sequence cleavage will occur. Common sequences known as *consensus sequences* are found at both the 5'-(GT) and 3'-(AG) ends of all introns (Fig. 2-1). Needless to say, the fidelity of precise splicing at exon–intron boundaries is of crucial importance. Improper splicing could lead to insertion or deletion of amino acids in the final protein product or to a frameshift in the translation of the RNA sequence, leading to an altered amino acid sequence. The reason for the presence of intervening sequences in eukaryotic genes has not been established. However, in some genes they serve to separate specific structural and/or functional domains of the resultant protein. This may facilitate evolution by allowing specific functional domains (for example an NADPH binding site) to be transferred from one gene to another by genetic recombination. When genes encoding the same protein are compared amongst different species, the nucleotide sequence within introns diverges much more rapidly than the nucleotide sequence within exons. This lack of nucleotide conservation within introns indicates that the precise sequence is less important. However, the precise positioning of exon–intron junctions is conserved across species lines, indicating that while the precise intronic sequences may not be important, their locations within the amino acid coding sequence are.

In addition to splicing, a second post-transcriptional event that occurs in the nucleus during the course of RNA maturation is the addition of a methylated structure at the 5'-end of the RNA molecule known as the RNA *cap*. The cap structure is formed by addition of a guanosine to the 5'-end of the mRNA through an unusual 5'-5'-triphosphate bond followed by methylation of this guanosine in the N-7 position and subsequent methylation of the adjacent nucleotide(s). This structure facilitates the binding of eukaryotic RNA to ribosomes and thus enhances the initiation of translation, and has also been proposed to render the RNA more stable. An additional modification of the primary transcript before the appearance of mature mRNA in the cytoplasm is the addition at the 3'-end of the RNA of a nucleotide sequence consisting of repeated adenylic acid residues known as the *poly(A) tail*. This step also occurs in the nucleus, and although the precise function of poly(A) tails is unknown, they are thought to play a role in RNA stability. Poly(A) segments generally range between 50 and 150 bases and are found 12–16 bases downstream from a consensus poly(A) addition site, AAUAAA or AUUAAA. Thus, the mature messenger RNA molecule (Fig. 2-1) that can be translated in the cytoplasm results from a series of post-transcriptional modifications of the original primary transcript of a gene. Within the nucleotide sequence of this mature transcript lies an open reading frame that encodes the protein beginning at an initiator methionine codon (AUG). An open reading frame is a stretch of triplets that encode the amino acid sequence of a protein, beginning with the initiator codon and ending with a stop codon (UAA or UAG). In the mature messenger RNA molecule, the open reading frame is flanked by both 5' and 3' untranslated regions. The cap structure is located at the 5'-end of the RNA and the poly(A) tail at the 3'-end of the RNA.

As indicated above, in many genes expression (transcription) is controlled by sequences located in the 5'-flanking region beyond the transcriptional unit. Collectively, the sequences that are required for regulating expression are known as the *cis-acting* elements of the gene. *Cis*-acting indicates that these elements (DNA sequences) are part of the same double-stranded DNA that contains the transcription unit. One such region is the site recognized by RNA polymerase as a binding site. In many genes this region includes a short nucleotide sequence known as the *Goldberg-Hogness* or *TATA box* (Fig. 2-1). In such genes, a consensus (TATAA) or related sequence lies approximately 30 bases upstream from the site at which transcription begins. Upstream (5'-ward) from the TATA box in many genes there is a second consensus sequence, CCAAT *(CAT box)*, that may play a role in the activity of the transcription process. A few genes (such as that encoding HMG CoA-reductase) do not contain either the TATA or CAT *cis*-acting elements, but rather have a *GC box* (GGGGCGGGGC) thought to be required for transcription. Thus, for a gene that is expressed constitutively throughout the life of the cell and that requires no additional regulation, these *cis*-acting elements (TATA box, CAT box, or GC box) may be sufficient to regulate expression. In such genes, the level of constitutive expression could be controlled by the binding constant of RNA polymerase for the sequence in this 5'-flanking region.

In many genes, however, expression is controlled by additional physiological or xenobiotic factors. For example, diet, circadian rhythms, or hormones each may regulate expression of a certain subset of genes. In these cases additional *cis*-acting elements are associated with the genes, which are usually located in the 5'-flanking region. These segments of the genes bind proteins, known as *trans-acting factors* (transcription factors), which can influence the binding of RNA polymerase at the region surrounding the TATA box and in this way, regulate the rate of transcription. One such class of regulatory elements is known as *enhancers*. Enhancers are DNA sequences that bind proteins and that can function when inserted in either orientation (5' → 3' or 3' → 5') or at a considerable distance from the site at which transcription begins. Furthermore, enhancers from one gene are found to function *in vitro* to regulate transcription of a heterologous gene. Enhancers can bind proteins that increase gene expression (stimulators) or that decrease gene expression (repressors). It is likely that competition between the binding of stimulator and repressor proteins plays an important role in the overall regulation of eukaryotic gene expression, just as has been described for prokaryotic gene expression. As noted earlier, in a few instances, enhancer sequences have been found within the transcription unit of genes in introns.

Certain groups of genes are known to have common *cis*-acting sequences and consequently are thought to be regulated by identical or very similar *trans*-acting factors. Several genes important to endocrine physiology fall into this category. For example, genes that are regulated by the steroid hormone cortisol contain a common sequence that binds a glucocorticoid receptor, which itself binds cortisol. The binding of the protein receptor to the genomic regulatory sequence is greatly enhanced by the binding of cortisol to the receptor (see Chapter 3). It is presumed that formation of this ligand–receptor complex alters the conformation of the receptor, thereby facilitating its binding to the consensus *cis*-acting element. A similar mechanism is thought to be involved in the regulation of genes by other steroid hormones, such as progesterone or estrogen. In these cases both the *cis*-acting element and the receptor (*trans*-acting factor) are related but contain features that are unique to the specific steroid hormone. Another type of protein that regulates expression of several different genes is the cAMP response element binding protein (CREB). Several eukaryotic genes whose expression is regulated by cAMP have been found to share a common *cis*-acting element (DNA sequence) in their 5′-untranscribed regions (TGACGTCA). CREB binds to this pallindromic *cis*-regulatory sequence as a homodimer (two identical subunits) to activate transcription. Phosphorylation of CREB by cAMP-dependent protein kinase is necessary for activation of transcription. Genes utilizing this mode of transcriptional activation include α-chorionic gonadotropin, somatostatin, vasoactive intestinal peptide, and proenkephalin. In this form of transcriptional regulation, cAMP, a second messenger whose synthesis is regulated by peptide hormones that bind to specific cell surface receptors (see Chapter 3), regulates gene expression through this transcription factor.

In addition to *trans*-acting factors that regulate multiple genes such as CREB and the glucocorticoid receptor, there very likely are a large number of proteins that regulate expression of only one or a very few genes. These will include both stimulators and repressors whose synthesis is regulated by diet, circadian rhythms, etc. In the former group the level of the ligand or second messenger (glucocorticoid or cAMP) is crucial to regulation of gene expression. In this latter case, the level of the *trans*-acting factor itself may also be crucial to the regulation of gene expression. Both tissue specificity of gene expression and developmental patterns of gene expression are controlled by such gene-specific *trans*-acting factors. Thus gene expression must be considered to be a dynamic process that is regulated by a complex group of protein factors working in concert to enhance and repress transcription, depending on the physiological needs of the organism. In fact, it is now clear that not only protein–DNA interactions (*cis*-elements and *trans*-acting factors) are important in the regulation of transcription, but also protein–protein interactions may play an important role. A *trans*-acting protein might bind to a *cis*-acting sequence, but this binding would not lead to an alteration of transcription of the associated gene. However, as a result of a conformational change produced by the binding of this *trans*-acting protein to the DNA, a second protein can bind to the bound protein, thereby leading to alteration of the gene's transcription.

Particularly intriguing aspects of gene expression include tissue specificity and developmental regulation. In the case of tissue specificity, specific genes are expressed in certain but not all tissues. This leads to different structure and enzymatic properties being associated with different organs or cell types. The vast majority of studies in endocrine physiology involve investigation of tissue-specific events. For example, among steroidogenic tissues only the adrenal cortex expresses the steroid 21-hydroxylase and 11β-hydroxylase enzymes that are required for glucocorticoid and mineralocorticoid production. The gonads express a different profile of steroid hydroxylase enzymes leading to sex hormone production. Finally, nonsteroidogenic tissues, such as heart, liver, and kidney, express none of the steroid hydroxylase genes. Although many of the details of tissue-specific gene expression remain to be elucidated, it can generally be imagined that it is the expression of certain *trans*-

acting factors that is tissue specific. If a gene requires binding of a *trans*-acting factor for expression, particularly if the factor overcomes repression by another factor, the gene will be expressed in the tissues in which the stimulating *trans*-acting factor is found. Thus, in this sense, it is the expression of *trans*-acting factors that is tissue specific. Presumably studies on the tissue-specific nature of gene expression will, by necessity, involve investigation of expression of genes encoding *trans*-acting factors. Developmental regulation also includes tissue-specific gene expression. However, development also involves time-dependent regulation of gene expression. In this case, it is imagined that regulation of certain *trans*-acting factors will occur in a time-dependent fashion so that gene expression may be turned on and off at the appropriate times in development. A detailed working hypothesis for regulation of gene expression involving *trans*-acting factors and *cis*-acting genomic elements must also encompass both tissue-specific and time-dependent (developmental) regulation.

The biochemical mechanisms by which transcription is activated or repressed by *trans*-acting factors binding to their specific DNA elements are not yet clearly understood. However, it has now been established in a few cases that transcription factors interact with the basal transcription machinery and in this way effect RNA polymerase binding. This protein–protein interaction can occur directly between an enhancer and one of the general transcription complexes of the basal transcription machinery such as transcription factor IID or with one of the TATA-box binding protein associated factors (TAFs). Alternatively, in the case of certain genes regulated by cAMP through CREB, the interaction between phosphorylated, DNA-bound CREB and the transcription machinery is mediated by a non-DNA binding protein known as *CREB binding protein* (CBP). Thus, in this case, it is not CREB which interacts with the basal transcription machinery but rather CBP which itself is bound to CREB.

In summary, the regulation of expression of genes encoding hormones as well as genes regulated by hormones involves complicated interactions between proteins and DNA. Perhaps a single general biochemical mechanism is involved in the regulation of expression of all genes. However, the multiplicity and diversity of *trans*-acting factors and *cis*-acting elements and the complexity of the basal transcription machinery provide the necessary diversity of this process. Many genes are likely to be regulated by a cassette of *cis*-acting elements that binds different proteins at different times, leading to regulation of gene expression in response to developmental, tissue-specific, and physiological events. Thus the second generation of experiments on regulation of gene expression will involve the study of a myriad of yet-to-be-identified genes that encode trans-acting factors.

POST-TRANSCRIPTIONAL REGULATION AND PROTEIN SYNTHESIS

The regulation of amounts of protein products can be controlled not only at the transcriptional level but also by alterations in RNA stability and/or protein stability. Different messenger RNA molecules display very different half-lives. Ornithine decarboxylase is an example of a messenger RNA (mRNA) that turns over rapidly (less than 1 hour), whereas adrenal steroid hydroxylase mRNAs turn over with half-lives of several hours and approaching 1 day in some instances. Also, in some cases, these half-lives are altered by their physiological environment. Hormones such as ACTH are known to alter the half-lives of certain RNA species, although the mechanism by which such a change occurs is unknown. The final product of gene expression, the resultant protein, can also be regulated by the physiological environment. It is well known that endogenous factors such as cAMP can alter the activity of certain enzymes by activating phosphorylation. In addition, the half-life of the resultant protein might be altered by post-translational modifications such as phosphorylation. Thus, in most cases, the regulation of gene expression as assayed by enzymatic activity will be the result of a combination of more than one mechanism. To fully understand hormonal action it is necessary to investigate the effects of transcriptional regulation, translational regulation, RNA turnover, protein turnover, and post-translational modification on enzymatic activity.

Most enzymes and other proteins are synthesized in the cells in which they function. Peptide hormones are an exception to this rule as they are synthesized in one cell type and secreted to be utilized by another cell type, often in another organ. Thus a specialized system has developed for synthesis and secretion of such proteins. Many proteins that traverse the intracellular secretory pathway are synthesized as higher molecular weight precursor forms that are proteolytically cleaved upon secretion from the cell. The additional presequences of these precursor proteins are usually located at their amino-terminal ends. Some peptide hormones are synthesized as *prehormones*, which contain at their amino terminus an amino acid sequence called a *signal peptide*, which is cleaved by a peptidase located on the cisternal face of the membrane of the endoplasmic reticulum. Synthesis of prehormones follows the signal hypothesis established by Blobel and his collaborators (Fig. 2-2). Translation of prehormone mRNA is initiated on free polysomes. As the newly synthesized N-terminal sequence of the prehormone emerges from the translation complex, it is recognized by a signal recognition particle, a complex structure consisting of both protein subunits and RNA molecules. Many such newly synthesized polypeptide chains contain sequences called *halt-transfer sequences* that serve to arrest translation upon binding of the signal recognition particle. The bound signal recognition particle then recognizes a receptor (docking protein) in the endoplasmic reticulum leading to attachment of the translation complex to this membrane. The remainder of the synthesis of the prehormone takes place on bound polysomes. Upon binding of the signal recognition complex to the docking protein, the signal recognition particle is released from the polypeptide chain, leading to release of the translation arrest and continuation of protein synthesis. The amino-terminal signal sequence that is hydrophobic serves to anchor the newly synthesized protein in the endoplasmic reticulum and protein synthesis continues with the polypeptide chain extruding into the cisternae of the endoplasmic reticulum. The signal peptidase then cleaves this precursor segment and the resultant mature hormone appears in the cisternae of the endoplasmic reticulum. The mature hormone moves through the endoplasmic reticulum to the Golgi apparatus where it is sorted and packaged into a secretory vesicle for transport from the cell.

In addition to those peptide hormones that are synthesized as prehormones, certain peptide hormones are synthesized as *preprohormones* (Fig. 2-3). Preprohormones are synthesized as described above for prehormones, however, the resultant product in the cisternae of the endoplasmic reticulum is a prohormone. The proteo-

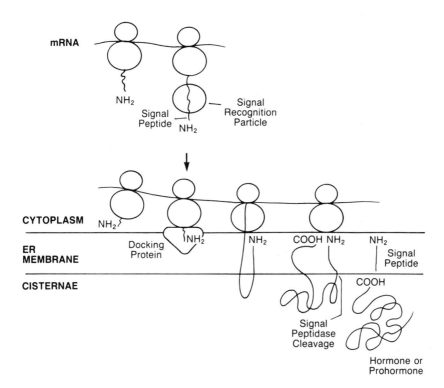

Fig. 2-2 Schematic illustration of the general process of protein synthesis required for the production of peptide hormones or prohormones that are subsequently secreted from the cell following passage through the cisternae of the endoplasmic reticulum to the Golgi structure.

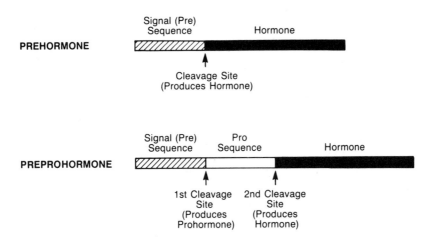

Fig. 2-3 Schematic representation of the structural difference between a prehormone and a preprohormone. Included is the designation of the cleavage site(s) required for production of the mature peptide hormones.

lytic removal of the prepiece from a preprohormone leads to the appearance of the prohormone in the cisternae of the endoplasmic reticulum. A prohormone then undergoes further proteolytic processing to remove the propiece, leading to production of a mature hormone. This second cleavable segment is at the amino terminus of the prohormone. The signal peptide or prepiece of either prehormones or preprohormones is a hydrophobic sequence of amino acids that is cleaved by an endopeptidase of unknown specificity in the cisternae of the endoplasmic reticulum. The aminoterminal propiece of preprohormones contains basic amino acids and is cleaved by endopeptidases having trypsin-like activity. It is not clear why certain secretory proteins such as growth hormone and prolactin are synthesized as prehormones while other secretory proteins such as insulin and parathyroid hormone are synthesized as preprohormones. It should also be noted that certain secretory proteins appear to contain no N-terminal extension and presumably do not undergo proteolytic processing during their maturation and secretion. Examples of such proteins are basic fibroblast growth factor, endothelial cell growth factor, and interleukin-1.

A few prohormones contain multiple peptide hormones within their primary sequence. The best studied and perhaps most complex example of this is proopiomelanocortin (POMC), which is synthesized in the pituitary and hypothalamus (Fig. 2-4, also see Chapter 6). The POMC protein contains 239 amino acids plus a signal peptide of 26 amino acids. In the center of the mature sequence of POMC are 39 amino acids encoding adrenocorticotropin (ACTH), which is a primary product derived from this protein in the anterior pituitary. In the intermediate lobe of the pituitary a portion of this ACTH sequence (1–13) is released that functions as α-melanotropin (α-MSH). Furthermore, in the anterior pituitary, the C-terminal fragment of POMC following ACTH formation is further cleaved by an endopeptidase to produce two additional peptide hormones, γ-lipotropin (γ-LPH) from the N-terminal portion of this fragment and β-endorphin from the C-terminal portion of this fragment. β-Melanotropin (β-MSH) is derived from the C-terminal region of the γ-LPH sequence in the intermediate lobe of the pituitary and γ-MSH is derived from the N-terminal fragment of POMC in the intermediate lobe of the pituitary. This is a fascinating example of the generation of biochemical diversity at the posttranslational level in a tissue-specific fashion.

Tissue-specific variation in the processing of preproglucagon between the pan-

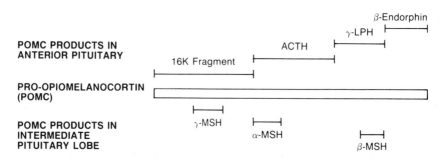

Fig. 2-4 Comparison of the peptide hormones produced from proopiomelanocortin (POMC) in the anterior pituitary and the intermediate pituitary lobe.

creas and intestine is another example of this phenomenon. In pancreatic islets the major product of preproglucagon processing is glucagon itself. In the intestine glicentin is produced, a predominantly proteolytic product which consists of glucagon plus both N-terminal and C-terminal extensions. Glicentin is further processed proteolytically into the peptide hormone oxyntomodulin, which participates in the regulation of gastric HCl secretion. Thus, while regulation of gene expression is the most prevalent mechanism by which the events associated with endocrine physiology are regulated, post-translational regulation, including chemical modification by phosphorylation or peptide bond cleavage, also plays an important role in these processes.

MOLECULAR BASIS OF GENETIC DISEASE

One of the exciting medical advances to result from the development of new biochemical techniques in molecular biology is our ability to determine at the molecular level the basis of genetic diseases. The hemoglobinopathies associated with globin chain synthesis (thalassemias) have taught us that it will not be surprising to encounter diseases associated with each step in the biosynthetic pathway leading to the production of a particular enzyme or protein. An obvious cause of genetic disease is *gene deletion*, which is defined as the absence of all or a portion of the genomic information encoding a particular protein in an individual. Gene deletion has been found to be one cause of congenital adrenal hyperplasia, a group of autosomal recessive genetic diseases resulting from aberrant glucocorticoid, mineralocorticoid, and sex hormone production (see also Chapter 8). By far the most common cause of congenital adrenal hyperplasia is steroid 21-hydroxylase deficiency. The steroid 21-hydroxylase is a steroid hydroxylase localized in the endoplasmic reticulum of the adrenal cortex. Although the number of individuals examined by DNA analysis to date is not great (several hundred), it appears that in at least 20% of the cases with a deficiency of this enzyme, gene deletion is the cause. The remainder of these cases result from a variety of other causes. For example, a point mutation in the coding sequence of the 21-hydroxylase gene that leads to an amino acid substitution in the primary sequence of the enzyme and subsequently, altered enzymatic activity, has been shown to be a cause of this deficiency state. However, we can also well imagine defects arising from point mutations (single base changes) occurring in enhancers that regulate transcription, at splice junctions producing aberrant mRNA molecules, and at start-and-stop codons producing aberrant proteins—all of these leading to reduced or absent steroid 21-hydroxylase activity. One of the important challenges that lies before those investigating the clinical aspects of endocrine physiology is to establish the molecular basis of the large number of genetic diseases associated with this area of medicine.

This task is made possible by our ability to utilize polymerase chain reaction (PCR) to generate DNA fragments from suspected mutant genes from individuals with genetic diseases and to sequence them directly to locate mutations. By comparing the nucleotide sequence of the normal gene with that of the mutant gene it is usually possible to establish the amino acid change that leads to the alteration of enzymatic activity. The complete gene does not usually have to be sequenced in such a study, rather, only the exons need be sequenced from PCR-amplified genomic

fragments, and this can be readily accomplished by using the Sanger dideoxynu-
cleotide sequencing method in which oligonucleotides corresponding to specific
regions of each exon act as primers for the sequencing reactions. It is also possible
in many instances to localize the mutant exon by electrophoretic comparison of PCR
fragments covering exons of mutant and normal genes, so that all exons need not
be sequenced. Techniques have also been developed that permit the incorporation
of cDNA inserts into expression vectors and the generation of the resultant protein
in specific cell lines. This technology, coupled with the ability to mutagenize a cDNA
insert such that its sequence corresponds to that found in a mutant gene, makes it
possible to express the mutant enzyme and confirm its altered activity as well as to
evaluate the detailed kinetic parameters of the mutant enzyme. The modern techniques of molecular biology have led to an increased understanding of the regulation
of gene expression at both the transcriptional and post-transcriptional levels. This
information, along with our ability to rapidly sequence amino acid coding and regulatory regions of mutant genes and to carry out site-specific mutagenesis studies
using expression vectors, provides the modern clinician–scientist with the necessary
tools to establish in detail the molecular basis of genetic diseases, including those
associated with aberrant endocrine physiology.

SUGGESTED READING

Donohoue PA, Parker K, and Migeon CJ: Congenital adrenal hyperplasia. In: *The Metabolic
    and Molecular Bases of Inherited Disease*, 7th ed., CR Scriver, AL Beaudet, WS Sly,
    and D Valle, eds., McGraw-Hill, New York, pp. 2929–2966, 1995.
Habener JF: Genetic control of hormone formation. In: *Williams Textbook of Endocrinology*,
    8th ed., JD Wilson and DW Foster, eds., Saunders, Philadelphia, pp. 9–33, 1992.
Montminy MR, Gonzelez GA, and Yamamoto K: Regulation of cAMP-inducible genes by
    CREB. Trends Neurol Sci 13:184–188, 1990.
Walter P, Gilmore R, and Blobel G: Protein translocation across an endoplasmic reticulum.
    Cell 38:5–8, 1984.

3

Mechanisms of Hormone Action

CAROLE R. MENDELSON

The capacity of a cell to respond to a particular hormone depends upon the presence of cellular receptors specific for that hormone. The steroid and thyroid hormones as well as retinoids and 1,25-dihydroxyvitamin D_3 are thought to diffuse freely through the lipophilic plasma membrane of the cell and to interact with receptors that are primarily within the nucleus. Hormones that are water soluble, such as the peptide hormones, catecholamines, and other neurotransmitters, interact with receptors in the plasma membrane.

RECEPTOR PROPERTIES

The function of a receptor is to recognize a particular hormone among all the molecules in the environment of the cell at a given time, and, after binding the hormone, to transmit a signal that ultimately results in a biological response. Hormones are normally present in the circulation in extremely low concentrations, from 10^{-11} to 10^{-9} M. The receptor must therefore have an *affinity* for the hormone that is of a magnitude appropriate to the circulating levels. The receptor must also bind the hormone with high *specificity*, so that it has a greater affinity for a particular biologically active molecule rather than a related but less biologically active species. The affinity of the receptor for a particular ligand in relation to its affinity for other related molecules determines the specificity of the hormone–receptor interaction.

The affinity of a hormone for its receptor results from *noncovalent* binding, primarily in the form of hydrophobic interactions that provide the driving force for the binding reaction, and from electrostatic interactions. The latter, which occur between oppositely charged groups on peptide hormones and their receptors, are important for hormone–receptor specificity. The affinity of a hormone–receptor interaction is defined in terms of the *equilibrium dissociation constant* (K_d). In a system in which there is a single class of binding sites with no interactions among receptors, the K_d is defined as the concentration of hormone, at equilibrium, that is required for binding to 50% of the receptor sites. The affinity can also be expressed as the *equilibrium association constant* (K_a), which is the reciprocal of the K_d.

The binding of hormone to receptor is a *saturable* process; there are a finite number of receptors for a given hormone on a target cell. In addition, the binding of hormone to receptor must either precede or accompany the biological response, and the magnitude of the biological response must be associated, in some manner, with receptor occupancy. Hormones or analogues that bind to receptors and elicit the same biological response as the naturally occurring hormone are termed *agonists*. Molecules that bind to receptors but fail to elicit the normal biological response are termed *competitive antagonists*, since they occupy the receptors and prevent the binding of the biologically active molecules. Molecules that bind to receptors, but are less biologically active than the native hormone, are termed *partial agonists*. The term *partial antagonist* also applies since partial agonists bind to receptors and prevent the binding of the fully biologically active native hormone.

RELATIONSHIP OF BINDING TO BIOLOGICAL RESPONSE

The biological response of a target cell to a hormone is determined by a number of factors, including the concentration of hormone, the concentration of receptors, and the affinity of the hormone–receptor interaction. Normally, the concentration of a particular hormone in the circulation is much lower than the K_d of the hormone–receptor interaction. Therefore, the receptors on a target cell are almost never saturated and an increase in the concentration of circulating hormone results in an increase in the number of occupied receptors.

In a number of examples of peptide hormone binding to cell surface receptors, the response of the target cell to the hormone is directly proportional to the number of receptor sites that are occupied, that is, the binding and biological response curves are superimposable over the entire range of hormone concentrations and a maximum biological response is achieved when 100% of the receptor sites are occupied (Fig. 3-1). If the number of cellular receptors is reduced without a change in the K_d, both the binding and biological response curves are reduced and remain superimposable.

In the majority of examples, however, the maximum biological response of a target cell is achieved at concentrations of hormone lower than those required to fully occupy all of the receptors on that cell. For example, in isolated adipocytes, maximum stimulation of glucose oxidation is achieved at concentrations of insulin that occupy only 2–3% of the cellular receptors. In Leydig cells, maximum stimulation of testosterone synthesis occurs at concentrations of gonadotropin that occupy only 1% of the cellular receptors. In these systems, >97% of the receptors are referred to as *spare receptors*. The term spare receptors does not imply that these receptors are not being utilized, but rather that a maximum biological response is achieved when all of the receptors on a target cell are occupied <3% of the time. The degree of spareness of receptors for a particular hormone on a target cell can vary from one cellular response to another, resulting in a different dose–response curve for each biological response of a cell to a given hormone.

Let us consider the hormone–binding (Fig. 3-2A) and biological response (Fig. 3-2B) curves for a hypothetical target cell that contains spare receptors for a particular biological response. The cell normally contains 20,000 receptors for the hormone. In this example, 75% of these receptors are considered to be spare receptors, since a maximum biological response is achieved at concentrations of hormone required to occupy only 5000 receptor sites per cell. When the number of cellular receptors is reduced by 50% and 75% without a change in receptor affinity, the

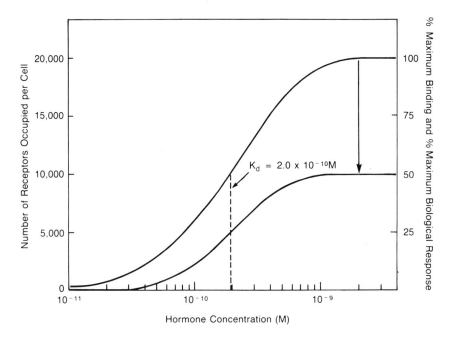

Fig. 3-1 Hormone binding and biological response curves when no spare receptors are present. In this example, the hormone binding and biological response curves are superimposed over the entire range of hormone concentrations. A 50% decrease in receptor number with no change in the K_d of the hormone–receptor interaction will result in an equivalent reduction in the maximum biological response.

maximum biological response remains unchanged; however, maximum response is achieved at progressively increased concentrations of hormone. When the number of cellular receptors is reduced further, by 88%, the maximum biological response is reduced proportionately. From this example, one can see that the greater the proportion of spare receptors for a particular biological response, the more sensitive is the target cell to the hormone, that is, the lower the concentration of hormone required to achieve half-maximum biological response. In addition to increasing the sensitivity of the target cell to the hormone, spare receptors also serve to prolong the biological response of a cell to short bursts of circulating hormone. The hormone–receptor complexes in excess of those required for the maximum biological response will maintain the response for longer periods as the concentration of hormone in the circulation declines.

PEPTIDE AND POLYPEPTIDE HORMONES, NEUROTRANSMITTERS, AND PROSTAGLANDINS

Receptor Structure and Function

Receptors for peptide and polypeptide hormones, catecholamines, and other neurotransmitters and prostaglandins are integral membrane proteins that are interspersed

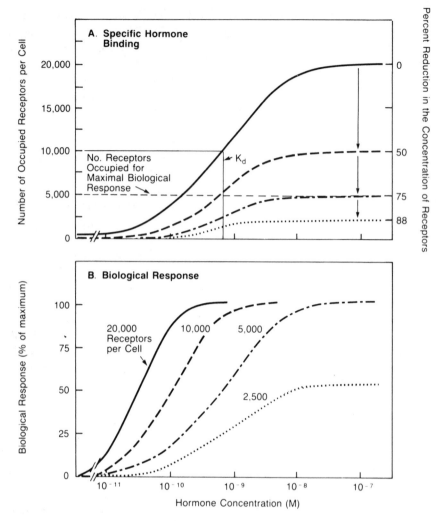

Fig. 3-2 Hormone binding (A) and biological response (B) curves when spare receptors are present. In this example, when the target cell contains its full complement of 20,000 receptors, a maximum biological response is achieved at concentrations of hormone required to occupy only 25% of the cellular receptors. When the number of cellular receptors is reduced to 10,000 or 5,000 without a change in the K_d, a maximum biological response can still be achieved, albeit at progressively increased concentrations of hormone. When the number of cellular receptors is decreased by 88%, the maximum biological response is reduced by 50%.

within the phospholipid bilayer of the plasma membranes of target cells. Because these receptors are soluble in aqueous media only in the presence of detergents, they have proved difficult to purify and, until recently, little was known of their structure. Most of the cellular components that are involved in the initial response of the target cell to such hormones are also present within the plasma membrane.

Cell surface receptors for some ligands, such as low-density lipoprotein (LDL), transferrin, the asialoglycoproteins, IgA/IgM, toxins, and viruses are also integral membrane proteins. The function of these receptors is to translocate the respective ligands to intracellular sites where the ligands themselves act to alter cellular function. For these systems, the information is contained within the ligand itself and the receptor serves merely to facilitate the delivery of the ligand to its site of action within the cell. An example of a ligand that operates through such a system is cholera toxin, which binds to ganglioside G_{M1} receptors on the plasma membrane. The receptor serves to concentrate the toxin molecule and to facilitate its translocation to sites within the plasma membrane where the toxin causes the activation of adenylyl cyclase (see below).

In the case of receptors for peptide and polypeptide hormones, prostaglandins, and neurotransmitters, the ligand merely activates the receptor, which in turn transmits various signals that result in altered cellular function. In fact, antibodies against the insulin and thyrotropin (thyroid-stimulating hormone, TSH) receptors can activate the respective receptors and mimic the actions of the hormones themselves. It is probable, therefore, that the hormone causes some conformational change in the receptor, which in turn transmits a signal that results in the generation of a biological response.

The primary structures of a large number of cell surface receptors for hormones and other ligands have recently been determined by the use of recombinant DNA techniques. Figure 3-3 diagrammatically shows the structures of the β-adrenergic receptor, and the receptors for epidermal growth factor (EGF), insulin, platelet-derived growth factor (PDGF), atrial natriuretic peptide (ANP), and growth hormone (GH). All of these receptors are glycoproteins that cross the plasma membrane either once (receptors for EGF, insulin, PDGF, ANP, and GH) or several times (β-adrenergic receptor).

Seven-Transmembrane Domain Receptors

The β-adrenergic receptor, a 418 amino acid polypeptide which mediates some of the actions of the catecholamines, epinephrine and norepinephrine, is a member of a large family of related receptor molecules all containing seven hydrophobic regions, each of which is capable of forming an α-helix of sufficient length to span the plasma membrane. These receptor molecules are therefore likely to cross the plasma membrane seven times (Fig. 3-3). This seven-transmembrane domain receptor family includes the β- and α-adrenergic receptors, the muscarinic cholinergic receptors, receptors for vasopressin, angiotensin II, serotonin, substance P, dopamine, luteinizing hormone (LH), follicle-stimulating hormone (FSH), thyroid-stimulating hormone (TSH), platelet-activating factor, prostaglandins, as well as the retinal rod outer segment protein rhodopsin, which serves as a ''receptor'' for light and mediates a complex signal transduction mechanism that culminates in the visual response. Each member of this receptor family interacts within the plasma membrane with a specific member of another family of proteins, the guanine nucleotide-binding proteins (G proteins). The G protein family includes G_s, which mediates adenylyl cyclase stimulation; G_i, which mediates adenylyl cyclase inhibition; G_q, which mediates phospholipase C activation; and G_t (transducin), a protein of the rod outer segment which mediates activation of cyclic GMP phosphodiesterase, resulting in

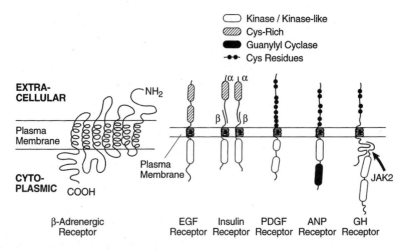

Fig. 3-3 Schematic representation of the proposed structures and plasma membrane orientations of the β-adrenergic, EGF, insulin, PDGF, ANP, and GH receptors. All six receptors are transmembrane proteins oriented so that their carboxy termini are in the cytoplasm. The β-adrenergic receptor crosses the plasma membrane seven times, whereas the other receptors cross the plasma membrane only once. The hatched areas represent regions of the EGF and insulin receptors enriched in cysteine residues. The open areas represent domains with a high degree of sequence similarity to tyrosine kinases of the *src* gene family. The filled-in region of the ANP receptor has homology to soluble guanylyl cyclase. Solid circles represent individual cysteine residues. The cytoplasmic domain of the GH receptor is shown in association with the Janus kinase JAK2, a cytoplasmic tyrosine kinase that associates with the ligand-bound receptor and mediates signal transduction.

the closure of Na^+ channels in the rod cell membranes to produce the visual response. All of the G proteins are heterotrimers composed of α-, β-, and γ-subunits. The α-subunit is unique to each G protein, whereas the β- and γ-subunits are similar. Some 15 different α-subunit proteins have now been characterized, as well as several related β- and γ-subunit proteins. The binding of the catecholamines epinephrine and norepinephrine to β-adrenergic receptors promotes interaction of the receptors with the stimulatory guanine nucleotide-binding protein of adenylyl cyclase (G_s). Interaction of the β-adrenergic receptor with G_s promotes the activation of adenylyl cyclase (see below).

Nicotinic Acetylcholine Receptors

In higher vertebrates, there are two basic subtypes of acetylcholine receptors, *muscarinic* and *nicotinic*. As mentioned above, muscarinic receptors are members of the seven-transmembrane domain receptor family, are localized in cells throughout the body, and interact with G proteins. Nicotinic receptors are present in the neuromuscular junctions. Acetylcholine, which is released from the presynaptic nerve endings upon depolarization of the neuron, binds to such receptors clustered in the motor endplates of skeletal muscle. The acetylcholine receptors of electric eels and electric fish are structurally and functionally homologous to nicotinic acetylcholine receptors and are probably the best characterized of the cell surface receptors because they are present in very high concentrations in the plasma membranes of electrocytes, which are specialized cells that develop from embryonic muscle cells and generate the electric potential that such species use to stun or kill their prey. The acetylcholine receptor is composed of five similar subunits each of 50,000–65,000 molecular weight, which are the products of four different but homologous genes: the five subunits surround an ion-conducting channel. The binding of acetylcholine to the receptor results in the opening of the ion-conducting channel created by a conformational change in the subunits of the receptor, which form the walls of the ion channel. The channel predominantly transports sodium ions, resulting in a depolarization of the plasma membrane of the postsynaptic cell.

Receptors with Intrinsic Protein Tyrosine Kinase Activity (Fig. 3-3)

EGF Receptor. The EGF receptor is a glycoprotein containing an intrinsic protein kinase that specifically catalyzes the phosphorylation of tyrosine residues on proteins. The tyrosine kinase activity of the receptor is stimulated upon the binding of EGF and is presumed to mediate some of its actions. The receptor is composed of a single polypeptide chain of 1186 amino acids. The protein can be divided into three domains: an N-terminal domain of 621 amino acids that contains the EGF binding site, a membrane-spanning domain of 26 hydrophobic amino acids, and a C-terminal cytoplasmic domain of 542 amino acids that shares sequence homology with other tyrosine-specific protein kinases. The N-terminal region contains many cysteine residues, which are clustered into two regions that may form an EGF-binding cleft. The EGF receptor is found to be overproduced in a number of tumor cell lines, suggesting that overexpression of the EGF receptor gene may contribute to the phenotype of cellular transformation. It is also of interest that the transmembrane and the cytoplasmic portion of the EGF receptor, which encodes the tyrosine kinase, have a very high degree of sequence homology with one of the transforming proteins of the avian erythroblastosis virus, the v-*erb-B* oncogene product. Viral oncogenes are derived from host cellular genes that are acquired during the course of infection. Apparently, the progenitor of the avian erythroblastosis virus acquired from host cells is the portion of the EGF receptor gene that encodes the transmembrane and cytoplasmic domain, but not the part that encodes the extracellular EGF-binding region. It has been suggested that the v-*erb-B* gene product induces cellular transformation because of the constitutive expression of the tyrosine kinase domain in the absence of expression of the regulatory EGF-binding domain.

Insulin Receptor. The polypeptide hormone, insulin, exerts a variety of metabolic and growth-promoting effects on its target cells that are initiated by its interaction with specific plasma membrane receptors. The insulin receptor is a high-molecular-weight glycoprotein that exhibits insulin-dependent, tyrosine-specific protein kinase activity. The receptor probably exists in the plasma membrane as a tetramer consisting of two disulfide-linked heterodimers $(\alpha\beta)_2$ (Fig. 3-3). The α- and β-subunits of the insulin receptor are synthesized as part of a single precursor polypeptide chain, which is subsequently glycosylated, proteolytically cleaved, and inserted into the plasma membrane. The α-subunit can be chemically cross-linked to ^{125}I-labeled insulin and, therefore, contains at least part of the hormone-binding site of the receptor. The β-subunits exhibit the insulin-dependent tyrosine kinase activity. As previously discussed, tyrosine protein kinase activity also is associated with the EGF receptor and with several other growth factor receptors, including those for insulin-like growth factor-I (IGF-I, closely related structurally to the insulin receptor), PDGF, the colony-stimulating factor CSF-1, fibroblast growth factors (FGF), HER-2/*neu* (closely related structurally to the EGF receptor), as well as a number of transforming retroviral oncogene products. Such findings are indicative of the role of tyrosine protein kinases in growth control (see Chapter 12). The α-subunit of the insulin receptor does not contain a hydrophobic membrane-spanning sequence and probably is localized exclusively on the outer face of the plasma membrane. Like the N-terminal extracellular domain of the EGF receptor, the α-chain is rich in cysteine residues.

Guanylyl Cyclase-Linked Receptors

Receptors for natriuretic peptides are membrane-associated guanylyl cyclases that are comprised of an extracellular ligand-binding domain, a single α-helical trans-membrane domain, and a cytosolic catalytic domain consisting of one region with homology to protein kinases and another with homology to soluble guanylyl cyclases and the catalytic domain of adenylyl cyclases (Fig. 3-3). Atrial natriuretic peptide, released from the atrium of the heart, binds to receptors on target cells, activates guanylyl cyclase, and promotes increased production of cyclic GMP. Members of this receptor family also include receptors for heat-stable enterotoxins from *E. coli* (guanylins) and sea urchin sperm receptors for a number of chemoattractant peptides produced by sea urchin eggs.

Growth Hormone and Prolactin Receptors Are Members of the Cytokine Receptor Superfamily

Receptors for growth hormone and prolactin belong to a large superfamily that includes receptors for the hematopoietic growth factors, which control the proliferation, differentiation, and activity of eight lineages of blood cells that are generated from an ancestral pool of stem cells. Hematopoietic growth factors include the interleukins (IL)-2-7, IL-9-11, colony-stimulating factors G-CSF and GM-CSF, leukemia-inhibitory factor (LIF), oncostatin M (OSM), erythropoietin, as well as ciliary neurotrophic factor (CNTF). The hematopoietic growth factors exert pleiotropic effects; i.e., most of the factors are active on cells of more than one lineage, and more than one factor controls growth and differentiation of cells in any one lineage. On the other hand, growth hormone has a variety of metabolic effects and stimulates skeletal growth through induction of IGF-I. Prolactin has numerous biological effects: in mammals, its best-known action is to stimulate growth and differentiation of the mammary gland, as well as lactogenesis. Interestingly, receptors for prolactin and growth hormone also are present on cells of the immune system. On the other hand, the CNTF receptor α-chain is restricted to the nervous system; CNTF supports the survival of chick embryonic ciliary ganglion cells and of sympathetic sensory and motor neurons.

The high-affinity binding forms of receptors in the cytokine superfamily are comprised of either a homodimeric complex of specific α-chains, or a heterodimeric complex of the specific α-chains with one or more β-subunits that also are members of this superfamily. The extracellular domains of these receptors share a number of common structural features, including two pairs of disulfide-linked cysteine residues in conserved positions (usually near the aminoterminal end, i.e., distal to the plasma membrane), and a WSXWS sequence (tryptophan, serine, any amino acid, followed by another tryptophan, serine) proximal to the plasma membrane. Receptors for the interferons (IFN)α/β, IFNγ, and IL-10 are structurally related to the cytokine receptors described above, but lack this WSXWS sequence in their extracellular domains.

Unlike the cytoplasmic domains of receptors for EGF, insulin, IGF-I, PDGF, and FGF, which have intrinsic protein tyrosine kinase activity, the cytoplasmic domains of members of the cytokine receptor superfamily lack homology to known protein kinases, and have no intrinsic kinase activity. Despite this absence of a kinase domain, binding of cytokines, growth hormone, and prolactin to their receptors is followed by a rapid increase in tyrosine phosphorylation of cellular proteins and of the receptor itself. Within the plasma membrane-proximal region of the cytoplasmic domain are two regions (box 1 and box 2) that are required for mediation of the growth-promoting effects of the cytokines. As will be discussed later in this chapter, signal transduction is mediated by the binding and activation of soluble, cytoplasmic tyrosine kinases (JAKs) to the membrane-proximal cytoplasmic region. A schematic of the growth hormone receptor and its associated kinase, JAK2, is shown in Fig. 3-3.

As mentioned above, in the case of all members of the cytokine receptor superfamily, high-affinity binding requires either homodimerization of the receptor α-chains (prolactin, growth hormone, G-CSF, and erythropoietin receptors), or heterodimerization with one or more β-subunits that also are members of this superfamily. The fact that some of these β-subunits are shared in common with different family members helps to explain the observed pleiotropic effects. As an example, the high-affinity binding forms of receptors for IL-3, IL-5, and GM-CSF are comprised of a heterodimeric complex of the specific α-chains with a common β-subunit. The α-chain receptors for IL-6, LIF/OSM, and CNTFR also share a common β-subunit (gp130).

Signal Transduction Mechanisms

It is generally accepted that polypeptide hormones and catecholamines exert their effects on cellular metabolism by binding to receptors on the surface of target cells and activating a membrane-associated enzyme that in turn elaborates an intracellular second messenger. This second messenger subsequently mediates the various biological effects of the hormone. This so-called second messenger hypothesis of hormone action was proposed in the early 1960s when it was discovered that the activation of glycogen phosphorylase by epinephrine and glucagon in liver slices was mediated by the formation of a heat-stable compound, which was identified as adenosine $3',5'$-monophosphate (cyclic AMP). Cyclic AMP is formed from $Mg^{2+} \cdot$ ATP by a membrane-associated enzyme, adenylyl cyclase. According to this concept, the hormone, or first messenger, carries information from its site of production to the target cell where it binds to specific receptors on the cell surface. This results in the activation of a membrane-bound enzyme or effector (e.g., adenylyl cyclase), which generates a soluble intracellular second messenger (e.g., cyclic AMP), which then transmits the information to the cellular machinery, resulting in a biological response. In recent years, a number of other effector/second messenger systems have been discovered that mediate the actions of a variety of hormones on cellular metabolism and function.

Hormone-Sensitive Adenylyl Cyclase

The hormone-sensitive adenylyl cyclase system has at least three components: the receptor (R_s or R_i), a form of guanine nucleotide-binding regulatory protein (G_s or G_i), and the catalytic component (C), which enzymatically converts $Mg^{2+} \cdot ATP$ to cyclic AMP (Fig. 3-4). The guanine nucleotide-binding regulatory protein, G_s, mediates the action of hormones that stimulate adenylyl cyclase activity, whereas G_i mediates the actions of those hormones that inhibit adenylyl cyclase. All of these components are integral plasma membrane proteins that can be independently isolated and recombined *in vitro* to reconstitute a functional hormone-sensitive adenylyl cyclase system. It is apparent that a number of different types of hormone receptors on a single cell can interact with the same pool of regulatory and catalytic components and that these combined interactions result in either a net stimulation or inhibition of adenylyl cyclase activity. G_s and G_i are heterotrimers comprised of a unique α-subunit (α_s or α_i) and similar β- and γ-subunits. The common $\beta\gamma$-subunits shared by G_s and G_i are essential for the integrated actions of stimulatory and inhibitory hormones on adenylyl cyclase activity. It has been proposed that the binding of hormone to receptor results in a conformational change that promotes the interaction or coupling of the hormone–receptor complex with the other components of hormone-sensitive adenylyl cyclase. This coupling may depend upon changes in

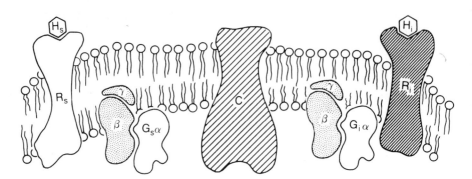

Fig. 3-4 Plasma membrane components of hormone-sensitive adenylyl cyclase. Hormone-sensitive adenylyl cyclase is composed of the following integral proteins of the plasma membrane: receptors for either stimulatory (R_s) or inhibitory (R_i) hormones, the guanine nucleotide-binding regulatory proteins G_s or G_i, and the catalytic component (C). G_s is composed of a unique α-subunit, $M_r \simeq 45,000$, and β- and γ-subunits, $M_r \simeq 35,000$ and $10,000$, respectively. G_i has a unique α-subunit, $M_r \simeq 41,000$, and β- and γ-subunits that are similar to those of G_s.

membrane fluidity that facilitate the lateral diffusion of the hormone–receptor complex through the plasma membrane.

Receptor-Mediated Stimulation of Adenylyl Cyclase. The proposed mechanism for the hormonal activation of adenylyl cyclase, which is based on the studies from a number of laboratories, is presented in Fig. 3-5. As discussed above, G_s has three subunits, α_s, β, and γ. In the inactive state, the guanine nucleotide that is bound to the α-subunit of G_s is GDP. The binding of hormone (H) to receptor (R) is believed to promote the formation of the ternary complex, $H \cdot R \cdot G_s$, which facilitates the dissociation of GDP and the binding of GTP to the $G_s\alpha$-subunit. The binding of GTP to $G_s\alpha$ results in the dissociation of the α_s from the $\beta\gamma$-subunits. The $G_s\alpha \cdot$ GTP then associates with the catalytic subunit (C) of the adenylyl cyclase to form the active holoenzyme ($G_s\alpha \cdot$ GTP $\cdot$ C). The activated catalytic subunit then converts $Mg^{2+} \cdot$ ATP to cyclic AMP. The activated $G_s\alpha$ contains a GTPase activity, which catalyzes the hydrolysis of GTP to GDP and terminates the cycle of adenylyl cyclase activation. Cholera toxin, which is produced by the bacterial organism *Vibrio cholera*, binds to gangliosides (complex glycolipids) on the cell surface and penetrates

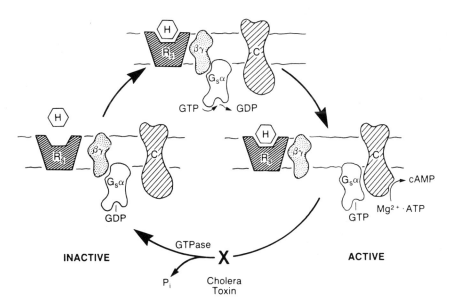

Fig. 3-5 Proposed mechanism for the hormonal activation of adenylyl cyclase, G_s is composed of three subunits, α_s, β, and γ (depicted here as $G_s\alpha$ and $\beta\gamma$). In the absence of the binding of a stimulatory hormone to its receptor, the guanine nucleotide GDP is bound to $G_s\alpha$ and adenylyl cyclase is in the inactive state. The binding of hormone to receptor facilitates the dissociation of GDP and the association of GTP with the α-subunit of G_s. This, in turn, results in the dissociation of $G_s\alpha$ from the $\beta\gamma$-subunits and its association with the catalytic component (C) resulting in adenylyl cyclase activation. The activated $G_s\alpha$ contains a GTPase, which hydrolyzes the bound GTP to GDP resulting in a return of adenylyl cyclase to an inactive state. Cholera toxin, which inhibits the GTPase activity, causes a persistent activation of adenylyl cyclase.

the cell membrane. Once within the cell membrane, the toxin catalyzes the ADP-ribosylation of an arginine residue (Arg201) in $G_s\alpha$, causing inhibition of the GTPase activity (Fig. 3-5). This change results in a persistent activation of adenylyl cyclase. In intestinal mucosal cells, the binding of cholera toxin and subsequent activation of adenylyl cyclase results in the stimulation of ion (primarily Cl^-) and water secretion across the intestinal brush border, causing massive diarrhea.

Receptor-Mediated Inhibition of Adenylyl Cyclase. A number of hormones inhibit adenylyl cyclase activity. They include catecholamines that bind to α_2-adrenergic receptors, muscarinic–cholinergic agonists, and opioids. These hormones bind to cell surface receptors that interact with the inhibitory guanine nucleotide-binding regulatory protein, G_i. G_i is similar in structure to G_s, having three subunits, α_i, β, and γ. The $\beta\gamma$-subunits of G_i are similar to those of G_s, whereas the α-subunit is distinct. $G_i\alpha$ contains a guanine nucleotide-binding site. The binding of these hormones to their receptors promotes the exchange of GTP for GDP on $G_i\alpha$. This results in the dissociation of α_i from $\beta\gamma$. The inhibition of adenylyl cyclase activity appears to be mediated primarily by the interaction of the free $\beta\gamma$-subunits of G_i with the α-subunit of G_s, reducing the concentration of free $G_s\alpha$. A direct interaction of $\alpha_i \cdot$ GTP with adenylyl cyclase also appears to play a role in its inhibition. Islet-activating protein, one of the toxins of *Bordetella pertussis*, prevents the dissociation of G_i, which results in adenylyl cyclase activation, since less free $\beta\gamma$ is available to interact with $G_s\alpha$. Thus, cholera toxin activates adenylyl cyclase by promoting the dissociation of G_s, whereas islet-activating protein activates adenylyl cyclase by inhibiting the dissociation of G_i.

Cyclic AMP Regulation of Cellular Function

The major mechanism by which cyclic AMP regulates cellular function is through its binding to cyclic AMP-dependent protein kinases (PK-A). The cyclic AMP-dependent protein kinase holoenzyme is comprised of a regulatory subunit (R) dimer and two catalytic subunits (C). The R subunit is a cyclic AMP-binding protein, while the C subunit, when free of R, expresses protein kinase activity. Each mole of R dimer binds 4 mol of cyclic AMP. The binding of cyclic AMP to R causes dissociation of the inactive holoenzyme to yield two active catalytic subunits. The reaction is generally represented as

$$R_2C_2 + 4 \text{ cAMP} \rightarrow R_2 \cdot 4 \text{ cAMP} + 2 \text{ C}$$
$$\text{(inactive)} \qquad\qquad\qquad \text{(active)}$$

The active free catalytic subunits can now catalyze the phosphorylation of a number of cellular proteins. The protein kinase catalyzes the transfer of a high-energy phosphate from ATP to the hydroxyl groups of serine and, to a much lesser extent, threonine residues on cellular proteins. The phosphorylation of enzymes may result in changes in enzyme activity. For example, phosphorylation of hormone-sensitive lipase, cholesteryl esterase, or glycogen phosphorylase results in enzyme activation. On the other hand, phosphorylation of glycogen synthase decreases enzyme activity. The specific responses of various cell types to an increase in cyclic AMP and the activation of cyclic AMP-dependent protein kinase is determined by the cellular phenotype and, therefore, by the enzymes and substrates available for regulation. Thus, a major response of a liver cell to an increase in cyclic AMP is glycogenolysis, because the liver cell expresses the enzymes to synthesize and to metabolize glycogen. In a fat cell, on the other hand, the primary response to an increase in cyclic AMP is lipolysis, because the fat cell expresses the enzymes for uptake of triglyceride precursors from the circulation, and for the synthesis and metabolism of triglycerides.

In addition to its PK-A-mediated actions to alter the activities of various metabolic enzymes, cyclic AMP also regulates the transcription of specific genes in eukaryotic cells. An effect of cyclic AMP at the level of gene transcription has been found for a number of eukaryotic genes, including phospho*enol*pyruvate carboxykinase (PEPCK), tyrosine aminotransferase (TAT), the human glycoprotein hormone α-subunit gene, preprosomatostatin, vasoactive intestinal polypeptide (VIP), and several forms of cytochrome *P*-450 involved in steroid hydroxylation.

Our understanding of the molecular mechanisms whereby cyclic AMP activates gene transcription in mammalian cells has been greatly advanced by the discovery of a DNA-binding protein or transcription factor, termed *cyclic AMP response element binding protein* (CREB), which is activated by cyclic AMP to regulate the transcriptional activity of a number of specific genes. CREB is a member of a family of related DNA-binding proteins, the activating transcription factors (ATF proteins), that bind to a common sequence of DNA, TGACGTCA, which is palindromic; that is, it possesses a 2-fold rotational axis of symmetry. Increased cellular levels of cyclic AMP result in an activation of CREB through its phosphorylation on serine residues by PK-A. The activated CREB appears to bind as a dimer to the palindromic cyclic AMP response element (CRE), TGACGTCA. The binding of activated CREB to the CRE results in the transcriptional activation of a number of eukaryotic genes, including those encoding PEPCK, VIP, TAT, and the human glycoprotein hormone α-subunit gene, to name a few. It should be noted, however, that the CRE does not appear to be present in the regulatory regions of all genes that are regulated by cyclic AMP. Therefore, additional cyclic AMP-responsive transcription factors and mechanisms may mediate AMP regulation of eukaryotic gene transcription.

Cyclic AMP is rapidly metabolized in cells by the cyclic nucleotide phosphodiesterases, which hydrolyze cyclic AMP to form the inactive 5'-AMP.

$$\text{cyclic AMP} \xrightarrow{\hspace{1cm} \text{phosphodiesterase} \hspace{1cm}} \text{5'-AMP}$$

The actions of a number of hormones, including insulin and catecholamines binding to α_1-adrenergic receptors, are mediated in part by an activation of phosphodiesterase and a subsequent decrease in cellular cyclic AMP levels.

Hormonal Induction of Phospholipid Turnover

A significant number of polypeptide hormones exert their actions on cellular metabolism and function by mechanisms that do not involve adenylyl cyclase activation and cyclic AMP. Table 3-1 lists hormones that act through cyclic AMP, as well as those that do not. Hormones that act through cyclic AMP-mediated mechanisms include adrenocorticotropic hormone (ACTH), glucagon, and the pituitary glycoprotein hormones, luteinizing hormone (LH), follicle-stimulating hormone (FSH), and

Table 3-1   Signal Transduction Mechanisms in Hormone Action

Hormones that act through cyclic AMP-mediated mechanisms	Hormones that do not act through cyclic AMP-mediated mechanisms
Epinephrine and norepinephrine (β-receptors)	Epinephrine and norepinephrine (α_1, α_2-receptors)
Glycoprotein hormones (LH, FSH, TSH, hCG)	Opioid peptides
Glucagon	Acetylcholine
ACTH	Insulin
Vasopressin (V_2 receptors)	Vasopressin (V_1 receptors)
	Growth factors (EGF, FGF, PDGF, IGF-I)
	Growth hormone

thyroid-stimulating hormone (TSH). The glycoprotein hormone of human placental origin, human chorionic gonadotropin (hCG), which is highly homologous to LH and which binds to LH receptors, also increases cyclic AMP formation.

Other hormones, such as insulin, growth hormone, various growth factors, and hypothalamic releasing hormones, act through mechanisms that are independent of cyclic AMP. Certain hormones may act through a cyclic AMP-mediated mechanism in one tissue and by a cyclic AMP-independent mechanism in another. For example, vasopressin binds to a specific subset of receptors (V_2) on cells of the kidney collecting tubules and loop of Henle to promote sodium and water reabsorption. These actions of vasopressin are mediated by increased cyclic AMP formation and cyclic AMP-dependent protein kinase activation. On the other hand, in the liver, vasopressin acts through another subset of receptors (V_1) to enhance glycogenolysis, and this effect is mediated by a cyclic AMP-independent pathway. The catecholamines, epinephrine and norepinephrine, bind to several subsets of receptors that are present in different relative amounts in various tissues. Binding of catecholamines to β-adrenergic receptors activates adenylyl cyclase and increases cyclic AMP formation. Interaction of catecholamines with α_2-adrenergic receptors, on the other hand, inhibits adenylyl cyclase, while binding to α_1-adrenergic receptors increases inositol phospholipid turnover with a resulting increase in the levels of free cytosolic calcium ion and an activation of protein kinase C. In liver cells, norepinephrine binds both to β- and to α_1-adrenergic receptors to increase glycogenolysis. Binding to β-receptors results in adenylyl cyclase activation, an increase in cyclic AMP, and activation of cyclic AMP-dependent protein kinase, which in turn catalyzes the phosphorylation and activation of phosphorylase kinase. The activated phosphorylase kinase catalyzes the phosphorylation and activation of phosphorylase, resulting in enhanced glycogenolysis. Phosphorylase kinase also catalyzes the phosphorylation of glycogen synthase, resulting in its inactivation. Norepinephrine, acting through α_1-receptors, promotes an increase in the levels of free cytosolic calcium ion, which causes, by mechanisms discussed below, the activation of phosphorylase kinase and the subsequent increase in glycogen breakdown.

Inositol Trisphosphate. As previously mentioned, a number of hormone–receptor interactions increase free cytosolic calcium ion and activate protein kinase C, which serve important roles in the regulation of cellular function. Hormone–receptor interactions that result in the formation of these second messengers include the binding of acetylcholine to muscarinic receptors, epinephrine and norepinephrine, to α_1-adrenergic receptors, vasopressin to V_1 receptors, and angiotensin II to receptors on liver cells. The proposed mechanism of action of such hormones is presented in Fig. 3-6. The binding of hormone to receptor results in the rapid activation of a plasma membrane-associated phospholipase C (PLC), which catalyzes the hydrolysis of a specific inositol phospholipid within the plasma membrane, phosphatidylinositol 4,5-bisphosphate (PIP_2), to form the second messengers diacylglycerol (DG) and inositol 1,4,5-trisphosphate (IP_3). At least three phospholipase C proteins have been characterized, PLC-β, PLC-γ, and PLC-δ. It appears that the specific isoform of this enzyme that is involved in catalyzing the hydrolysis of PIP_2 is PLC-β. The hormonal activation of PLC-β appears to be mediated by the interaction of the receptor with a specific G protein (G_q).

The hydrolysis of PIP_2 and the formation of IP_3 are specifically associated with an increase in the levels of free cytosolic calcium ion and the subsequent physiological response. Furthermore, incubation of permeabilized cells with IP_3 results in a profound increase in the release of calcium ion from intracellular stores, primarily the endoplasmic reticulum. An endoplasmic reticulum-associated receptor for IP_3 has recently been cloned and characterized. The IP_3 receptor has an apparent molecular weight of 260,000 and contains at least four membrane-spanning domains. It is postulated that these membrane-spanning domains form the calcium channel. In addition to mediating IP_3-stimulated calcium release from the endoplasmic reticulum, this receptor may act in conjunction with the dihydropyridine (DHP)-gated calcium channels to facilitate the influx of extracellular calcium into the cell. Once formed, IP_3 is rapidly hydrolyzed to IP_2, IP, and inositol by the actions of specific phosphomonoesterases. The action of the esterase that hydrolyzes IP to inositol is inhibited by lithium ion.

Calcium as a Second Messenger. The levels of free calcium ion in the cytoplasm are normally quite low ($\approx 10^{-7}$ M) as compared to the levels of calcium ion in the extracellular fluid ($\approx 10^{-3}$ M). The influx of calcium into most cells at rest is minimal, despite a large electrochemical gradient, since the plasma membranes of most cells at rest are relatively impermeable to calcium ion. Within the cell, calcium is stored in relatively high concentrations in the mitochondrial membrane, the endoplasmic reticulum (sarcoplasmic reticulum of muscle cells), and the plasma membrane. An increase in the levels of free cytosolic calcium ion can have a variety of effects on the cell, including changes in cell motility, contraction of muscle cells, increased release of secretory proteins, and activation of a number of regulatory enzymes. Calcium ion exerts most of these effects in cells by binding to specific calcium-binding proteins, such as *calmodulin*, in nonmuscle cells, and *troponin C* in striated muscle cells. The binding of calcium results in the activation of these calcium-binding proteins. Calmodulin has no intrinsic activity of its own; after binding calcium ion (each molecule of calmodulin has four calcium-binding sites), it is

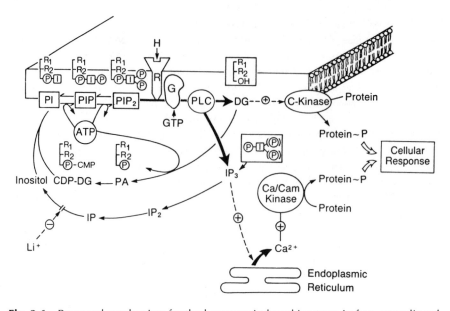

Fig. 3-6 Proposed mechanism for the hormone-induced increase in free cytosolic calcium and activation of protein kinase C. The polyphosphoinositides phosphatidylinositol 4-phosphate (PIP) and phosphatidylinositol 4,5-bisphosphate (PIP_2) are formed within the plasma membrane from phosphatidylinositol (PI) and ATP in reactions catalyzed by kinases. The binding of hormone to receptor results in the activation of a specific phospholipase C (PLC) within the plasma membrane, which catalyzes the hydrolysis of PIP_2 to form the putative second messengers, inositol trisphosphate (IP_3) and diacylglycerol (DG). The hormone–receptor activation of the PLC is believed to be mediated by a guanine nucleotide-binding regulatory protein (G). The water-soluble IP_3 diffuses into the cytoplasm and stimulates the release of calcium from storage sites, primarily within the endoplasmic reticulum. The increased free cytosolic calcium exerts most of its effects on cellular metabolism by binding to calmodulin (Cam). The calcium–calmodulin complex binds to various enzyme or effector proteins, causing changes in their activities. One such protein is phosphorylase kinase, which is activated by an increase in cytoplasmic calcium ion. The IP_3 is subsequently hydrolyzed by specific phosphatases to form the inactive IP_2, IP, and inositol. The DG remains within the plasma membrane, where it facilitates the activation of protein kinase C (C-Kinase) by calcium ion and phospholipid. The DG and inositol are utilized for the resynthesis of PI.

activated and binds to various enzymes or effector molecules, causing a change in their activities. Two enzymes that are activated by the calcium–calmodulin complex are phosphorylase kinase and cyclic AMP phosphodiesterase. The activation of phosphorylase kinase by cyclic AMP-dependent protein kinase is dependent upon calcium. Phosphorylase kinase has four subunits, α, β, γ, and δ. The δ-subunit is calmodulin, and the γ-subunit is the catalytic component of the enzyme. The α- and β-subunits are phosphorylated by cyclic AMP-dependent protein kinase. The phosphorylation of the α- and β-subunits in the presence of calcium ion activates the γ-

subunit and increases the affinity of the δ-subunit (calmodulin) for calcium. In the presence of calcium ion, the enzyme binds a second molecule of calmodulin (δ'). This in turn activates the dephosphorylated form of the enzyme but has little effect to promote the further activation of the phosphorylated enzyme. Phosphorylase kinase is an example of an enzyme that is activated by an increase either in intracellular calcium ion or cyclic AMP, or both. This provides an example of a system in which the actions of cyclic AMP and calcium occur in the same direction. There are other examples in which the actions of calcium antagonize those of cyclic AMP; cyclic AMP phosphodiesterase, the enzyme that catalyzes the metabolism of cyclic AMP to the inactive $5'$-AMP, is activated by the calcium–calmodulin complex. An increase in free cytosolic calcium ion, therefore, can cause a decrease in the levels of cyclic AMP and of cyclic AMP-dependent protein kinase activity and a change in the activities of the enzymes that are substrates for this kinase.

Protein Kinase C. The other product of the hydrolysis of inositol phospholipids, diacylglycerol, is also believed to serve as a second messenger by acting within the cell membrane to activate protein kinase C. Protein kinase C is a phospholipid- and calcium-dependent enzyme that catalyzes the phosphorylation of serine and threonine residues on a number of cellular proteins. Diacylglycerol dramatically increases the affinity of the enzyme for calcium ion and for phospholipid and therefore promotes an increase in enzyme activity at resting levels of intracellular calcium. The diacylglycerol-mediated hormonal activation of protein kinase C can be mimicked by incubating cells with tumor-promoting phorbol esters, which interact with the enzyme at the same site as diacylglycerol. Since phorbol esters are not rapidly degraded, these agents cause long-term activation of protein kinase C.

In addition to phosphorylating serine and threonine residues on a number of enzyme proteins, protein kinase C can alter the expression of a number of specific genes. This action of protein kinase C is mediated by the activation of a DNA-binding protein complex or transcription factor known as *AP-1*. AP-1 is a heterodimer of two related proteins, c-*fos* and c-*jun*, which are recognized as protooncogenes. AP-1 binds to a specific DNA sequence $TGAC/_GTCA$, which is quite similar to that recognized by the cyclic AMP-activated transcription factor CREB. In contrast to CREB, which is activated by increased phosphorylation, AP-1 binding and transcriptional activity appear to be increased by dephosphorylation of constiutively phosphorylated amino acid residues near the carboxyterminus and phosphorylation of other residues near the aminoterminus. Thus, it is likely that activated protein kinase C activates both a specific phosphatase and a specific kinase.

It is thus apparent that the hormonal activation of phosphatidyinositol turnover results in the elaboration of two important mediators, IP_3 and diacylglycerol, which in turn promote an increase in intracellular calcium and activation of protein kinase C, respectively. The result is a variety of cellular responses that are dependent upon cellular phenotype and include alterations in enzyme activity and in transcription factor activation.

Phosphatidylcholine Turnover. Although numerous hormones and neurotransmitters exert their effects through PIP_2 hydrolysis, a significant number have been found to stimulate the hydrolysis of phosphatidylcholine (PC). Phospholipase C-mediated hydrolysis of PC yields diacylglycerol and phosphorylcholine. The diacylglycerol that is formed has the capacity to activate protein kinase C as described above. Since PC is present in much greater abundance in cell membranes than is PIP_2, the former may serve as an important source of diacylglycerol for long-term activation of protein kinase C. Another phospholipase, phospholipase D, has been implicated in the actions of a number of hormones that act through G protein linked receptors. In addition to stimulating phospholipase C-mediated PIP_2 hydrolysis, the binding of vasopressin to V_1 receptors on hepatocytes and of acetylcholine to muscarinic receptors in the brain promotes the activation of phospholipase D, which catalyzes the hydrolysis of PC to form choline and phosphatidate. Phosphatidate may serve as a calcium ionophore by promoting calcium entry into cells and the mobilization of intracellular calcium from membrane stores. The phosphatidate also may be hydrolyzed by phosphatidate phosphohydrolase to form diacylglycerol, which in turn can activate protein kinase C.

The Sphingomyelin Cycle—Ceramide as a Second Messenger. Sphingomyelins are the only phospholipids in cell membranes that do not have a glycerol backbone. Rather, the backbone of sphingomyelin is sphingosine, an amino alcohol that contains an 18 carbon chain with one double bond. The amino group of sphingosine is linked to a long-chain fatty acid by an amide bond, and the primary hydroxyl group is in ester linkage with phosphocholine (Fig. 3-7). Although it once was believed that sphingomyelin served merely as a structural component of plasma membranes and myelin sheaths of nerves, it recently has become apparent that sphingomyelin is actively metabolized by a cytoplasmic neutral *sphingomyelinase* and serves as a reservoir of second messengers in the action of a number of hormones and cytokines. Sphingomyelinase-induced hydrolysis of sphingomyelin to form ceramide and phosphocholine mediates the action of tumor necrosis factor-α (TNF-α) to stimulate apoptosis (programmed cell death) of lymphoid and myeloid cells. The ceramide that is formed causes the activation of a ceramide-activated protein phosphatase (CAPP), a serine–threonine phosphatase related to protein phosphatases 2A. The finding that exogenously added ceramide activates CAPP and mimics the action of TNF-α to induce apoptosis, and that the phosphatase inhibitor okadaic acid inhibits TNF-α action to stimulate apoptosis, supports the role of CAPP as a mediator of TNF-α action. Ceramide action also may be mediated in part by ceramide-activated protein kinases; however, the role of these kinases in mediating the biological effects of ceramide is unclear.

The active form of vitamin D_3, 1,25-dihydroxyvitamin D_3, promotes differentiation of HL60 leukemia cells. Sphingomyelin hydrolysis and ceramide generation are among the earliest biochemical effects of 1,25-dihydroxyvitamin D_3 on these cells and appear to mediate some of the effects of the vitamin to promote cellular differentiation. The effect of 1,25-dihydroxyvitamin D_3 to stimulate sphingomyelin hydrolysis and ceramide formation is apparently initiated at the plasma membrane and does not involve its nuclear receptor, a member of the steroid receptor superfamily, which will be discussed later in this chapter. It therefore appears that ceramide blocks cell proliferation and promotes cell differentiation and apoptosis, as opposed to diacylglycerol, which is formed by the action of phospholipase C on

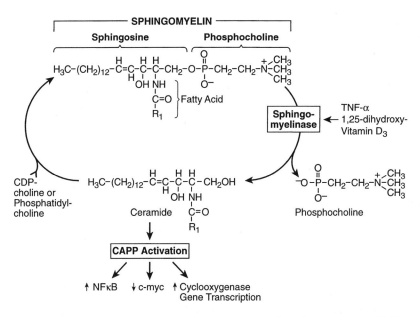

Fig. 3-7 The sphingomyelin cycle. Upon binding to cell surface receptors, TNF-α activates a neutral sphingomyelinase by unknown coupling mechanisms. The sphinomyelinase hydrolyzes sphingomyelin to form ceramide and phosphocholine. The ceramide causes activation of a ceramide-activated protein phosphatase (CAPP), a serine–threonine phosphatase, which alters the activities and/or levels of a number of cellular proteins and transcription factors (e.g., an induction of cyclooxygenase gene expression resulting in an increase in prostaglandin biosynthesis, an activation of the transcription factor nuclear factor κB [NFκB], and decreased activity of the transcription factor c-myc). The ceramide can then react with cytidine diphosphocholine (CDP-choline) or with phosphatidylcholine to form sphingomyelin and complete the cycle.

membrane glycerophospholipids, and which activates protein kinase C and induces cell proliferation. It has been suggested that ceramide may function as a tumor suppressor lipid, as compared to diacylglycerol which may function to stimulate cell proliferation and tumorigenesis.

Signal Transduction by Receptors That Contain Tyrosine Kinase Domains
As discussed above, cell surface receptors for insulin and other growth factors are transmembrane proteins that contain a cytoplasmic domain that is a tyrosine kinase. More than 50 receptor tyrosine kinases have been identified which belong to 14 different families. Binding of ligands to these receptors and subsequent activation of the tyrosine kinases results in changes in cell proliferation, differentiation, shape, and migration. Listed below are four of the families of receptor tyrosine kinases.

1. *The EGF, HER-2/*neu *family:* the structure of the EGF receptor has already been described. HER-2/*neu*, which is homologous to the EGF receptor and to the *neu* oncogene, has been found to be amplified in many adenocarcinomas and is overexpressed in breast tumors of a significant number of

breast cancer patients. The identification of a ligand for this receptor, produced by human breast cancer cells in culture, suggests a role of HER-2/ *neu* in autocrine growth regulation of breast tumors.

2. *The insulin/IGF-I receptor family:* like the insulin receptor, discussed above, the IGF-I receptor is a tetramer consisting of two disulfide-linked heterodimers $(\alpha\beta)_2$. The IGF-I receptor differs from the insulin receptor in its extracellular cysteine-rich region, which is believed to be responsible for ligand binding, and in the C-terminus downstream of the tyrosine kinase domain.

3. *The PDGF, CSF-1, and c-kit receptor family:* PDGF is released from platelets when they adhere to injured vessels and acts to stimulate proliferation of mesenchymal cells. CSF-1 acts on hematopoietic precursor cells and promotes their differentiation into monocytes and macrophages. The c-*kit* protooncogene appears to encode a receptor for a mast cell growth factor. The extracellular regions of these receptors contain 5 repeats of an immunoglobulin-like domain containing cysteine residues. In contrast to the EGF and insulin receptors, the tyrosine kinase domains of this group of receptors contain a 70–100 amino acid insertion (Fig. 3-3). PDGF and CSF-1 are disulfide-linked heterodimers. PDGF-induced activation of its receptor involves receptor dimerization and interaction with specific cytoplasmic effector molecules.

4. *The FGF receptor family:* This family includes the receptors of acidic and basic FGF, as well as a transmembrane protein encoded by the *int*-2 protooncogene. FGFs appear to serve an important role in wound healing and angiogenesis. The extracellular domain of these receptors contains three immunoglobulin-like domains with two cysteines in each, like the PDGF receptor family. The tyrosine kinase domain also is divided into two regions by an inserted stretch of only 14 amino acids.

Most, if not all of the actions of the tyrosine kinase family of receptors are mediated by ligand-induced receptor dimerization/oligomerization, tyrosine kinase activation, and autophosphorylation. With the exception of the insulin receptor (which already exists as a dimer of two disulfide-linked α,β-heterodimers [see above]), all receptor tyrosine kinases undergo dimerization upon ligand binding. Receptor dimerization serves an essential role in tyrosine kinase activation, autophosphorylation, and transmembrane signaling. The tyrosine kinase of one receptor polypeptide chain catalyzes the phosphorylation of tyrosines on the other receptor chain in the dimer. Autophosphorylation of the receptor on tyrosine residues increases tyrosine kinase activity and enhances its capacity to phosphorylate other cellular target proteins. This occurs because the phosphorylated tyrosines on the receptor serve as binding or docking sites for target proteins that contain a sequence of amino acids referred to as a *src-homology 2* (SH2) domain. The SH2 domain is so named because of its similarity to a region present in Src, a cellular protooncogene product involved in cell growth control. Src, which lacks an extracellular ligand-binding domain, is activated by binding to receptor tyrosine kinases via its SH2 region. The binding of Src or other cellular target proteins to activated receptor tyrosine kinases through their SH2 domains orients these proteins for phosphorylation by the activated receptor tyrosine kinase.

Many of the SH2-containing cellular proteins that bind to activated receptor tyrosine kinases are components of downstream signaling pathways which involve other second messenger systems. In this manner, the tyrosine kinase receptor family interacts with other signal transduction systems. As an example, PDGF binding to its receptor results in an increase in phosphatidylinositol turnover and in cytosolic calcium. This action of PDGF appears to be mediated in part by the association and activation of a specific isoform of phospholipase C, PLC-γ, which interacts via its SH2 domain with the phosphotyrosines in the cytoplasmic domain of the activated PDGF receptor and consequently, is phosphorylated on tyrosine residues. The activated phospholipase C, in turn, catalyzes the hydrolysis of PIP_2 resulting in the elaboration of IP_3 with associated increase in cytoplasmic calcium. PDGF activation of phospholipase C also catalyzes the hydrolysis of PC. It has been postulated that increased diacylglycerol formed through PC hydrolysis and the associated protein kinase C activation provides an important component of the mitogenic action of PDGF. Autophosphorylation sites of the activated, PDGF receptor also interact with the adaptor protein, growth factor receptor-bound protein 2 (Grb2), which then recruits guanine nucleotide exchange factors that promote the binding of GTP to Ras, which in turn binds and activates the serine–threonine protein kinase Raf. This ultimately results in the activation of mitogen-activated protein kinases (MAP kinases), which phosphorylate a number of cellular proteins, including the transcription factor AP-1.

It is therefore apparent that the tyrosine kinase family of cell surface receptors acts in parallel as well as in a cross-talking manner with other receptor-regulated effector systems that involve G proteins. Some of these interactions are additive or synergistic, whereas others are antagonistic. The net result is the integrated control of cellular function, homeostasis, and growth.

Signal Transduction through Guanylyl Cyclase Receptors

In the early 1970s, there was a great deal of excitement concerning the possibility that cyclic GMP might serve as a second messenger that mediates the actions of hormones that are antagonistic to those acting through cyclic AMP. Cyclic GMP is formed from GTP in a reaction that is catalyzed by guanylyl cyclase. In contrast to adenylyl cyclase, which is exclusively localized to the plasma membrane, two forms of guanylyl cyclase have been well characterized. One is a heme-containing soluble enzyme comprised of two subunits and the other is a plasma membrane-associated protein. A third form of guanylyl cyclase, which is associated with structural elements, is present in retinal rod cells and is involved in the synthesis of cyclic GMP involved in the visual cycle. The plasma membrane-associated form was intensively studied since atrial natriuretic peptides were found to activate guanylyl cyclase and promote increased synthesis of cyclic GMP in target cells. The actions of ANP, which include vasodilation, natriuresis, and diuresis (see also Chapter 7), appear to be mediated either entirely or in part by the binding to and activation of the ANP receptor, a membrane-bound guanylyl cyclase (Fig. 3-3) resulting in increased production of the second messenger, cyclic GMP. Oligomerization of the receptor appears to be required for guanylyl cyclase activation, as is binding of ATP to the protein kinase homology domain. The mechanism(s) whereby cyclic GMP acts to mediate the actions of ANP have yet to be elucidated; however, it is apparent from

studies in retinal rod cells that cyclic GMP serves to maintain cation channels in an open conformation. This action may, in part, explain the natriuretic action of ANP. Alternatively, this natriuretic action of ANP may be mediated by its effect to inhibit aldosterone biosynthesis by glomerulosa cells of the adrenal cortex. In addition, cyclic GMP may function to activate a phosphodiesterase, which, in turn, lowers cyclic AMP levels. In this manner, cyclic GMP may act to antagonize the actions of cyclic AMP in a number of cell types. It is apparent that much remains to be learned of the mechanisms whereby cyclic GMP regulates cellular function.

Signal Transduction by Members of the Cytokine Receptor Family

As discussed above, receptors for growth hormone, prolactin, and other members of the cytokine receptor superfamily do not have kinase domains. Furthermore, the cytoplasmic domains of superfamily members have only limited sequence similarity; this is confined to a membrane proximal region (box 1 and box 2 motifs), which is functionally required for mitogenic responses. In spite of the absence of a kinase domain, ligand binding results in the rapid phosphorylation of cellular proteins, as well as the receptor itself, on tyrosine residues. The box 1 and box 2 regions of the receptor are required for coupling of ligand binding to tyrosine phosphorylation. It has recently been found that tyrosine phosphorylation is mediated by a family of cytoplasmic tyrosine kinases, the Janus (JAK) kinases, which physically associate with the box1–box2 domains of the ligand-bound receptor, leading to auto-

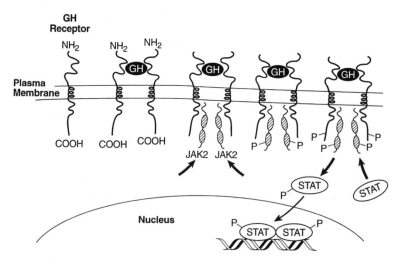

Fig. 3-8 Signal transduction mechanisms associated with binding of growth hormone (GH) to its receptor. The binding of a single molecule of GH promotes receptor dimerization. This in turn results in an increased affinity of the membrane-proximal region of the cytoplasmic domain of the receptor for the cytoplasmic tyrosine kinase, JAK2, a member of the Janus kinase family, which undergoes autophosphorylation and further activation. The activated JAK2 catalyzes phosphorylation of the receptor, as well as other cellular proteins including a STAT (signal transducers and activators of transcription) transcription factor, which is activated and binds as a dimeric complex to a specific enhancer element in the regulatory region of a GH-inducible gene, i.e., IGF-I, resulting in increased transcription initiation.

phosphorylation on tyrosine residues, further activation of the JAK kinase, and phosphorylation of the receptor, as well as other cellular proteins that are recruited to the receptor–kinase complex. The major cellular targets for JAK kinases are members of a family of transcription factors called *signal transducers and activators of transcription* (STATs). Phosphorylation of STATs on tyrosine residues results in their activation and binding to specific response elements within the regulatory regions of genes. It is apparent that receptor dimerization or oligomerization, which occurs as a result of ligand binding, is essential to initiation of the signal transduction cascade. The receptor oligomerization causes an increased affinity for JAKs and results in their increased concentration, facilitating crossphosphorylation of their autophosphorylation sites. A hypothetical model of the signal transduction pathway mediated by the binding of growth hormone to its receptor is represented in schematic form in Fig. 3-8.

To date, four JAK kinase family members have been identified: JAK1, JAK2, JAK3, and tyk2; these have specificity for different ligand-receptor complexes. For example, JAK2 has been found associated with activated growth hormone, prolactin, and erythropoietin receptors; activated interferon-α/β receptors associate with JAK1 and tyk2; interferon-γ receptors associate with JAKs 1 and 2; and activated interleukin-2 receptors associate with JAK 3. Further complexity results from the identification of at least four different STAT proteins which appear to be substrates for the different JAKs. It should also be noted that binding of EGF and PDGF to their receptors, which have intrinsic tyrosine kinase activity and are autophosphorylated, promotes tyrosine phosphorylation of JAK1 and activation of a STAT1 transcription factor that binds as part of a complex to an enhancer in the c-*fos* promoter.

STEROID HORMONES

Steroid hormones exert long-term effects on their target cells; they stimulate cell growth and differentiation and regulate the synthesis of specific proteins. Steroid hormones exert these actions on target cells after binding to specific receptors, which are localized primarily within the nucleus. The steroid–receptor complex regulates the synthesis of specific proteins primarily by altering the rate of transcription of specific genes. The first clear indication that steroid hormones could interact with specific genes and alter their activity was the finding in 1960 that the insect steroid ecdysone rapidly induced the formation of puffs in specific regions of the salivary gland polytene chromosomes of the midge *Chironomus*. It has subsequently been shown that the puffs correspond to active genes; several of these genes have been found to encode salivary proteins. Although the ecdysteroids were the first of the steroid hormones shown to regulate gene activity, studies of the receptors and mechanisms of action of steroid hormones in vertebrates rapidly progressed and outpaced those concerning the ecdysteroids. This was primarily due to the synthesis of radiolabeled vertebrate steroid hormones of high specific activity and the subsequent discovery of receptor proteins in cells of vertebrates.

Over the past two decades, studies from a large number of laboratories have led to a generally accepted hypothesis for the mechanism of action of steroid hormones (Fig. 3-9). Steroids travel in the circulation predominantly bound to several classes of serum proteins. Estrogens and androgens are transported in the circulation

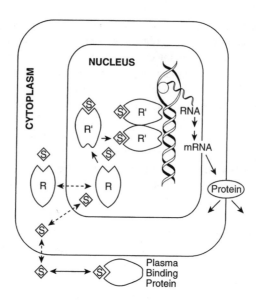

Fig. 3-9 Schematic representation of a steroid hormone responsive cell. Steroid hormones (S) in the circulation are predominantly bound to specific proteins. Only a small proportion of the circulating steroid is free; it is the unbound steroid that has the capacity to enter cells by free diffusion. Within target cells, the steroids bind with high affinity to specific receptors that are primarily localized within the nucleus. After binding steroid, the receptor (R) undergoes a process of "transformation," which results in an increase in the affinity of the receptor for specific sequences of DNA proximate to steroid-regulated genes. The binding of the receptor as a dimer to these genomic regions can result in increases or decreases in the rates of gene expression. The steroid–receptor complex is believed to increase the rate of transcription of specific genes by causing the formation of a stable preinitiation complex and facilitating the binding of RNA polymerase II to the promoter regions of such genes. This results in the synthesis of RNA molecules that are processed within the nucleus to mRNAs that enter the cytoplasm and are translated into specific proteins.

bound to testosterone-binding globulin (TeBG), which binds estradiol-17β and testosterone with relatively high affinity ($K_d \simeq 10^{-9}\ M$). These steroids are also weakly bound to serum albumin. Glucocorticoids and progesterone are bound in the circulation to corticosteroid-binding globulin (CBG), also referred to as transcortin. It is generally believed that only the free steroid can enter cells (see Chapter 5). Presumably, steroids can freely diffuse across the plasma membranes of all cells but are sequestered only within cells that contain specific intracellular receptors. The steroid binds to its receptor with an affinity ($K_d \simeq 10^{-10}\ M$) that is at least 10-fold greater than the affinity with which it binds to the serum globulins.

Steroid Hormone Receptors

Cellular Localization

An issue that has been a source of controversy over the years is the subcellular localization of the unoccupied or so-called free steroid receptor. There has always been general agreement that the occupied or "bound" steroid receptor has increased affinity for chromatin and is localized to the nucleus. It was believed for some time, however, that the free steroid receptor was a cytoplasmic protein that moved to the nucleus only after binding hormone and undergoing an undefined process termed *transformation* or *activation*. This concept was derived from studies in which cells not previously exposed to a given steroid were homogenized and cytosolic and nuclear fractions were prepared by differential centrifugation. When these fractions were incubated with the radiolabeled steroid, the receptor was usually found to be present predominantly in the cytosol. On the other hand, if the cells were first incubated with the steroid hormone, then homogenized and subcellular fractions prepared, the receptor was found to be present almost exclusively in the nuclear fraction. With the availability of monoclonal antibodies specific for a number of steroid receptors and the technique of immunocytochemistry, it has generally been found that both the free and occupied steroid receptors are localized to the nucleus. Receptors for thyroid hormones, retinoic acid, and 1,25-dihydroxyvitamin D_3 are present within the nucleus, bound to their respective *hormone responsive elements* (HREs) in the absence or presence of bound hormone. By contrast, receptors for glucocorticoids (GR), progesterone (PR), estrogen (ER), androgens (AR), and mineralocorticoids (MR) in the absence of hormone binding are not bound to their respective HREs. Rather, these "free" receptors are present within the nucleus (PR, ER, AR, MR) or cytoplasm (GR) bound to a complex of *heat shock proteins* (hsps). Upon hormone binding, these receptors are released from the hsp complex and are activated to a DNA-binding, transcription-modulating state (see below). These activated receptors are now found within the nucleus bound to their respective HREs which usually lie upstream of target genes.

Receptor Structure

The primary structures of the receptors for all the major classes of steroid hormones including 1,25-dihydroxyvitamin D_3 have been determined by the use of recombinant DNA technology. The steroid hormone receptors belong to a superfamily of nuclear receptors that also includes several tissue-specific forms of receptors for thyroid hormones and retinoic acid, as well as a number of nuclear proteins for which ligands have not as yet been identified. It is apparent that these molecules are members of a large superfamily of ligand-activated transcription factors that bind to specific sequences of DNA as dimers and regulate the transcriptional activity of specific genes.

All of these receptors have the same general structure (Fig. 3-10) as well as a high degree of homology in their hormone-binding and DNA-binding regions. By comparison of the nucleotide sequences of the mRNAs as well as the deduced amino acid sequences of the receptor proteins, it has been found that there are three regions that are highly conserved among all members of this superfamily. The region of greatest homology is the DNA-binding domain (DBD), which is composed of 66–68 amino acids and contains nine perfectly conserved cysteine residues. This domain contains two repeated units enriched in the amino acids Cys, Lys, and Arg.

Such repeated units have been identified in a number of nucleic acid-binding proteins. Each of these units is folded into a finger-like structure containing four cysteines that coordinate one zinc ion. The loop of the so-called zinc finger is made

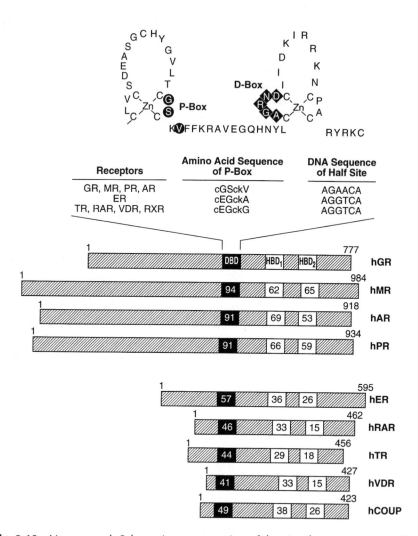

Fig. 3-10 Upper panel: Schematic representation of the zinc finger–containing DNA-binding domain of the glucocorticoid receptor. Amino acids in the P-box (important in DNA sequence recognition) and the D-box (important in receptor dimerization and required spacing of the two half-sites of the response element) are highlighted. The amino acids of the P-box that are important for half-site recognition of three different groups of receptors and the sequences of the half-sites to which they bind, are indicated. Lower panel: Domain structures of members of the steroid receptor superfamily, including human receptors for glucocorticoid (hGR), mineralocorticoid (hMR), androgen (hAR), progesterone (hPR), estrogen (hER), retinoic acid (hRAR), thyroid hormone (hTR), 1,25-dihydroxyvitamin D_3 (hVDR), and the orphan receptor, hCOUP. Three highly conserved regions are indicated, the DNA-binding domain (DBD) and two regions involved in hormone binding (HBD$_1$ and HBD$_2$). The numbers within these regions indicate the percent amino acid similarity with hGR. The hypervariable region of the receptors extends from the amino terminus (indicated by the number 1) to the DBD. The number at the right end of each schematic indicates the number of amino acids in each receptor protein.

up of 12–13 amino acids; a spacer region of 15–17 amino acids is present between the two fingers (Fig. 3-10). These DNA-binding "fingers" have the capacity to insert into a half-turn of DNA. Such repeated units have been identified in a number of nuclear proteins that are known to act as transcription factors. Near the carboxy-terminal end of the receptor molecules are two conserved regions (HBD_1 and HBD_2) of 42 and 22 hydrophobic amino acids, respectively, which comprise the hormone-binding domain. The carboxy-terminal regions of the members of this superfamily of receptors have a high degree of sequence similarity with the v-*erb-A* protein of the oncogenic avian erythroblastosis virus. The v-*erb-A* protein, which is not on-cogenic by itself, increases the oncogenic potential of another viral protein, v-*erb-B* (which has a high degree of sequence homology with the tyrosine kinase domain of the EGF receptor, as previously discussed). Estrogens stimulate the growth of estrogen receptor-positive mammary tumors, and have been implicated as contrib-utory factors in the pathogenesis of human breast cancer. These effects could be mediated through the induction of synthesis of growth factors or through a direct action on growth factor receptors. It has been postulated that the estrogen receptor may potentiate the action of an oncogene in breast cancer in much the same way as v-*erb-A* potentiates the oncogenic potential of v-*erb-B*.

The region of these receptors between the aminoterminus and the DNA-binding domain is hypervariable (HV) both in size and in amino acid sequence. For example, the HV region of the glucocorticoid receptor is 421 amino acids in size, whereas the HV region of the receptor for 1,25-dihydroxyvitamin D_3 consists of only 24 amino acids. There is evidence that this region contributes to the transcription modulating activity of the glucocorticoid receptor. On the basis of sequence similarities among the DNA-binding domains of the various receptors, these molecules can be grouped into two subfamilies. The first is composed of the receptors for glucocorticoids, mineralocorticoids, progesterone, and androgens. The second family consists of the receptors for estrogen, thyroid hormones, vitamin D_3, and retinoic acid. This family also includes the so-called orphan receptors, which are related proteins for which ligands have not as yet been identified. The members of the first subfamily contain HV regions of 420–603 amino acids, whereas receptors in the second subfamily contain HV regions of only 24–185 amino acids. Interestingly, the transcription factor COUP, which is involved in the transcriptional activation of a number of genes, including those for chicken ovalbumin, mammalian insulin, and proopiome-lanocortin, is a member of the steroid receptor superfamily of proteins based on structural homology. Like the thyroid hormone (TR), retinoic acid (RAR), and 1,25-dihydroxyvitamin D_3 (VDR) receptors, COUP binds to DNA as a heterodimer with the retinoid-X-receptor (RXR); however, the existence of a ligand for COUP remains to be determined. The finding that the carboxyterminal region of COUP has signif-icant sequence similarity to other members of the steroid receptor superfamily, and that this region of COUP is evolutionarily conserved, is suggestive of the existence of such a ligand. It has been suggested that COUP may be activated by a ligand produced within the same cell in which it is synthesized. It is postulated that the steroid receptor superfamily originated from a primordial receptor gene in primitive organisms in which it may have served as a receptor for environmental nutrients and other factors. The different classes of receptors for steroids and other regulatory molecules within the present superfamily of receptors and related molecules evolved in response to the increasing regulatory requirements imposed by cellular specialization of higher eukaryotes.

Molecular Forms and Receptor Transformation (Activation)

"Free" glucocorticoid, mineralocorticoid, androgen, progesterone, and estrogen receptors are believed to exist in the cell as monomers that are associated with a complex of proteins that includes two molecules of a 90,000 dalton phosphoprotein, hsp90, one molecule of hsp70, and a number of other proteins (Fig. 3-11). In this conformation the receptor is incapable of binding DNA or regulating gene transcription. The binding of hormone to the receptor results in the dissociation of receptor from this complex, which forms a homodimer with another receptor molecule. This receptor homodimer has a greatly increased affinity for binding to HREs in DNA. Phosphorylation of the receptor homodimer, either just before or after binding to DNA, is believed also to serve a role in the activation of the receptor to a transcription modulatory state. By contrast, "free" receptors for thyroid hormone, retinoic acid, and 1,25-dihydroxyvitamin D_3 are not bound to hsps. Rather, they are bound to their HREs either as homodimers or as heterodimers with a molecule of a member of the retinoic acid receptor subfamily, termed *retinoid-X receptor* (RXR). In their "free" state, receptors in this subgroup act to inhibit transcription of their respective target genes. Upon hormone binding, there is a conformational change in these receptors which results in target gene activation.

Regulation of Gene Expression by Members of the Steroid Receptor Superfamily

The primary action of steroid, retinoid, and thyroid hormones is to regulate the rate of transcription of specific genes and thus to alter the synthesis of specific proteins. As discussed above, members of the steroid receptor superfamily are transcription factors that modulate gene transcription. This group of transcription factors interacts with specific genomic sequences, the so-called HREs, which are predominantly localized in the 5'-flanking regions of target genes, usually within several hundred base pairs upstream of the transcription initiation sites. The activated forms of these receptors bind to their HREs as homo- (GR, MR, AR, PR, ER) or hetero- (TR, VDR, RAR) dimers.

Hormone-Responsive Elements and Regulation of Gene Transcription

After hormone binding and receptor activation, the GR, PR, MR, AR, and ER bind as homodimers to their respective HREs. The *GR, PR, MR*, and *AR* bind to an identical HRE, which consists of a double-stranded inverted repeat or palindrome of six nucleotides (half-site) separated by a three nucleotide spacer:

AGAACAnnnTGTTCT
TCTTGTnnnACAAGA

The fact that more than one hormone receptor binds to the same sequence of DNA implies that specificity is dependent upon the type of hormone receptor(s) present

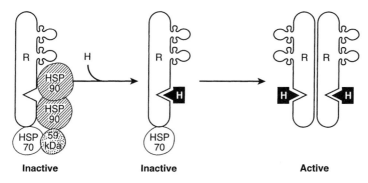

Inactive	Inactive	Active

Fig. 3-11 Proposed mechanism for hormone-induced activation of the glucocorticoid receptor. The inactive "free" glucocorticoid receptor (R) exists in the cytoplasm as a monomer in association with a protein complex comprised of two molecules of heat shock protein (hsp)90, one molecule of hsp70, and a number of other proteins. After hormone (H) binding, the receptor dissociates from the hsp complex and forms a homodimer with another molecule of glucocorticoid receptor. This process is required for transformation of the receptor to a DNA-binding, transcription modulating state.

in the target cell. Alternatively, DNA sequences outside of the HRE, as well as interaction with other transcription factors bound to *cis*-acting enhancer elements, may be responsible for differential activation of target genes by structurally related hormone receptors.

The *ER* does not bind to HRE recognized by GR, MR, AR, and PR. Rather, the ER binds specifically to a HRE that consists of a double-stranded inverted repeat of similar sequence, also with a three nucleotide spacer:

$$\text{AGGTCAnnnTGACCT}$$
$$\text{TCCAGTnnnACTGGA}$$

Specificity for the primary sequence of the DNA half-site (e.g. AGAACA vs. AGGTCA) is determined by 3 amino acids in the shoulder of the first zinc finger of the receptors, the so-called *P-box* (Fig. 3-10). In the GR, MR, PR, and AR, these amino acids are glycine, serine, and valine. By contrast, in the estrogen receptor, these amino acids are glutamate, glycine, and alanine. Within the shoulder of the second zinc finger is the *D-box* or "dimerization" box (Fig. 3-10). This region of the receptor is important for dimerization of the two receptor monomers and establishes orientation of the monomers for appropriate binding to differentially spaced half-sites.

Receptors for *thyroid hormones, retinoids*, and *1,25-dihydroxyvitamin D$_3$* usually bind as heterodimers with RXR to a double-stranded, direct hexameric repeat separated by a spacer of 1–5 nucleotides. Although the sequence of the half-site is the same as that of the estrogen-responsive element ([ERE] AGGTCA/TGACCT), binding specificity is determined by the presence of a direct repeat, rather than a palindrome (as is the case for the ERE), and by the number of nucleotides in the spacer. As an example, the VDR–RXR heterodimer preferentially binds to a direct repeat with a 3-nucleotide spacer, the TR–RXR heterodimer binds preferentially to a direct repeat with a 4-nucleotide spacer, and the RAR–RXR heterodimer binds preferentially to a direct repeat with a 5-nucleotide spacer, as shown below.

$$AGGTCAn_{1-5}AGGTCA$$
$$TCCAGTn_{1-5}TCCAGT$$

Glucocorticoid, Mineralocorticoid, Progesterone, and Androgen Receptor Regulation of Gene Expression. A model of the proposed mechanism whereby steroid hormones such as glucocorticoids, mineralocorticoids, progesterone, and androgens regulate eukaryotic gene expression is shown in Fig. 3-12. After binding a molecule of hormone, the receptor is released from the complex with hsp90 and other heat shock proteins and forms a homodimer with another receptor molecule. The receptor dimer is phosphorylated and binds with high affinity to a HRE, which may exist

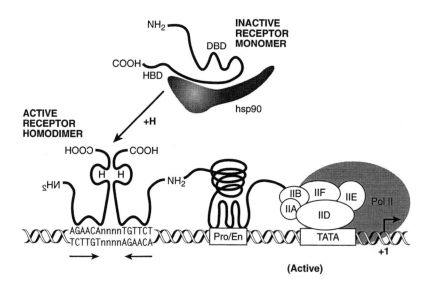

Fig. 3-12 Proposed mechanism for regulation of gene transcription by glucocorticoid, progesterone, mineralocorticoid, and androgen receptors. After binding hormone (H), the receptor dissociates from the complex with heat shock protein 90 (hsp90) and other proteins, and binds to its hormone response element (HRE) upstream of the target gene as a homodimer. The activated receptor dimer can now interact with other receptor dimers bound to their HREs, with other transcription factors bound to promoter or enhancer elements (Pro/En), and/or directly with the preinitiation complex of proteins bound to the TATA box. This results in the stabilization of the preinitiation complex which, in turn, facilitates the binding of RNA polymerase II (Pol II) and activation of transcription initiation. It recently has been found that a number of receptors in the steroid receptor superfamily can interact directly with transcription factor IIB (TFIIB). Since binding of TFIIB is rate limiting in preinitiation complex assembly, it is suggested that the steroid receptor may facilitate preinitiation complex assembly through its interaction with TFIIB.

several hundred base pairs upstream of the target gene. The receptor dimer undergoes further phosphorylation and can now interact with other receptor dimers bound to HREs, with transcription factors bound to other promoter/enhancer (Pro/En) elements, and to the preinitiation complex of proteins bound to the TATA box. These cooperative interactions result in stabilization of the TATA binding protein (preinitiation) complex and the binding of RNA polymerase II (Pol II), resulting in activation of transcription initiation. The formation of the preinitiation complex is a sequential process; the binding of transcription factor IIB (TFIIB) is the rate-limiting step. It recently has been found that a number of receptors in the steroid receptor superfamily can bind directly to TFIIB. This suggests that the steroid receptor may facilitate the interaction of TFIIB with the preinitiation complex.

Thyroid, Retinoid, and Vitamin D_3 Receptor Regulation of Gene Expression. As discussed above, receptors for retinoids, thyroid hormones, and 1,25-dihydroxyvitamin D_3, when bound to their HREs in the absence of ligand binding, cause silencing of their target genes. The proposed mechanism whereby the active thyroid hormone triiodothyronine (T_3) regulates gene expression is shown in Fig. 3-13. In the absence of hormone binding, the thyroid hormone receptor (T_3R) is bound to its response element (a direct repeat of the sequence AGGTCA with a 4-nucleotide spacer) as a homodimer or heterodimer with RXR. The carboxyterminus of the T_3R interacts with TFIIB, and together with a co-repressor, silences transcription. The binding of T_3 to the receptor favors heterodimerization of T_3R with RXR (which binds 9-*cis*-retinoic acid) and causes dissociation of the co-repressor and a decrease in the interaction of the carboxyterminus of T_3R with TFIIB. This may allow the receptor aminoterminus, which in other members of the steroid receptor superfamily contains a transcription activation domain, to interact with TFIIB. Transcription of the thyroid hormone-responsive gene is now activated.

Tissue-Specific Expression of Hormonally Regulated Genes

The phenotype of a given cell is the result of the complement of genes that are expressed; the expression of a number of these cellular genes will be subject to hormonal regulation. A hormone acting through its cellular receptors can regulate the expression of different genes in different tissues. For example, estrogen regulates the expression of the ovalbumin gene in the chick oviduct and the vitellogenin gene in chick liver. Since the estrogen receptors in oviduct and liver tissues are apparently identical, and since the DNA sequences within and surrounding eukaryotic genes should be essentially the same in all cell types of a given organism, there must be other structural features that determine which genes are expressible and available to hormonal regulation. The complement of genes that are expressed in a given cell type is determined at the time of cellular differentiation. It is apparent that such expressible genes exist in so-called DNase I-sensitive regions; that is, they reside in regions of the chromatin that are more readily digested *in vitro* by DNase I than is the bulk of chromosomal DNA. The DNase I-sensitive regions define the structural framework in a given cell for the genes that are available for expression. Most of the cellular genes are not expressed and are tightly packaged with histone proteins in higher order chromosomal structures, rendering them DNase I resistant.

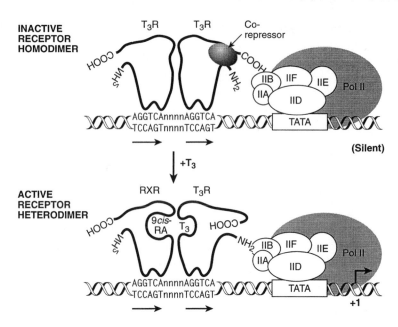

Fig. 3-13 Proposed mechanism for regulation of gene transcription by thyroid hormones, retinoids, and 1,25-dihydroxyvitamin D_3. In its "free" state, the thyroid hormone receptor (T_3R) binds to its hormone response element as homodimer (shown), or as a heterodimer with a molecule of retinoid-X receptor (RXR). The carboxyterminus of T_3R interacts with transcription factor IIB (TFIIB) and, together with a co-repressor, silences transcription. The interaction of the receptor carboxy-terminus with TFIIB may prevent the association of TFIIB with the other factors bound to the TATA box and prevent formation of a stable preinitiation complex. Upon binding of triiodothyronine (T_3), the receptor undergoes a conformational change resulting in the dissociation of the co-repressor, a decreased interaction of the T_3R carboxyterminus with TFIIB, and an increased interaction of the T_3R aminoterminus with TFIIB. This in turn facilitates TFIIB binding and assembly of a stable preinitiation complex, the binding of RNA polymerase II (Pol II), and the activation of transcription initiation.

ENDOCRINE DISORDERS DUE TO ALTERATIONS IN RECEPTOR NUMBER AND FUNCTION

As discussed above, changes in the concentration of cellular receptors for a specific hormone can markedly alter the sensitivity of the target cell to that hormone; a decrease in the concentration of receptors can decrease sensitivity to the hormone, whereas an increase in receptor concentration can increase the sensitivity of the target cell to the hormone. It is apparent that the concentration of cellular receptors for a specific hormone can vary considerably with the physiologic state and that the concentration of the hormone itself can regulate the concentration of its own receptors on target cells. Most commonly, an increase in the level of a specific hormone will cause a decrease in the available number of its cellular receptors. This decrease in available receptors can be due either to receptor phosphorylation by a specific kinase and sequestration of receptors away from the cell surface, or to an actual disappearance or loss of receptors from the cell. This hormonally induced negative regulation of receptors is termed *homologous "down-regulation"* or *"desensitization."* Studies of the down-regulation of receptors for insulin and EGF by the homologous hormones indicate that down-regulation is caused, at least in part, by a clustering of hormone–receptor complexes in coated pits on the cell surface, internalization within coated vesicles, and degradation by lysosomal enzymes. Coated pits and coated vesicles are so named because they contain a single protein, clathrin, that forms a "coat" on their cytoplasmic surfaces. There is little doubt that receptor internalization provides an important homeostatic mechanism that serves to protect the organism from the potential toxic effects of hormone excess.

Desensitization is defined as a decrease in the responsiveness of a cell to a constant level of hormone or factor upon prolonged exposure. Homologous desensitization can also result from a hormone-induced alteration in the receptor, which uncouples it from some component of the signal transduction pathway. Another form of desensitization, *heterologous desensitization*, occurs when incubation with one agonist reduces the responsiveness of a cell to a number of other agonists that act through different receptors. This phenomenon is most commonly observed with receptors that act through the adenylyl cyclase system. Heterologous desensitization reflects a broad pattern of refractoriness that has a slower onset than homologous desensitization.

In certain instances, endocrine dysfunction arises from pathologic alterations in receptor levels and/or function. Specific examples of such endocrine dysfunction are discussed below.

Endocrine Dysfunction Caused by Homologous and Heterologous Receptor Regulation

The process of homologous down-regulation of receptors may be a contributing factor to a number of clinical states. Probably the most common receptor defect in man is the down-regulation of insulin receptors that is caused by the chronic hyperinsulinemia of *obesity*. Obesity is generally characterized by hyperinsulinemia, variable degrees of glucose intolerance, and resistance to both endogenous and exogenously administered insulin. The insulin resistance associated with obesity is caused, at least in part, by a decrease in the concentration of available insulin receptors on target cells because of increased insulin–receptor endocytosis and degradation; postreceptor defects may play some role in the reduced sensitivity to insulin. When obese individuals are maintained on calorie-restricted diets for several weeks, there is a marked reduction in the circulating insulin concentration, an increase in the concentration of insulin receptors on target cells, and a decline in the insulin resistance.

Another example of endocrine dysfunction resulting from homologous desensitization is the reduced sensitivity to both endogenous and exogenous catecholamines that results from the use of β-adrenergic agonists as bronchodilators in the treatment of asthma. This reduced sensitivity to catecholamines is caused by the β-adrenergic agonist-induced reduction in the concentration of β-adrenergic receptors on the cell surface and to a reduced coupling of receptors to adenylyl cyclase. Homologous desensitization of β-adrenergic receptors is associated with phosphorylation of the receptors by a β-adrenergic receptor kinase. The phosphorylation of the receptors either promotes the sequestration of receptor away from the cell surface or renders it less effective in activating adenylyl cyclase, or both. The sequestered receptors are then either recycled to the cell surface or targeted for degradation within lysosomes.

There are a number of examples of one hormone acting through its own receptors to alter the sensitivity of target cells to another hormone. Thyroid hormones can enhance the sensitivity of cardiac muscle cells to catecholamines by increasing the number of β-adrenergic receptors. This can result in an increased heart rate, atrial fibrillation, and congestive failure. Frequently, an alteration in cardiac function is the only manifestation of an otherwise masked hyperthyroidism.

Endocrine Dysfunction Caused by Autoimmune Disease Involving Antireceptor Antibodies

The presence of antibodies to receptors appears to be of central importance in several disease states.

Acetylcholine Receptors

In myasthenia gravis, a disease characterized by muscle weakness and flaccid paralysis, there are circulating antibodies to the nicotinic acetylcholine receptors localized in the neuromuscular junctions. These antibodies do not block the binding of acetylcholine to the receptor; however, they enhance the rate of receptor degradation and thus render the skeletal muscle cells less sensitive to endogenous acetylcholine.

Thyroid-Stimulating Hormone (Thyrotropin, TSH)

Individuals with hyperthyroidism associated with diffuse goiter (Graves' disease) have circulating antibodies to the receptor for TSH, a polypeptide hormone produced by the anterior pituitary that acts on the thyroid gland to stimulate the production of the thyroid hormones (see Chapter 13). One of the effects of the thyroid hormones is to act on the hypothalamus and pituitary to inhibit the production of TSH. These autoantibodies bind to the TSH receptors on thyroid cells and mimic the actions of TSH itself. Since such antibodies are not regulated by negative feedback of the thyroid hormones as is TSH, the thyroid gland is persistently stimulated to produce elevated levels of circulating thyroid hormones (see Chapter 13).

Insulin Receptors

In patients who have an unusual form of diabetes associated with insulin resistance, there are circulating autoantibodies to the insulin receptor. The antibodies bind to insulin receptors, block insulin binding, and mimic insulin action. This usually results in desensitization to endogenous insulin with associated hyperinsulinemia and extreme insulin resistance. In rare instances, receptor desensitization does not occur and the patient manifests severe hypoglycemia.

Disorders Due to Receptor and Postreceptor Defects

Disorders Due to Defects in G Protein-Coupled Signal Transduction

As previously mentioned, the family of guanine nucleotide-binding (G) proteins mediate the actions of a wide variety of hormones and factors in tissues throughout the body. The massive secretory diarrhea associated with *Vibrio cholerae* infection was the first disorder found to be caused by the modification of a G protein. As discussed above, the cholera exotoxin catalyzes ADP-ribosylation of arginine 201 of $G_s\alpha$ in intestinal epithelial cells. This causes a marked inactivation of the GTPase activity associated with this subunit resulting in persistent activation of adenylyl cyclase and of cyclic AMP formation. It now is apparent that there are a variety of somatic mutations of G proteins that result in activation and inactivation of G proteins. Two endocrine disorders associated with activation and inhibition of G_s function are described below.

McCune-Albright Syndrome. This disorder is caused by a missense mutation of arginine 201 of $G_s\alpha$ resulting in decreased GTPase activity and constitutive activation of adenylyl cyclase in one or more endocrine glands, including the adrenal cortex, gonads, thyroid, and pituitary thyrotrophs. This results in elevated levels of cyclic AMP formation, abnormal cell proliferation, and hyperfunction of the affected endocrine tissues. The symptoms can include acromegaly, precocious puberty, hyperthyroidism, hyperparathyroidism, Cushing's syndrome, fibrous dysplasia of bone that is polyostotic (involving more than one bone), and café au lait skin pigmentation.

PTH-Resistant Hypoparathyroidism. Subjects with this disorder manifest a variety of missense mutations in the gene encoding $G_s\alpha$ that include loss of function mutations, and mutations that prevent coupling of $G_s\alpha$ with activating receptors. Most affected individuals manifest a 50% decrease in $G_s\alpha$ activity. This results in pseudohypoparathyroidism (see Chapter 15) accompanied by variable resistance to a variety of hormones that act through adenylyl cyclase activation, including thyroid-stimulating hormone, gonadotropins, and other hormones that act through G_s-coupled receptors.

Androgen Resistance

Testosterone, the major steroid product of the Leydig cells of the testis and the principal circulating androgen, serves as a prohormone in most androgen target tissues for dihydrotestosterone (DHT), which is formed from testosterone in these tissues by the action of the enzyme 5α-reductase (see Chapter 10). During embryogenesis, androgens produced by the fetal testis play a critical role in the differentiation of the male internal and external genitalia (see Chapter 8).

Androgen resistance can be caused by (1) complete absence, diminished amounts, or qualitative abnormalities of androgen receptors in target tissues, or (2) defects in the signal transduction pathway despite apparently normal androgen binding. The gene that encodes the androgen receptor is X linked; therefore, mutations are expressed in the hemizygous 46,XY state. A syndrome of complete androgen resistance in humans, termed *testicular feminization* (see Chapter 8), is commonly due to decreased amounts of functional androgen receptor. Testicular feminization has been found to be associated with a variety of point mutations scattered throughout the androgen receptor gene. Such mutations can result either in the insertion of a premature translational ''stop codon'' so that a truncated nonfunctional protein is synthesized, or in a loss of functional activity of the hormone-binding or DNA-binding domains. Affected genotypic males are born with female external genitalia and a blind vaginal pouch. Wolffian duct structures are absent or vestigial; müllerian duct structures are absent and the external genitalia are female. Testes are present in the labia, inguinal canals, or abdomen. At puberty, these individuals develop female secondary sex characteristics. Since the negative feedback of testosterone at the hypothalamic–pituitary level is defective, circulating levels of luteinizing hormone and testosterone are elevated. Increased levels of luteinizing hormone cause enhanced secretion of estrogen by the testis, and the increased estrogen production contributes to the development of female secondary sex characteristics at puberty.

Vitamin D Resistance

Vitamin D-dependent rickets types I and II are inherited disorders in the pathway of vitamin D action. Both are inherited as autosomal recessive traits and are associated with childhood rickets and adult abnormalities of bone mineralization termed *osteomalacia*. Vitamin D-dependent rickets type I appears to arise from a deficiency in the activity of renal 1α-hydroxylase, the enzyme that converts 25-(OH)D$_3$ into 1,25-(OH)$_2$D$_3$, the active form of the vitamin. The defect can be treated by administering physiologic amounts of 1,25-(OH)$_2$D$_3$. On the other hand, vitamin D-dependent rickets type II appears to be due to a defect in the vitamin D receptor, or at the postreceptor level. Children suffering from vitamin D-dependent rickets, type II in two unrelated families were found to have single point mutations in the tip of one of the zinc fingers of the DNA-binding domain of the vitamin D receptor. These mutant receptors are apparently defective in their ability to interact with and transcriptionally regulate vitamin D-responsive genes. Almost all individuals afflicted with this disorder present within the first year of life with hypocalcemia, osteomalacia, or rickets, and secondary hyperparathyroidism. An unexplained feature that is frequently associated with this disorder is total alopecia. Almost all afflicted individuals are resistant to pharmacologic doses of vitamin D and its metabolites in the healing of bone lesions and normalization of serum calcium levels.

SUGGESTED READING

Benovic JL, Bouvier M, Caron MG, and Lefkowitz RJ: Regulation of adenylyl cyclase-coupled β-adrenergic receptors. Annu Rev Cell Biol 4:405–428, 1988.

Berridge MJ: Inositol trisphosphate and calcium signalling. Nature 361:315–325, 1993.

Cobb MH, and Goldsmith EJ: How MAP kinases are regulated. J Biol Chem 270:14843–14846, 1995.

Cohen P: Signal integration at the level of protein kinases, protein phosphatases and their substrates. Trends Biochem Sci 17:408–413, 1992.

Davis RJ: The mitogen-activated protein kinase signal transduction pathway. J Biol Chem 268:14553–14556, 1993.

Dohlman HG, Thorner J, Caron MG, and Lefkowitz RJ: Model systems for the study of seven transmembrane segment receptors. Annu Rev Biochem 60:653–688, 1991.

Exton JH: Signalling through phosphatidylcholine breakdown. J Biol Chem 265:1–4, 1990.

Garbers DL, and Lowe DG: Guanylyl cyclase receptors. J Biol Chem 269:30741–30744, 1994.

Gilman AG: G proteins and the regulation of adenylyl cyclase. J Am Med Assoc 262:1819–1825, 1989.

Glass CK: Differential recognition of target genes by nuclear receptor monomers, dimers, and heterodimers. Endocr Rev 15:391–407, 1994.

Griffin JE, McPhaul MJ, Russell DW, and Wilson JD: The androgen resistance syndromes: Steroid 5α-reductase 2 deficiency, testicular feminization, and related disorders. In: *The*

Metabolic and Molecular Bases of Inherited Disease, 7th ed., Vol. 2, CR Scriver, AL Beaudet, WS Sly, and D Valle, eds., McGraw-Hill, New York, Chap. 95, pp 2967–2998, 1995.

Hannun YA: The sphingomyelin cycle and the second messenger function of ceramide. J Biol Chem 269:3125–3128, 1994.

Ihle JN, Witthuhn BA, Quelle FW, Yamamoto K, Thierfelder WE, Kreider B, and Silvennoinen O: Signaling by the cytokine receptor superfamily: JAKs and STATs. Trends Biochem Sci 19:222–227, 1994.

Kahn CR, Smith RJ, and Chin WW: Mechanism of action of hormones that act at the cell surface. In: *Williams Textbook of Endocrinology*, 8th ed., JD Wilson and DW Foster, eds., Saunders, Philadelphia, pp. 91–134, 1992.

Kitamura T, Ogorochi T, and Miyajima A: Multimeric cytokine receptors. Trends Endocrinol Metab 5:8–14, 1994.

Lemmon MA, and Schlessinger J: Regulation of signal transduction and signal diversity by receptor oligomerization. Trends Biochem Sci 19:459–464, 1994.

Linder ME, and Gilman AG: G proteins. Sci Am 267:56–65, 1992.

Meyer TE, and Habener JF: Cyclic adenosine 3′,5′-monophosphate response element binding protein (CREB) and related transcription-activating deoxyribonucleic acid-binding proteins. Endocr Rev 14:269–290, 1993.

Nishizuka Y: Intracellular signaling by hydrolysis of phospholipids and activation of protein kinase C. Science 258:607–614, 1992.

O'Malley BW: The steroid receptor superfamily: More excitement predicted for the future. Mol Endocrinol 4:363–369, 1990.

Pratt WB: Interaction of hsp90 with steroid receptors: Organizing some diverse observations and presenting the newest concepts. Mol Cell Endocrinol 74:C69–C76, 1990.

Pratt WB: The role of heat shock proteins in regulating the function, folding, and trafficking of the glucocorticoid receptor. J Biol Chem 268:21455–21458, 1993.

Quigley CA, De Bellis A, Marschke KB, El-Awady MK, Wilson EM, and French F: Androgen receptor defects: Historical, clinical and molecular perspectives. Endocr Rev 16:271–321, 1995.

Rhee SG, and Choi KD: Regulation of inositol phospholipid-specific phospholipase C enzymes. J Biol Chem 267:12393–12396, 1992.

Schlessinger J: How receptor tyrosine kinases activate Ras. Trends Biochem Sci 18:273–275, 1993.

Spiegel AM, Weinstein LS, and Shenker A: Abnormalities in G protein-coupled signal transduction pathways in human disease. J Clin Invest 92:1119–1125, 1993.

Strader CD, Sigal IS, and Dixon RAF: Structural basis of β-adrenergic receptor function. FASEB J 3:1825–1832, 1989.

Stroud RM, and Finer-Moore J: Acetylcholine receptor structure, function, and evolution. Annu Rev Cell Biol 1:317–351, 1985.

Tsai M-J, and O'Malley BW: Molecular mechanisms of action of steroid/thyroid receptor superfamily members. Annu Rev Biochem 63:451–486, 1994.

Ullrich A, and Schlessinger J: Signal transduction by receptors with tyrosine kinase activity. Cell 61:203–212, 1990.

Williams LT: Signal transduction by the platelet-derived growth factor receptor. Science 243:1564–1570, 1989.

4

Cytokines and Immune–Endocrine Interactions

M. LINETTE CASEY

It has been suspected, since the end of the 19th century, that there is a close interaction between the endocrine and immune systems. This relationship was first recognized with the observation that thymic size increased in response to prepubertal castration of male rabbits. Later, it was observed that challenges which evoke an inflammatory response also lead to changes in the endocrine system. These include a striking increase in the rate of secretion of cortisol by the adrenal cortex. Indeed, a variety of stressful challenges lead to alterations in the production of steroid hormones by the adrenal cortex. With severe thermal injuries, for example, there is an increase in the rate of cortisol formation and, paradoxically, an equally dramatic reduction in the rate of formation of adrenal C_{19}-steroids, most notably dehydroepiandrosterone and dehydroepiandrosterone sulfate.

Several other observations support the concept of interaction between the endocrine and immune systems: (1) the dimorphic nature of the immune response of the two genders; (2) alterations in the immune response after gonadectomy or sex steroid hormone treatment; (3) modification of the immune response during pregnancy; and, (4) the identification of steroid hormone receptors in cells of the immune system. It is now clear that hormones affect cells of the immune system and that cytokines (the mediators of the immune response) affect the neuroendocrine system and the actions of hormones. During the past two decades, cytokines have been identified in appreciable numbers and their structure, function, and mechanism(s) of action have been elucidated in part.

CYTOKINES: DEFINITIONS AND CATEGORIZATION

The term *cytokine* was originally coined, by Stanley Cohen in 1974, to refer to a soluble product produced by nonimmune or immune cells that mediated an immune response. With time, the definition of cytokine has broadened so that the term now includes a variety of growth factors and other regulatory proteins that affect the function of immune or nonimmune cells. (The term *immune cells* refers to cells of bone marrow origin that have a central role in the immune response, namely, lymphocytes, granulocytes, monocytes, macrophages, neutrophils, eosinophils, and mast cells.) *Cytokine* is used to describe proteins produced by various cell types in response to stimuli arising from various physiologic and pathophysiological states. Cytokines act in target cells through plasma membrane receptors to provoke intracellular signals that eventuate in biological effects (Fig. 4-1). The range of biological processes affected by cytokines is broad and, in recent years, has been expanded to include nonimmune processes. The cytokines are very potent agents that can elicit massive changes in cells; a number of mechanisms have evolved to modulate these actions, including the presence of naturally occurring antagonists (Fig. 4-1).

The cytokine family of proteins includes the interleukins (presently interleukin-1 [α and β] through interleukin-16); tumor necrosis factor-α and -β; the hematopoietic colony-stimulating factors (CSF) (macrophage CSF, granulocyte CSF, granulocyte-macrophage CSF); the interferons (-α[s], -β, and -γ); transforming growth factors-β (and the related proteins müllerian-inhibiting hormone, activin, and inhibin [Chapters 9 and 10]); chemokines (which include interleukin-6, interleukin-8, interleukin-11, monocyte chemoattractant proteins, oncostatin-M, leukemia-inhibitory factor, and others); and a number of growth factors (including epidermal growth factor and transforming growth factor-α, insulin-like growth factors [Chapter 12], and platelet-derived growth factor). Various classification systems have been used to categorize cytokines on the basis of cell type of origin, function, or structure. It is noteworthy that the usefulness of these classification systems is limited by the overlapping and multifunctional nature of many cytokines. For example, the terms *lymphokine* and *monokine* were coined to denote classes of cytokines produced by lymphocytes (especially T cells) or monocytes, respectively. But the *lymphokine* and *monokine* are used less frequently as it was discovered that the origin of some cytokines placed in these categories is not restricted to lymphocytes or monocytes. In another classification scheme, cytokines are categorized as inflammatory cytokines, immunomodulatory cytokines, chemokines, and growth factors (Table 4-1). *Chemokine* denotes a class of cytokines that attract specific cells of the immune system.

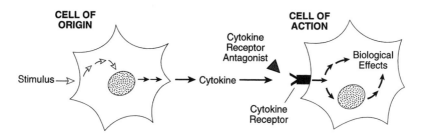

Fig. 4-1 Cells of many types respond to a stimulus, such as bacterial or viral toxin, stress, or inflammatory cytokines, to produce an intracellular signal that leads to the synthesis and secretion of cytokines. These potent proteinaceous agents act in nearby cells (or in some cases, enter blood, Fig. 4-2) through specific, plasma membrane–bound receptors to effect a myriad of profound biological effects. Naturally occuring cytokine receptor antagonists may be present to limit the effects of cytokines by blocking interaction of the cytokine with the specific receptor.

Table 1 Functional Categorization of Cytokines

Function	Cytokine
Inflammatory cytokines	Interleukin (IL)-1β Tumor necrosis factor (TNF)-α IL-6
Immunomodulatory cytokines	IL-2 IL-9 IL-15 IL-3 IL-10 IL-4 IL-12 IL-5 IL-13 IL-7 IL-14

	Colony stimulating factors (CSF)
	  M-CSF macrophage CSF
	  G-CSF granulocyte CSF
	  GM-CSF granulocyte-macrophage CSF
	Monocyte chemotactic proteins (MCPs)
	  MCP-1
	  MCP-2
	  MCP-3
	Interferons (IFN)
	  IFN-α
	  IFN-β
	  IFN-γ

Chemokines	IL-6 IL-8 IL-11 MCP-1 Oncostatin M RANTES Leukemia inhibitory factor (LIF) Ciliary neurotrophic factor (CNTF)
Growth factors and other mediators	Transforming growth factor-β (TGF-β) Platelet-derived growth factor (PDGF) Epidermal growth factor (EGF) Insulin-like growth factors (IGFs) Interferons IL-6 IL-1α

Cytokines act in a variety of cell types through autocrine, paracrine, endocrine, and juxtacrine mechanisms (Fig. 4-2). *Autocrine* refers to a mechanism by which a cytokine (or other agent) acts in the same cell from which it was produced and secreted. *Paracrine* refers to a mechanism by which a cytokine (or other agent) produced in one cell acts on a neighboring cell, access to which is by diffusion through the intercellular spaces rather than through the blood. Paracrine mechanisms are quantitatively the most important mechanisms of cytokine action. *Endocrine* refers to a mechanism by which a cytokine (or hormone) enters blood and reaches a target cell through the circulation. *Juxtacrine* refers to a mechanism whereby a cytokine that is embedded in, bound to, or associated with the plasma membrane of one cell interacts with a specific receptor in a juxtaposed cell. Cells of many different types and in all tissues have specific receptors for multiple cytokines.

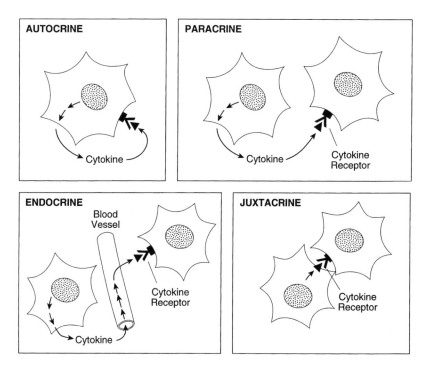

Fig. 4-2 Mechanisms of intercellular communication and action of cytokines, hormones, and other agents. Specific illustrations are given for cytokines. *Autocrine* refers to a mechanism by which a cytokine acts in its cell of origin. *Paracrine* refers to a mechanism by which a cytokine produced in one cell acts on a neighboring cell, access to which is by diffusion through the intercellular spaces rather than through the blood. Paracrine mechanisms are quantitatively the most important mechanisms of cytokine action. *Endocrine* refers to a mechanism by which a cytokine (or hormone) enters the blood and reaches a target cell through the circulation. *Juxtacrine* refers to a mechanism whereby a cytokine that is embedded in, bound to, or associated with the plasma membrane of one cell, interacts with a specific receptor in a juxtaposed cell.

Cytokines effect multiple diverse functions, physiologically and pathophysiologically: (1) inflammatory cytokines are produced in response to infection and immunologic challenge; (2) some cytokines serve as normal growth or regulatory factors; and (3) some cytokines act to facilitate the maintenance of normal immune function. Each of these may modulate the secretions and function of the neuroendocrine system and each may be modulated by the actions of hormones.

CYTOKINES AND NEUROENDOCRINE FUNCTION

Cytokines may modulate endocrine function in a number of ways. They act in the hypothalamus to modify the formation of releasing hormones and in the anterior pituitary to modulate the formation of hypophyseal hormones, e.g., corticotropin (ACTH), gonadotropins, thyroid-stimulating hormone (TSH), prolactin, and growth hormone (GH). They also act in endocrine glands to modify responsiveness to tro-

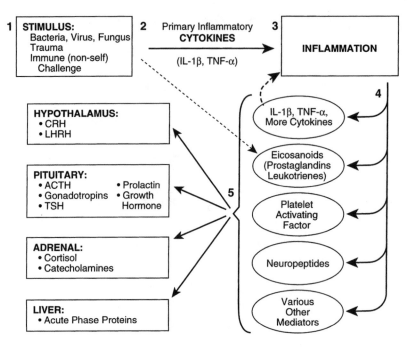

Fig. 4-3 The inflammatory response and consequences thereof. Some of the effects of the inflammatory response on the neuroendocrine system are illustrated. 1: A stimulus such as trauma, stress, immune challenge, or bacterial, viral, or fungal toxin acts to provoke the inflammatory response in mononuclear phagocytes. 2: Monocytes (mononuclear phagocytes in blood) respond rapidly with a massive outpouring of primary inflammatory cytokines, namely, IL-1β and TNF-α. 3: A profound process of inflammation ensues and propagates itself with the autoinduction of IL-1β by IL-1β. 4: Inflammation includes the production of IL-1β as well as TNF-α, IL-6, MCP-1, and many other cytokines (chemoattractants, mitogens, and hemopoietic factors), and the formation of eicosanoids, platelet-activating factor, nitric oxide, neuropeptides (by immune cells), and other bioactive agents. 5: These agents, in particular, the inflammatory cytokines IL-1β, TNF-α, and IL-6, act either directly or indirectly to increase the production of releasing hormones in the hypothalamus, pituitary hormones, cortisol, and catecholamines. In addition, the liver is stimulated to produce acute-phase proteins.

phic hormones or else to regulate the rate of synthesis of hormones (steroid or protein) by the target endocrine gland. These actions of cytokines are illustrated in Fig. 4-3. In addition, cytokines are produced by and act in hormonally responsive tissues in hormonally modulated fashions.

CYTOKINE PRODUCTION IN RESPONSE TO IMMUNE CHALLENGE OR INFECTION

The production of cytokines is a dynamic process that is critical for normal function and well-being. The influence of cytokines on the neuroendocrine system can be appreciated by considering first the central role of interleukin-1β in the immune response.

The term *interleukin* (IL) was coined to describe proteins that are produced by and act on leukocytes. Since that time it has been discovered that some interleukins also are produced by non-immune cells and that many interleukins act in cells other than leukocytes. At the time of writing this chapter, 17 interleukins have been identified in the human: IL-1α, IL-1β, and IL-2 through IL-16, numbered consecutively. In addition, a naturally occuring IL-1 receptor antagonist, which is a member of the IL-1 family, acts to block the actions of IL-1α and IL-1β.

Interleukin-1β: The Primary Cytokine

Interleukin-1β (IL-1β) is referred to as the *primary* cytokine. It is synthesized and secreted by mononuclear phagocytes in response to inflammatory stimuli such as infection or immunological challenge. It is primary because (1) it is produced very rapidly (Fig. 4-4) in response to challenge, and (2) it stimulates the production of many other cytokines.

IL-1β Synthesis, Processing, and Secretion

Bioactive IL-1β is a 17,000-dalton protein that is produced by proteolytic processing of pro-IL-1β, a 34,000-dalton precursor. Although pro-IL-1β is synthesized by many cell types, monocytes and macrophages are by far the principal source of bioactive IL-1β. This is because pro-IL-1β must be processed intracellularly by proteolytic cleavage to the active 17,000-dalton form before it can be secreted. The proteolytic cleavage of pro-IL-1β is catalyzed by a specific enzyme called *IL-1β converting enzyme* (Fig. 4-5), which is found in very few cell types, namely, mononuclear phagocytes (monocytes and macrophages) and neutrophils, and possibly placental

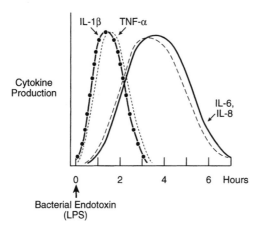

Fig. 4-4 Production of selected cytokines by monocytes in response to bacterial endotoxin (lipopolysaccharide, LPS). IL-1β, the primary cytokine, and TNF-α are produced rapidly and transiently. These cytokines stimulate the production of IL-6 and IL-8 by a variety of cell types.

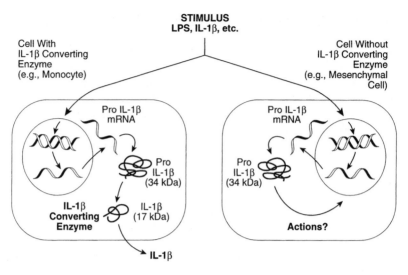

Fig. 4-5   Cell specificity of pro-IL-1β processing and active IL-1β secretion. The secretion of active IL-1β (17 kDa) in response to a stimulus is dependent on the processing of the precursor pro-IL-1β (34 kDa) by IL-1β–converting enzyme. This enzyme is expressed in a limited number of cell types, notable among which are monocytes, macrophages, and neutrophils. Other cell types that do not express IL-1β–converting enzyme respond to similar stimuli with an increase in pro-IL-1β synthesis as well. But the pro-IL-1β is not processed and therefore, cannot be secreted. Rather, it accumulates in the cell and may exert other, yet undefined actions. LPS, lipopolysaccharide.

trophoblasts. Many other cell types respond to immune challenge by a rapid, transient increase in the transcription of the IL-1β gene, an increase in the level of pro-IL-1β mRNA, and an increase in the synthesis of pro-IL-1β. But in cells that lack IL-1β-converting enzyme, the pro-IL-1β accumulates intracellularly and is not secreted. It is postulated that the accumulated pro-IL-1β has a specific, but yet undefined, function in these cells. In addition, the pro-IL-1β converting enzyme is believed to forestall apoptosis.

IL-1β Action
The secreted, bioactive form of IL-1β (17,000 daltons) acts locally in a paracrine or autocrine manner to stimulate the generation of IL-1β itself (a process called *autoinduction*), which serves to propagate the immune response. IL-1β also acts to stimulate the production of other interleukins, tumor necrosis factors, colony-stimulating factors, and interferons. IL-1β acts through two specific receptors to effect many biological responses. The effects of IL-1β are amplified by actions of each of these cytokines through specific receptors to cause the production of yet other potent agents. These include prostaglandins and other arachidonic acid metabolites, platelet-activating factor (PAF), nitric oxide, and other potent substances. For example, IL-1β acts in a variety of cells (commonly the mesenchymal cells of a given tissue) to increase the production of IL-6. IL-6, acting in an endocrine fashion by circulating to the liver, stimulates the synthesis and secretion of *acute-phase proteins*. These

proteins include gammaglobulin and C-reactive protein and others, all of which function to protect the host and repair wounds. This is one component of the *acute phase response* to infection or antigenic and immune challenge. IL-6 also acts in the liver to inhibit the synthesis of albumin and transferrin. IL-1β promotes hematopoiesis and stimulates the production of IL-2, which is a potent mitogen for T cells. IL-1β stimulates replication of fibroblasts, a process that facilitates wound repair.

Generally, IL-1β does not enter blood in a bioactive form. Rather, IL-1β in plasma is bound to α2-macroglobulin, an abundant plasma protein. When bound to α2-macroglobulin, IL-1β is inactive. Therefore, it is unlikely that IL-1β can act in an endocrine fashion; rather, IL-1β acts principally in a paracrine or autocrine manner.

IL-1α

IL-1α acts by way of the same receptors as IL-1β, yet the amino acid sequences of IL-1α and IL-1β are only 26% homologous. Whereas the actions of IL-1α and IL-1β are virtually indistinguishable, the amount of IL-1β secreted exceeds that of IL-1α by a factor of 10 under most circumstances.

The Inflammatory Response

The response of mononuclear phagocytes to infection and immunological challenges is extraordinarily rapid. It involves the local activation and chemoattraction of immune cells and the stimulation of antibody production by lymphocytes.

Tumor necrosis factor (TNF)-α, like IL-1β, is produced very rapidly in vastly increased amounts in response to antigenic challenge or infection. For example, in response to bacterial endotoxin (lipopolysaccharide), IL-1β and TNF-α are produced rapidly by mononuclear phagocytes (Fig. 4-4). The maximal production of IL-1β and TNF-α by mononuclear phagocytes occurs within approximately 2 hours of exposure to bacterial endotoxin (Fig. 4-4). The production of IL-6, IL-8, and other cytokines, which is stimulated by IL-1β and TNF-α, occurs later. IL-1β further promotes the formation and secretion of TNF-α. Additionally, IL-1β autoinduction leads to the production of more IL-1β. Thus the immune response is propagated in this manner. The cytokines IL-1β, TNF-α, and IL-6 are generally referred to collectively as *inflammatory cytokines*; IL-8, MCP-1, and IFN-γ are sometimes included in the group of inflammatory cytokines.

While the rapid and massive response of the cytokine system to infection and other challenges is essential to the successful containment of infection, the profound effects of this system, if left unchecked, are lethal. Thus, because of the potency of cytokines and the wide-ranging actions, a necessary, indeed crucial, component of the cytokine response is self-containment, involving mechanisms for limiting or even curtailing the potent actions of the inflammatory cytokines.

By way of an example, consider these effects of bacterial infection in blood. Bacterial endotoxin (lipopolysaccharide) acts to increase the production of IL-1β and TNF-α by monocytes (Fig. 4-3). The autoinduction of IL-1β leads to more IL-1β and thus, more TNF-α. IL-1β acts to promote the production of chemokines (such as IL-8, MCP-1, and others), which in turn act to recruit additional immune cells, including monocytes, neutrophils, and lymphocytes, to the site of infection. IL-1β

and TNF-α stimulate the formation of IL-6, which circulates to the liver and elicits the production of acute-phase proteins (Fig. 4-3). TNF-α and IL-1β act in the hypothalamus to stimulate the production of prostaglandins, especially PGE$_2$, which causes an increase in body temperature. TNF-α acts on endothelial cells of blood vessels to promote the formation of prostacylin, a potent vasodilator (Fig. 4-3). The rapid, excessive formation of prostacylin leads to vasodilatation, which is characteristic of the vascular collapse of endotoxic shock with infections by gram-negative microorganisms.

A number of mechanisms can inhibit the continued stimulation of the cytokine system and the action of selected cytokines once the inflammatory response has been initiated. Among them are the rapid degradation of cytokine mRNAs and the production of cytokine inhibitory factors, such as cytokine receptor antagonists (Fig. 4-1), e.g., the IL-1 receptor antagonist and proteins that bind cytokines with high affinity. The latter include a family of "soluble" receptors for cytokines (such as TNF-α soluble receptors). These proteins are partial or truncated forms of the plasma membrane–bound cytokine receptors that are present in plasma and bind the cognate receptors with high affinity. Two forms of TNF-α soluble receptors that correspond to the extracellular portions of each of two TNF-α plasma membrane receptors have been identified. These soluble receptors limit the effectiveness of cytokines in tissues that are removed from the site of infection, wound, or antigenic challenge. In addition, the production of IL-10, an inhibitory cytokine, is stimulated by IL-1β. IL-10 acts to limit the inflammatory response to infection by decreasing the activity and replication of T cells.

EFFECTS OF CYTOKINES ON THE HYPOTHALAMIC–PITUITARY–ADRENAL AXIS

The most familiar relationship between the immune and endocrine systems is the potent immunosuppressive action of glucocorticoids. That there are bidirectional responses between the endocrine and immune systems was a more recent observation. Cytokines, as mediators of the immune response, occupy a central role in these bidirectional responses. In fact, cytokines modulate functions of the hypothalamic–pituitary–adrenal axis, the hypothalamic–pituitary–gonadal axis (in women and men), the hypothalamic–pituitary–thyroid axis, and in other functions of the pituitary gland also, including the production of prolactin, GH, and proopiomelanocortin (POMC)-derived proteins (for example, β-endorphin, α-melanocyte-stimulating hormone (MSH), in addition to ACTH). These relationships between the immune and endocrine systems are summarized below.

The Origin of Cytokines That Act in the Hypothalamus

There is clear-cut evidence that cytokines, in particular IL-1β, IL-6, and TNF-α, act in the hypothalamus to increase the production of releasing hormones, including corticotropin-releasing hormone (CRH), thyrotropin-releasing hormone (TRH), and luteinizing hormone-releasing hormone (LHRH). If these actions were mediated by systemic cytokines (acting in an endocrine fashion by delivery through the circulation), permeation of the blood–brain barrier by cytokines would be required. The mechanism by which this could occur is not known. It has been postulated that these cytokines may gain access to responsive areas of the brain where the blood–brain barrier is weak or nonfunctional, for example, the median eminence, the organum vasculosum of lamina terminalis of the hypothalamus, the subfornical organ, the choriod plexus, and the area postrema at the base of the fourth ventricle. On the other hand, it also has been hypothesized that these cytokines may arise in cells in the brain. One mechanism by which this may occur is by transfer of a signalling agent that is permeable to the blood–brain barrier. Additionally, monocytes, lymphocytes, and macrophages traverse the blood–brain barrier and take up residence in various areas of the brain. Resident monocytes and macrophages may secrete IL-1β and TNF-α in response to bacterial toxin or antigenic challenge. Another potential source of IL-1β in the brain are microglia, which are embryologically and functionally related to macrophages and therefore might be expected to express IL-1β-converting enzyme, which would allow for the processing of pro-IL-1β and the secretion of active IL-1β. In addition, IL-2, IL-4, IL-6, and TNF-α are produced by glial cells. IL-1 receptors, IL-1 receptor antagonists, and receptors for a number of other cytokines are present in the hypothalamus.

Cytokine Actions in the Hypothalamic–Pituitary–Adrenal Axis

IL-1β acts in the hypothalamus to increase CRH production. TNF-α and IL-6 also increase hypothalamic CRH production. CRH, produced in increased amounts, acts in the pituitary to stimulate the production of ACTH and other POMC-derived proteins (including β-endorphin and α-MSH). The action of IL-1β to increase ACTH production appears to be dependent strictly on an increase in CRH production. Stated differently, IL-1β does not act directly in the pituitary to increase ACTH production but rather through its action to increase CRH production. ACTH, in turn, acts in the adrenal cortex to stimulate the production of cortisol, the principal glucocorticoid in the human. IL-6 and TNF-α also act in the pituitary to stimulate the production of ACTH and thus, cortisol.

Effects of Cortisol on the Immune System

The increase in cortisol secretion by the adrenal cortex in response to the actions of IL-1β, IL-6, and TNF-α on the hypothalamus and pituitary provokes many profound effects on multiple organ systems. These changes also are effected by the administration of synthetic glucocorticoids such as dexamethasone and betamethasone. In the immune system, cortisol has profound suppressive effects on immune cell trafficking and function. Cortisol inhibits the chemoattraction and accumulation of immune cells at the site of infection or inflammation. Cortisol acts to decrease the production of IL-1β, IL-6, and TNF-α by immune cells and other cells. These actions of cortisol are effected by a decrease in the transcription of the genes for these cytokines and several chemokines (including a neutrophil chemoattractant protein, IL-8) as well. Cortisol also promotes instability of mRNAs for some cytokines, thereby limiting the levels of cytokines that are attained.

The effects of cytokines also are limited by cortisol because it inhibits the expression of plasma membrane receptors for cytokines and membrane-associated adhesion molecules that are critical for the actions of several cytokines. The effects of inflammatory cytokines to stimulate the production of other mediators of the inflammatory response, such as prostaglandins, platelet-activating factor, nitric oxide, and others, are markedly attenuated by cortisol.

Cortisol also acts to inhibit the production and action of several growth factors that stimulate the proliferation of selected immune cell populations. Cortisol is a potent immunosuppressor of Th1 cells, a subset of T cells called *type 1 helper cells*. Some T cells and eosinophils respond to cortisol by apoptosis (a sequence of specific events in the life of a cell that eventuates in cell death). Cortisol acts to decrease the production of antibodies by B lymphocytes.

Other Effects of Inflammatory Cytokines on Adrenal Steroidogenesis

Whereas TNF-α, together with IL-1 and IL-6, acts to increase cortisol secretion by the adrenal cortex, it also acts to inhibit angiotensin II-induced aldosterone synthesis.

IL1-β Stimulates the Production of POMC-Derived Proteins

In addition to ACTH, the levels of other POMC-derived proteins and peptides are increased in response to IL-1β. Notable among these are β-endorphin and α-MSH. β-Endorphin, an endogenous opioid, effects analgesia. α-MSH has antipyretic and anti-inflammatory actions. Both α-MSH and ACTH act to oppose the effects of inflammatory cytokines by inhibiting the release of CRH in response to IL-6.

Physiologic and Pathophysiologic Importance of Interactions between the Immune System and Hypothalamic–Pituitary–Adrenal Axis

Interactions between the immune system and the hypothalamic–pituitary–adrenal axis comprise an important negative feedback mechanism. Specifically, the cytokine-mediated stimulation of the hypothalamic–pituitary–adrenal axis leads to the production of glucocorticoid (cortisol in the human) in large amounts, which in turn suppresses the inflammatory response and most components of the immune system. An intriguing model system illustrates these relationships clearly. The Lewis rat is characterized by failure of CRH production in response to antigenic challenge or infection. As a result, these rats do not produce glucocorticoid (corticosterone in the rat) in response to infection or immune challenge. Acute arthritis develops in Lewis rats that are treated with bacterial toxins. This inflammatory disease is suppressed by the administration of synthetic glucocorticoids (such as dexamethasone or betamethasone). These observations are indicative of the physiologic and pathologic responses of the adrenal to the action of cytokines and the ability to recapitulate these relationships by administration of exogenous glucocorticoid. The normal Fischer rat, in contrast, responds to the administration of bacterial toxin by the development of a normal inflammatory response. Adrenalectomy of the Fischer rat or suppression of adrenal function by the administration of a glucocorticoid antagonist leads to the development of an enhanced susceptibility to inflammatory disease, analogous to that in the Lewis rat. It is postulated that the development of rheumatoid arthritis in humans may be related to decreased cortisol production (or abnormal diurnal rhythm of cortisol secretion), which may be a consequence of an inappropriate response of the hypothalamus to infection or immune challenge.

IMMUNE CELL PRODUCTION OF HYPOTHALAMIC RELEASING HORMONES AND PITUITARY HORMONES

ACTH, β-endorphin, and CRH were once considered to be exclusive peptide markers of the hypothalamic–pituitary–adrenal axis. Each of these factors is now known to be produced by cells of the immune system and in the human placenta. POMC-derived peptides (ACTH and β-endorphin) were the first neuroendocrine peptides to be identified as products of macrophages and lymphocytes. CRH production by mononuclear phagocytes was discovered soon thereafter. Many skeptics were silenced by the demonstration that lymphocyte-derived and adrenal-derived POMC and ACTH have identical amino acid sequences. Similar demonstrations for lymphocyte-derived GH and LHRH had convinced even the most doubtful investigators that cells of the immune system produce these agents.

Now more than 20 different ''neuroendocrine'' factors have been identified as products of T or B lymphocytes, macrophages, mast cells, neutrophils, megakaryocytes, splenocytes, or thymocytes. These include CRH and LHRH, ACTH, TSH, follicle-stimulating hormone (FSH), luteinizing hormone (LH), prolactin, endorphins, [Met]enkephalin, arginine vasopressin (AVP), oxytocin, somatostatin, vasoactive intestinal (VIP) polypeptide, substance P, and neuropeptide Y. Moreover, the regulation of neuroendocrine peptide production by immune cells is similar to that in the hypothalamus and the pituitary, as pointed out by Blalock. One similarity is that the production of trophic factors is in response to the action of releasing hormones. Another similarity is that the production of these hormones is regulated by negative feedback mechanisms similar to those observed in the hypothalamus and pituitary. But there also are several differences: Recall that the production of POMC-derived peptides in the pituitary is induced by CRH and inhibited by glucocorticoids. The same is true of lymphocyte-derived POMC-derived peptides, except that the actions of CRH and glucocorticoid in lymphocytes is indirect. CRH stimulates the production of IL-1β by macrophages; IL-1β induces B-cell production of POMC. Glucocorticoids inhibit the production of IL-1β by macrophages, thereby blocking POMC production indirectly.

The functions and regulation of secretion of these immune cell–derived releasing hormones, trophic factors, and protein hormones have been investigated. Our current understanding of the regulation of production and secretion of these factors is based in part on *in vitro* studies conducted with cells in culture. A number of different potential paracrine actions of these neuroendocrine factors have been discovered, as the following examples illustrate:

1. CRH increases natural killer (NK) cell activity, likely by stimulating POMC, and thus, β-endorphin production in B lymphocytes.
2. TRH acts on T cells to promote the synthesis and release of TSH, which increases antibody production by B lymphocytes.

3. β-Endorphin, produced by macrophages, lymphocytes, and plasma cells, acts in a paracrine fashion on the opioid receptors on peripheral sensory neuron terminals as an analgesic.
4. T lymphocytes synthesize and release prolactin. These cells also respond to prolactin, which serves to facilitate the proliferative action of IL-2 in these cells.

The physiological and pathophysiological relevance of these findings are intriguing areas of study. Again, as stated by Blalock, who would have believed that an opiate antagonist, acting to block the action of lymphocyte-derived endorphins would be more effective in prolonging allograft survival than an immunosuppressive drug like cyclosporin? Or that a dopamine agonist that blocks prolactin release would be effective in treating experimental allergic encephalomyelitis? Or that GH would facilitate human lymphocyte reconstitution of mice with severe combined immunodeficiency? Or that the inflammatory actions of IL-1β would be attenuated markedly by α-MSH? There is little doubt that the practice of medicine will be profoundly affected as the bidirectional responses between the endocrine and immune systems are deciphered.

EFFECTS OF CYTOKINES ON THE HYPOTHALAMIC–PITUITARY–THYROID AXIS

Like hypothalamic–pituitary–adrenal function, hypothalamic–pituitary–thyroid function is suppressed by inflammatory cytokines. IL-1β and TNF-α act to inhibit the production of hypothalamic TRH and pituitary TSH, leading to decreases in the plasma levels of thyroxine (T_4) and triiodothyronine (T_3). IL-1β also stimulates the production of somatostatin by the hypothalamus, which inhibits TSH secretion by the pituitary. The principal action of IL-1β to inhibit thyroid function is in the hypothalamus, with minimal direct actions on the pituitary. Additionally, TNF-α inhibits glycosylation of TSH, which decreases its potency in the thyroid gland.

Both IL-1β and TNF-α inhibit iodide uptake by the thyroid. IL-1β also reduces the synthesis of thyroglobulin. TNF-α desensitizes the thyroid to TSH and, consequently, the production of T_4 and T_3 in response to the TSH is attenuated markedly. IFN-γ acts in the thyroid gland to block TSH action and to increase the expression of major histocompatibility (MHC) class I and II antigens.

Thus, IL-1β, TNF-α, and IFN-γ act to suppress thyroid function by affecting the hypothalamus, the pituitary, and the thyroid gland. It is possible, but not proven, that these direct actions of cytokines contribute to the development of *euthyroid sick syndrome*, a description of disorders characterized by decreased plasma levels of T_3 (see Chapter 13).

Cytokines, the Thyroid Gland, and Autoimmune Disease

Another postulated action of cytokines in the thyroid relates to the development of autoimmune disorders that affect this gland. Thyroid autoimmune disease, caused by the production of antibodies against self antigens expressed on cells of the thyroid gland, is characterized by impaired thyroid function. Normally, only non-self antigens are recognized by T cells as targets for immune response. The production of anti-thyroid autoantibodies is induced in response to thyroid infiltration of T lymphocytes, which produce IFN-γ. The expression of MHC class II antigens is induced by IFN-γ. Because the expression of MHC antigens in the thyroid is usually rather low, the high level of expression in response to IFN-γ is believed to cause a breakdown in recognition of and distinction between self and non-self. Additional lymphocytes are recruited to the site and commence antibody production against antigens presented in association with these MHC molecules.

CYTOKINES AND THE HYPOTHALAMIC–PITUITARY–OVARIAN AXIS

In women with acute or chronic severe illness or infection, delays in ovulation and even chronic anovulation resulting in amenorrhea or oligoamenorrhea are common. This process is associated in part with decreased gonadotropin secretion and thus, hypogonadotropic hypogonadism. The involvement of cytokines in this process also is implied from the observation that the altered state of ovarian function is temporary; namely, cyclic ovulation returns once the illness is resolved. A variety of studies conducted in experimental animals indicate that cytokines may modulate the normal function of the anterior pituitary by way of a direct action of the cytokine on pituitary function. Ovarian estradiol-17β, secreted in large amounts by the mature graffian follicle, serves as the stimulus for the release of LH (the LH surge) that eventuates in the induction of ovulation by the mature follicle. IL-1β acts to inhibit the estrogen-induced surge in LH release, thereby supporting the hypothesis that cytokines are involved in causing the anovulation observed in women with severe illness.

Effects of Cytokines on Hypophyseal LH and FSH

In pituitary cells in culture, IL-1β induces the release of LH and FSH, although this finding is not readily reproducible *in vivo*. Pituitary cells also produce cytokines, notably IL-6. This observation has led to the suggestion that IL-6 may serve as an intrapituitary releasing factor. IL-1β acts upon pituitary cells to cause increased synthesis and release of IL-6; and acting directly upon pituitary cells, IL-1β and IL-6 promote the release of FSH and LH in amounts similar to those evoked in response to gonadotropin-releasing hormone (GnRH). Other cytokines, namely, activin and inhibin (members of the TGF-β family) of gonadal origin, act to modulate the release of FSH from the pituitary (for details see Chapter 9).

Cytokine Action in the Ovary

There is appreciable evidence that cytokines may modify gonadal function in a physiologic and pathophysiologic sense. Pro-IL-β mRNA levels increase in the macrophage-enriched ovarian theca interna at about the time of ovulation; this response appears to be specific because an increase in the level of this mRNA is not identified in peripheral blood monocytes during the same time of the ovarian cycle. Granulosa cells of the ovary also produce IL-6 and the production of IL-6 by granulosa cells can be increased by treatment of these cells with IL-1β. There is a growing body of evidence that IL-6 and other chemokines modulate the rate of formation of steroid hormones in the gonads and adrenal cortex and in extraglandular tissues as well. For example, the rate of aromatization of C_{19}-steroids (estrogen formation) in adipose tissue stromal cells is increased strikingly by treatment with IL-11 and oncostatin-M.

IL-6 inhibits the gonadotropin-dependent formation of progesterone in granulosa cells. TGF-β1 and TGF-β2 are produced by the granulosa cells and the rate of formation of TGF-β2 is increased specifically by treatment with FSH. In addition, it has been shown that TGF-β acts in a number of steroid-producing endocrine cells (including the adrenal cortex) to inhibit steroidogenesis. This action of TGF-β is attributable in part to inhibition of the transcription of the gene that encodes steroid 17α-hydroxylase/17,20-desmolase.

It has been stated often that there is a delicate endocrine balance involving the brain, pituitary, and ovary that is essential for optimal ovarian function, which culminates in the rupture of a single mature follicle with ovulation. This premise, however, is almost certainly incorrect. The regularity of cyclic ovulation in women, despite rather wide differences in the peripheral levels of gonadotropins and ovarian steroids, is testimony to the fact that a delicate endocrine balance is not essential to normalcy of ovarian function. Moreover, a variety of environmental, psychological, and pathophysiological processes may come to bear without affecting the regular cyclic ovarian function of most women. These observations suggest that there is, in addition to the brain–pituitary–ovarian axis, an intraovarian control system that modulates external stimuli in such a manner as to favor normal ovarian function with ovulation despite wide variations in the environment and in bodily function. It is quite possible that the cytokines, produced in the ovary *in situ*, contribute to this intraovarian control. But in addition to this, the formation of cytokines in response to acute or severe illness may act at the level of the brain or pituitary to forestall ovulation and thereby prevent pregnancy during times of environmental or physical danger. Thus it can be envisioned that the cytokine family of proteins acts in a manner that modulates ovarian function—normalizing ovarian steroidogenesis and ovulation during mild fluctuations in external forces, but temporarily halting ovulation during times of impending disaster.

Cytokine Action in the Testis

Infection and illness also are associated with impaired testicular function. As in the ovary, IL-1β acts to inhibit steroidogenesis in the Leydig cells of the testis. TNF-α potentiates this inhibitory action of IL-1β. IL-6 is produced in the Sertoli cells in response to FSH. Interferons (IFNs), which are produced in the gonads, also act in an autocrine and paracrine manner to inhibit steroidogenesis.

CYTOKINES AND A STEROID-RESPONSIVE TISSUE: THE ENDOMETRIUM

The *endometrium* is a tissue that serves an essential role in reproduction and it is a classic example of sex steroid hormone responsiveness (Chapter 9). Normal function of the hypothalamic–pituitary–gonadal axes in women and men, sperm fertilization of oocyte, and blastocyst development are futile in the absence of the endometrium. Under the influence of estradiol-17β, the endometrium grows at a very rapid rate that is characterized primarily by an increase in volume due to a rapid increase in the glandular epithelium of this tissue. After ovulation, with the formation of progesterone by the corpus luteum, the endometrium is modified to a ''secretory'' type that also is associated with very specific changes in the stromal component of this tissue. The stromal changes include an increase in the extracellular fluid and extracellular matrix. The mesenchymal cells of the endometrium—those that form the stroma—are modified in a process referred to as *decidualization*. Presumably, the decidualized endometrium is ideal for blastocyst implantation and placentation.

Among all human tissues, the endometrium is one of the most distinctive. It is the only tissue that undergoes rapid growth at cyclic intervals, only to be purposefully targeted for death by hypoxia and necrosis at the time of menstruation and endometrial shedding during nonfertile cycles.

The endometrium is shed in most women no fewer than 450 times, i.e., with each nonfertile cycle. The process of endometrial growth and desquamation is unique to only a few primates, viz., humans, Old World monkeys, and the great apes. This is the case because these are the only species in which the endometrium is supplied with blood by a peculiar set of arteries referred to as the *spiral* or *curling arteries*. The response of these vessels to the endocrine events of the ovarian cycle and to modifications in the volume of the endometrium gives rise to a series of vascular changes that support endometrial growth, endometrial tissue modifications for blastocyst implantation, and (in the absence of implantation) tissue hypoxia, necrosis, and desquamation with hemorrhage.

Bone Marrow–derived Cells of the Endometrium

During the course of each endometrial cycle, there is a modification in the cellular components of the endometrium due to striking, but predictable, changes in the number and type of bone marrow–derived cells.

One of the more intriguing cells of the endometrium is the so called large granular lymphocyte which is present in secretory endometrium and in decidua in very large numbers. These cells are a very particular type of lymphocyte and are present in peripheral blood in very small numbers. It is postulated, therefore, that these cells originated in bone marrow, but in response to the endocrine changes of the ovarian cycle, replicate *in situ* in the endometrium and decidua. A role for prolactin in the modulation of function of these cells is especially attractive. These cells, in turn, are believed to serve as regulating trophoblast invasion into the endometrium during the time of blastocyst implantation and placentation. The cytotrophoblasts of the invading placenta do not express MHC-type II antigens; thus it is believed that this is one mechanism whereby the invading allogeneic tissue of the fetus is accepted. But in human pregnancy with the formation of the hemochorioendothelial type of placenta, the invasiveness of the trophoblasts is extensive, involving penetration of the endometrial epithelium, stroma, and the direct invasion and destruction of the walls of the spiral arteries that supply the endometrium with blood. Therefore, some mechanism must be in place to halt the trophoblastic invasion; otherwise, each pregnancy would constitute a case of choriocarcinoma, or malignancy of the trophoblast.

It is believed that the large granular cells act to permit and thus inhibit the invasiveness of the trophoblast by way of the expression of an MHC-type I antigen, called *HLA-G*, that is monomeric, i.e., it is identical in all persons. It has been established that the immune cells recognize not only foreign antigens (non-self antigens) but also recognize the lack of self (or "missing self") in an adjacent cell.

Cytokines, Hormones, and Blastocyst Implantation

In humans and other catarrhine primates (specifically, great apes and Old World monkeys), blastocyst implantation includes trophoblastic invasion of the endometrium and endometrial blood vessels to establish a hemochorioendothelial type of placentation. With this type of placental formation, maternal blood comes into direct contact with the trophoblasts, which are of fetal origin. Thus the maternal blood is in direct contact with an allogeneic tissue graft since paternal genes are expressed in fetal tissues. Immunologists and reproductive biologists alike have been fascinated by the fact that the foreign tissue graft of the fetus is accepted by the maternal tissues without evidence of immunological rejection. The involvement of cytokines in this process of maternal acceptance of the fetal graft is almost certain. Moreover, the action of cytokines, produced by one of several cells of the potential implantation site is virtually assured in the regulation of this process. It has been demonstrated that a host of cytokines are produced by the endometrium and the decidua. Some of these cytokines are produced primarily, if not exclusively, by the bone marrow–derived cells normally present in the endometrium and decidua, and others are formed by the decidual cells per se.

Cytokines, Hormones, and Endometrium as an Endocrine Organ

In addition to the responsiveness of the endometrium to sex steroid hormones, the decidua (the specialized endometrium of pregnancy) also is a source of endocrine secretions; namely, the decidua produces prolactin in enormous quantities. The prolactin produced by the decidua does not enter the peripheral circulation of the mother but rather is transported exclusively to amniotic fluid. Moreover, the factors that regulate the release of prolactin by the anterior pituitary (dopamine, TRH) are not operative in the decidual cell to modulate the rate of prolactin formation in this tissue. The role of the prolactin produced in the endometrium has not been defined. Several proposals have been made. The most common is that prolactin acts on the fetal membranes, which are physically contiguous with the decidua during pregnancy, to modulate the rate of water and solute transport to and from the amniotic fluid. Direct evidence for this proposition has not been presented, however. The intriguing relationship between prolactin and lymphocytes may be important in the maintenance of pregnancy and the development of the fetus. Prolactin acts on lymphocytes to increase immunological responsiveness; the formation of prolactin by lymphocytes has been demonstrated.

TGF-β, Steroid Hormones, Vasoactive Peptides, and Endometrial Function

It is also suspected that TGF-β serves an important role in the modulation of immune and endocrine functions in the endometrium. Menstruation is ultimately the result of the withdrawal of steroid hormones produced by the corpus luteum, viz., progesterone and estradiol-17β. This process comes about as follows: the endometrium, as stated above, undergoes rapid growth during the follicular phase of the ovarian cycle—this is referred to as the proliferative phase of the endometrial cycle. The spiral arteries, which are so named because of an extraordinary propensity of these vessels to curl and twist, also grow; indeed, these vessels lengthen at a rate far greater than the rate of increase in endometrial tissue volume. This disparity in growth between the two tissues results in even greater curling and spiraling. It has been suggested that the rate of blood flow to the endometrium may be regulated in a paracrine manner by vasoactive peptides produced in the endometrial stromal cells—namely, parathyroid hormone–related protein (PTH-rP; Chapter 15), a potent vasodilator, is produced in the endometrial stromal cells, as is endothelin-1 (ET-1; Chapter 7), a potent vasoconstrictor. PTH-rP formation is increased in response to progesterone treatment of the endometrial stromal cell, whereas the expression of prepro-ET-1 is inhibited by progesterone treatment. TGF-β acts in the stromal cells to counteract the action of progesterone. Thus TGF-β acts as an antiprogestin by way of a progesterone receptor–independent process in a gene-specific manner to modulate the function of endometrial cells. The expression of TGF-β in the endometrium increases at the time of decidualization and at the time of blastocyst implantation. Possibly, the increase in TGF-β expression first observed in the endometrium is the result of the early stages of hypoxia that come about because of the further coiling of the spiral vessels as progesterone levels begin to decline during the late secretory phase of the endometrial cycle. This comes about as follows: with the great lengthening of the endometrial spiral arteries that accompanies endometrial tissue growth, there is even greater curling of the vessels. With progesterone withdrawal, there is a reduction in endometrial tissue volume without a reduction in the spiral artery length. This creates further coiling of these vessels, which becomes so severe as to impede blood flow. With the reduction in blood flow, hypoxia begins. As the volume of the endometrium is reduced even further, the coiling of the spiral arteries becomes so severe that blood flow is almost completely impeded. Hypoxia becomes severe, resulting in necrosis of this tissue. Just prior to the onset of bleeding that accompanies menstruation, however, the spiral arteries enter into a phase of severe vasoconstriction. This vasoconstrictive phenomenon is believed to be the consequence of ET-1 action via the adventitial surface of the spiral arteries; the ET-1 originates in the contiguous stromal cells of the endometrium, which also is the site of TGF-β formation. Until recently, it was believed that progesterone withdrawal was the cause of the onset of menstruation. It is now clear that this concept was an overly simplistic view of the physiological processes involved. In retrospect, it is easy to understand that a process as physiologically important as menstruation and preparation for a renewed opportunity for pregnancy could not be left to the uncertainties of progesterone withdrawal. Rather, the regulation of endometrial blood flow by locally produced vasoactive peptides is a much more reasonable explanation for the lack of excessive blood loss in most women with menstruation; namely, ET-1 acts as the tourniquet of the endometrium. By promoting severe vasoconstriction of these vessels at the time of menstruation, blood loss is limited.

In the endometrium, as in other tissues, TGF-β is a potent inducer of ET-1 formation. Thus TGF-β is poised to participate in the process of implantation by promoting decidualization of the endometrium or in the process of menstruation by promoting the formation of the vasoconstrictor, ET-1. TGF-β acts to promote its own formation, i.e., autoinduction. In this manner, TGF-β can be produced in the endometrium in response to generalized changes in the endometrial environment (for example, sex steroid hormone modifications or hypoxia) or to the secretions of the implanting blastocyst, which includes the trophoblastic formation of TGF-β. Indeed, with implantation there is a wave-like propagation of the TGF-β immunoreactivity in the endometrium that begins at the site of implantation and proceeds outward to involve the entire endometrium in a decidual response.

Interestingly, PTH-rP is produced in many normal, nonmalignant tissues, and there appears to be an especially high production of PTH-rP in tissues of reproduction. Very large amounts of PTH-rP are produced in the mammary tissue of lactating women and experimental animals and PTH-rP enters the milk in very large quantities. PTH-rP is also produced in the myometrium during pregnancy and in the decidua of women. PTH-rP is produced in the placenta and in other fetal organs including the kidney and parathyroid gland. In fact, it is now believed that PTH-rP is the parathormone of the fetus. It is envisioned that the formation of PTH-rP in the endometrium serves not only to increase vasodilatation and endometrial and decidual blood flow but also to facilitate the transfer of calcium across the trophoblast after implantation of the blastocyst. TGF-β also acts to increase the rate of formation of PTH-rP in the endometrium and particularly in the endometrial stromal cell, the progenitor cell of the decidua. PTH-rP, but not PTH, will effect a transfer of calcium across the trophoblast from mother to fetus.

The endometrium, therefore, is virtually a prototype tissue for an examination of the relationships between immune and endocrine functions. The endometrium also acts to control its own cytokine destiny; it has been demonstrated that chemoattractant proteins are produced by the mesenchymal cells of the endometrium. IL-8, which is a potent chemoattractant for neutrophils and T lymphocytes as well as monocyte chemoattractant protein-1 (MCP-1), is produced in the endometrial stromal cells and the rate of formation of these chemoattractants is modulated by the sex steroid hormones of the ovarian cycle. It is likely, therefore, that the endometrium, in response to the changing sex steroid hormone milieu of the cycle, is able to attract the proper bone marrow–derived cells into this tissue to further modulate the function of the endometrium. At the same time, the endometrium is poised to respond rapidly to infection. As a consequence of the anatomic reality of the female genital tract, the endometrium is potentially accessible to the outside environment via the patent cervical and vaginal canals. Infection in the endometrium during pregnancy would represent a great risk for septic abortion or preterm labor. The containment of infection by the efficient immune system of the endometrium undoubtedly contributes to the high success rate of pregnancies, despite the finding of microorganisms in large quantities in the vaginal fluid of all women.

SUGGESTED READING

Blalock JE: The syntax of immune–neuroendocrine communication. Immunol Today 15:504–511, 1994.

Chrousos GP: The hypothalamic–pituitary–adrenal axis and immune-related inflammation. N Engl J Med 332:1351–1362, 1995.

Clemens MJ: *Cytokines*. Information Press Ltd, Oxford, U.K., pp. 1–121, 1991.

Fukata J, Imura H, and Nakao K: Cytokines as mediators in the regulation of the hypothalamic–pituitary–adrenocortical function. J Clin Invest 16:141–155, 1993.

Gaillard RC: Neuroendocrine–immune system interactions: The immune–hypothalamo–pituitary–adrenal axis. Trends Endocrinol Metab 5:303–309, 1994.

Hamblin AS: *Cytokines and Cytokine Receptors*, IRL Press, Oxford University Press, New York, 1993.

Reichlin S: Neuroendocrine–immune interactions. N Engl J Med 329:1246–1253, 1993.

Tabibzadeh S: Cytokines and the hypothalamic–pituitary–ovarian–endometrial axis. Hum Reprod Update 9:947–967, 1994.

5

Assessment of Endocrine Function

JAMES E. GRIFFIN

The first step in the laboratory assessment of endocrine function is usually to measure hormone levels. The ability to measure hormone concentrations accurately in plasma and urine has greatly enhanced our understanding of normal endocrine function. Perhaps the single most important factor in the development of modern endocrinology has been the discovery and use of the radioimmunoassay to measure the small quantities of hormones that circulate in the blood. Because static hormone measurements may not provide sufficient insight into the status of a given endocrine system, a variety of dynamic endocrine tests have been developed. These dynamic endocrine tests usually take advantage of a known action of a hormone or of the mechanisms that control its secretion in order to probe for evidence of an abnormality of endocrine function. And as we have gained understanding of the receptor mechanisms of hormone action, assessing the number and binding affinity of a hormone's receptors in certain target tissues has become increasingly important. This chapter deals with these various ways of assessing endocrine function.

PLASMA HORMONE LEVELS

Types of Assays

Hormones circulating in the plasma were first detected by *in vivo* bioassays in which plasma or extracts of plasma were injected into animals and a biological response was measured. Unfortunately, most *in vivo* bioassays lack the precision, sensitivity, and specificity to measure the low concentrations in plasma and are inconvenient to perform. Real progress in measuring plasma hormone levels came over 30 years ago with the use of competitive binding and subsequently the development of the radioimmunoassay. In both these assays the unknown concentration of a hormone in plasma is estimated by the ability of the hormone to compete for binding to a specific binding protein (antibody in the case of radioimmunoassays) with a labeled hormone. By comparing the competition of a known amount of hormone in a reference standard with the plasma or plasma extract, the amount of hormone in the plasma can be estimated. The observation that almost any hormone could be used as an antigen to develop a specific antibody meant that the technique could be applied to hormones without known binding proteins. In most such immunoassays, the hormone is labeled with an isotope (radioimmunoassay); however, nonisotopic immunoassays have been developed for some hormones in which an enzyme or fluorescent tag is used instead of an isotope. In some instances, these nonisotopic assays are quite sensitive and readily automated. Since assays that do not involve radioactivity result in less equipment costs and no potential risk of exposing personnel to radioactivity, it is likely that they will receive greater attention in the future. At present, the radioimmunoassay is the most widely used method of assessing hormone concentrations, and radioimmunoassays have been developed for all polypeptide, steroid, and thyroid hormones.

For some hormones, the radioimmunoassay is being replaced by an another form of immunoassay termed the *immunometric assay*. These immunometric assays often have better sensitivity and specificity than the radioimmunoassay and can be accomplished in a shorter period of time (see below).

While radioimmunoassays provide information about immunoreactive hormone in the plasma, they do not necessarily correlate with biological activity of the measured hormone. The site in the molecule to which the antibody is directed may not be the site involved in receptor binding (see Chapter 3). Thus an attempt was made to utilize the interaction of the hormone with its specific receptor in radioreceptor assays and in certain *in vitro* bioassays. In a radioreceptor assay the specific receptor sites on cells or cellular components are substituted for the antibody as a binding protein for the hormone (or ligand). Thus the specificity of binding is determined not by the antibody determinants but by the biological receptor that mediates hormone action. Such radioreceptor assays can add to the information obtained from radioimmunoassays by assessing in some sense ''bioactive'' hormone, as when there is heterogeneity of a polypeptide hormone (e.g., insulin or growth hormone) in plasma. The radioreceptor assay may also be useful in assessing total activity in plasma related to a class of compounds interacting with one type of receptor, such as competition for a labeled glucocorticoid binding to the glucocorticoid receptor. Likewise, potential drug interference with androgen receptor binding has been assessed by competition of such compounds with labeled androgen for binding to the androgen receptor. Radioreceptor assays are also useful in assessing the presence of autoantibodies to hormone receptors in the plasma of some subjects with endocrine disorders, for example, antibodies to the thyrotropin or insulin receptors.

Although radioreceptor assays give some clue to the biological activity of a hormone, they measure only the first step in hormone action, the binding of hormone to its receptor. They do not assess the effects of hormone–receptor interactions in generating second messenger(s) or in stimulating a specific response. Thus, the assessment of plasma hormone levels has come full circle with the development of *in vitro* bioassays. These assays involve the incubation of plasma or plasma extracts with endocrine tissues, membrane preparations, or cultured cells. Perhaps the most extensively used of these *in vitro* bioassays is the dispersed rat Leydig cell assay for luteinizing hormone (LH) and human chorionic gonadotropin (hCG) in which the testosterone response to gonadotropic stimulation is measured. This assay has greater sensitivity than the radioimmunoassay or radioreceptor assays. It permits the detection of bioactive LH when immunoreactive LH is unmeasurable, and has made it possible to demonstrate physiological regulation of the qualitative aspects of LH secretion in the form of changing bioactive to immunoactive ratios of LH in the plasma. The exact chemical nature of these qualitative differences in circulating LH is not fully understood.

Principles of Radioimmunoassay

The principle underlying radioimmunoassay of hormones is the competitive inhibition of binding of a radiolabeled antigen (Ag*) (hormone) to antibody (Ab) by unlabeled antigen (Ag) present in a reference hormone standard or in a plasma sample (the unknown):

$$Ag^* + Ab \underset{K_d}{\overset{K_a}{\rightleftharpoons}} Ag^* - Ab$$

$$Ag + Ab \underset{K_d}{\overset{K_a}{\rightleftharpoons}} Ag - Ab$$

The antigen–antibody reaction is reversible, as shown by the arrows, with an asscciation rate constant of K_a and dissociation rate constant of K_d. The rate of dissociation is slower than the rate of association, and at equilibrium the observed amount of antigen bound to antibody is a result of the combined rates of association and dissociation. Following an appropriate period of incubation the antibody-bound antigen (B) is separated from the unbound or free antigen (F) by one of several different methods. Since a fixed amount of antibody is present in each assay tube, the amount of labeled hormone bound to antibody (the measured radioactivity) is a function of the concentration of unlabeled hormone in either the standard or the unknown sample. The higher the concentration of unlabeled hormone, the lower the quantity of radioactive hormone bound to the fixed amount of antibody. The results of a radioimmunoassay can be plotted several different ways. The plot of the ratio of B/F labeled hormone, or the ratio of the amount bound in the presence of standard (B) to the amount bound in the absence of any unlabeled hormone (B_0), as a function of amount of standard is curvilinear (Fig. 5-1). In order to facilitate automated calculation of results, logarithmic transformation of the data expressed as B/B_0 that usually results in a straight line is often used. However, one should keep in mind that there are limitations to linear transformation of radioimmunoassay curves in that they obscure the lack of sensitivity at both ends of the standard curve. As shown in the curvilinear plot of Fig. 5-1, the ability to determine reliably the unknown hormone concentration when the B/B_0 ratio falls on the portions of the curve with the least slope (e.g., 5–10 ng and 500–1000 ng) is quite limited despite the ability of logarithmic transformation of the data to give calculated values. A requirement for validation of a radioimmunoassay is that a dilution curve of the reference standard and the hormone to be measured must exhibit parallel dose–response lines as evidence of similar interaction with the antibody.

Antibodies suitable for radioimmunoassays are usually of the IgG class of immunoglobulins. Partially purified hormones are injected into laboratory animals

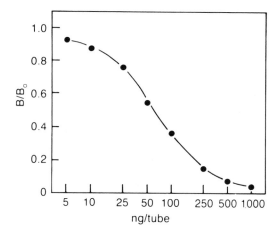

Fig. 5-1 Standard curve for radioimmunoassay of plasma testosterone. The ratio of the amount of radioactive testosterone bound in the presence of standard (*B*) to the amount bound in the absence of any unlabeled hormone (B_0) is plotted as a function of the amount of standard (unlabeled hormone) in each assay tube. Note that it is a semilogarithmic plot and that assay sensitivity is low at both extremes of hormone concentration.

to induce synthesis of immunoglobulins with sufficiently high sensitivity and specificity. If the hormone has a molecular size less than 1000 Da, it may be necessary to couple it to a carrier protein such as albumin to make it sufficiently immunogenic. In general, small peptides, steroids, and thyroid hormones need to be coupled to carrier proteins. Antibodies produced in this manner are polyclonal. It is also possible to produce monoclonal antibodies to hormones. For routine use, however, these monoclonal antibodies have not proved superior to traditional polyclonal antibodies because in general they are not of the highest affinity that might be found in a population of polyclonal antibodies.

Peptide and thyroid hormones are labeled with [125]I. For peptide hormones iodination is performed using $Na^{125}I$ to incorporate the label into tyrosyl or histidyl residues in the hormone. With small polypeptides lacking tyrosine or histidine in their structure, it is necessary to add a tyrosyl residue or substitute it for another acid in a manner to minimally alter the conformation of the polypeptide. For optimal stability and immunoreactivity only one atom ^{125}I is incorporated per mole of hormone. With steroid hormones ^{3}H is incorporated into the structure with a variable number of ^{3}H atoms per molecule to achieve the desired specific activity. The presence of ^{3}H in the structure does not result in any difference in immunoreactivity compared with unlabeled hormone.

Antibody-bound and free hormone can be separated by several different methods. The most widely used method for polypeptide hormone assays is the precipitation of the antigen–antibody complex with a second antibody. Thus, if the first antibody (to the hormone) was generated in a rabbit, an antibody to rabbit γ-globulin generated in another species is used to precipitate the original antigen–antibody complex. The most common means of separating antibody-bound and free hormone in steroid radioimmunoassays is by using dextran-coated charcoal. The unbound steroid is adsorbed to the surface of the charcoal and centrifuged to the bottom of the tube. The soluble antigen–antibody complex remains in the supernatant.

Problems in the Interpretation of Radioimmunoassays

In selecting an antisera for use in a given assay one must consider its specificity to avoid problems in the interpretation of assay results. The glycoprotein hormones LH, follicle-stimulating hormone (FSH), thyroid-stimulating hormone (thyrotropin, TSH), and hCG are examples of a group of related hormones that may present problems in assay specificity. Since they contain a common α-subunit, antisera generated to the intact hormone may cross-react with other members of the group. Since it is the β-subunit of each of these hormones that confers immunological and biological specificity, it is necessary to select antisera that recognize this component of the molecule if the assay is to have appropriate specificity.

Furthermore, there may be problems in assay specificity due to heterogeneity in the forms of a given peptide hormone that circulates in the blood. Many peptide hormones have glandular precursors (see Chapter 2) with different biological activities that circulate to a greater or lesser extent depending on the clinical situation. Likewise, some peptide hormones have inactive metabolites that normally circulate but may be present in greater concentrations in given disease states. Thus proinsulin may at times be present in the circulation with full immunoreactivity but less biological activity than insulin itself. And in the case of hormone metabolites, it is now recognized that a C-terminal portion of the parathyroid hormone (PTH) molecule is a metabolite that has a longer half-life than the intact hormone and is measured with some antisera in addition to intact PTH. The increased levels of the C-terminal fragment that accumulate in states of diminished renal function can result in inaccurate assessments of PTH. Fortunately, other antisera are available that recognize the intact molecule primarily by being directed to determinants on the N-terminal portion of the molecule (see Chapter 15).

Some peptide hormones are quite sensitive to proteases present in serum. The degradation of hormone before assay may result in artifactually low levels of hormone being measured since the degraded hormone may not react with antibody if its structure is sufficiently altered. In such instances, protease inhibitors may be added to the tube before the blood sample is collected, or the blood may be quickly refrigerated to inhibit enzyme activity. Corticotropin (ACTH), for example, requires such special attention to avoid degradation.

Circulating endogenous antibodies to hormones provide an additional problem in the interpretation of radioimmunoassays. Fortunately, such antibodies rarely occur spontaneously; however, antibodies to insulin are common in diabetic patients who are receiving exogenous insulin. The effect of a circulating antibody to the hormone on the usual double antibody radioimmunoassay is a spuriously high estimated concentration of the hormone in the plasma sample. The human circulating antibody will bind labeled hormone in addition to the assay antibody obtained from an animal species. However, the addition of a second antibody will only precipitate the labeled hormone bound to the animal antibody, resulting in apparently low bound radioactivity that is interpreted as high concentrations of unlabeled hormone in the unknown sample. When circulating antihormone antibodies are suspected, their presence can be detected by omitting the assay antibody and incubating the labeled hormone with the patient's plasma and then using agents such as ammonium sulfate to precipitate the hormone–antibody complex.

Characteristics and Advantages of Immunometric Assays

To achieve optimal sensitivity in the classical radioimmunoassay, the concentration of the labeled ligand must be critically adjusted, the antibody must be quite dilute, and the incubation must often be conducted for more than 24 hours. The immunometric assay method differs from the radioimmunoassay in several ways and does not have the requirements listed above for achieving high sensitivity. First, there is no labeled hormone or ligand in the assay. Instead, one of two monoclonal antibodies, each offering high specificity, is either labeled with radioiodine or another detection system. The second antibody is usually directed at another antigenic site on the hormone of interest and is coupled to a solid phase (e.g., a bead) for separation. Either antibody may be reacted with the ligand first. However, typically the labeled antibody is involved in the first incubation, and a ''sandwich'' reaction results when the second antibody is added, i.e., the ligand forms a bridge between the two antibody molecules. Second, since the sensitivity is not dependent on competition of hormone with labeled ligand for antibody, the antibodies are generally used in excess to allow all the hormone in a sample to be bound by the antibodies. This condition of antibody excess allows the binding reactions (and the assay) to be completed in only a few hours. Third, in the absence of the competition for antibody binding inherit in radioimmunoassays, the working range of assays is not so limited and may actually span all anticipated values without requiring repeat assays with dilution. Finally, since there is no labeled ligand in the assays, the problem of circulating endogenous antibodies to hormones (see above) is eliminated.

Immunometric assays have a variety of names based on the detection method or label attached to one of the antibodies. The label may be ^{125}I as in immunoradiometric assays, or a nonradioactive molecule may serve as signal such as a fluorophor in immunofluorometric assays, an enzyme in immunoenzymometric assays, or a chemiluminescent molecule in immunochemiluminometric assays. As discussed in Chapter 13, TSH is one of the hormones for which the development of immunometric assays has changed clinical assessment dramatically. There are different immunometric assays with each of the above signal systems developed for TSH. The chemiluminscent detection system may be the most sensitive because of a high signal-to-noise ratio. Immunometric assays are now becoming standard for parathyroid hormone, and it is likely they will have wide application.

Sampling the Plasma

The best way to sample the plasma in order to measure a hormone concentration depends on the given hormone. In the case of hormones whose plasma levels are relatively constant (e.g., thyroxine), the use of an isolated plasma sample provides a reliable estimate of hormonal status under usual circumstances. To assess hormone levels accurately for those hormones with pulsatile secretion (e.g., LH, testosterone), a single plasma sample may or may not be representative of mean plasma levels. With these hormones it is necessary to draw three or more samples at 20- to 30-minute intervals and pool aliquots of each for a single determination. For hormones with significant diurnal or sleep-related secretion the time of sampling may also be important. Because of the circadian rhythm of the ACTH–cortisol axis, for example, one expects higher plasma levels in the early morning hours. And the characteristic timing of an episode of growth hormone secretion soon after an individual falls asleep may be used to guide sampling for diagnostic purposes. In addition, one must recognize that sleep-related secretion may occur only during a certain period of life, as in LH secretion during mid-puberty. In women the plasma levels of gonadotropins, estradiol, and progesterone must be interpreted in light of the phase of the menstrual cycle.

Effect of Plasma-Binding Proteins

Whereas polypeptide hormones circulate in the plasma unbound to other plasma constituents, steroid and thyroid hormones circulate largely bound to albumin and specific binding proteins. Less than 1% of thyroxine (T_4) and triiodothyroxine (T_3) and only about 1–3% of androgens, estrogens, and glucocorticoids in the plasma are in the unbound or free form. The free hormone concentration of these hormones can be determined *in vitro* by a tedious procedure termed *equilibrium dialysis*. For some time, the free fraction was regarded as the biologically active portion of the circulating hormone concentration since it was thought to be the only fraction available for entry into cells and interaction with receptors.

It is now clear that the transport of steroid and thyroid hormones into cells is more complicated than free fraction estimated *in vitro* as being the only component of the hormone in the circulation available to enter cells. Dissociation of protein-bound hormone can occur within a capillary bed so that the active fraction can be larger than the free fraction measured under equilibrium conditions *in vitro. In vivo* methods (Fig. 5-2) are needed to estimate the capillary exchangeable hormone since interaction of binding proteins, such as albumin, thyroid hormone-binding globulin (TBG), or testosterone-binding globulin (TeBG), with the components on the microcirculation surface (e.g., endothelial glycocalyx) may lead to conformational changes at the hormone-binding site on the plasma protein. Such conformational changes may result in markedly increased rates of hormone dissociation from albumin and/or TeBG in the microcirculation that are not measured *in vitro.* Such *in vivo* methods have been developed using arterial injection of animals. Although there are differences for specific hormones, in general it now appears that it is the free *plus* the albumin-bound hormone that is available for entry into most tissues. In the case of plasma testosterone, this free plus albumin-bound component is about half of the total circulating hormone. The delivery of the free plus albumin-bound hormone to target tissues results in an amount of hormone in the tissue that would be necessary for half-saturation of steroid and thyroid hormone receptors at normal plasma concentrations of total hormone.

The laboratory measurement of free plus albumin-bound steroid and thyroid hormones is technically easier than the measurement of free hormone. The free plus albumin-bound fraction can be measured with the ammonium sulfate precipitation

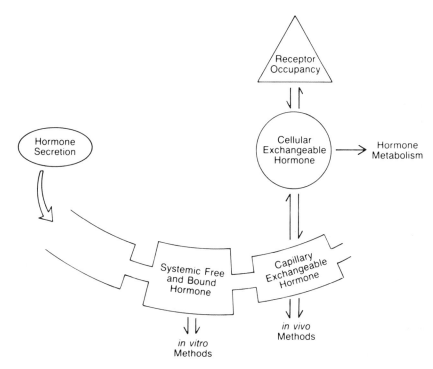

Fig. 5-2   Diagram depicting the concept of the difference between "free" hormone in the circulation as assessed by *in vitro* methods (e.g., equilibrium dialysis) and capillary exchangeable (or tissue available) hormone as assessed by *in vivo* methods. The capillary exchangeable fraction is usually much greater than the systemic free fraction. (Redrawn from Pardridge WM, unpublished.)

technique. This is not widely done at present; instead, total steroid and thyroid hormone concentrations are usually measured. When the level of specific binding proteins is likely to be increased (TBG and TeBG in pregnancy) or decreased (TBG in debilitating illness), some assessment of the amount of binding protein is important in interpreting the total hormone concentrations. TBG and TeBG may be quantitated directly and the amount of thyroid hormone-binding sites may be inferred by the T_3 resin uptake test (see Chapter 13).

Interpretation of "Normal Ranges"

Caution is needed in the interpretation of "normal ranges" for plasma levels of most hormones. Since the stated normal range is often quite broad, extending to a 2- to 3-fold variation (e.g., thyroxine, testosterone), it is possible for the level in an individual to be halved or doubled and still be within the so-called normal range (although abnormal for that individual). Since almost all hormone systems are under some regulatory feedback control, measuring both members of a "hormone pair" (e.g., thyroxine and TSH, testosterone and LH) may provide insights not available from individual values (Fig. 5-3). Although this is classically thought of in relation

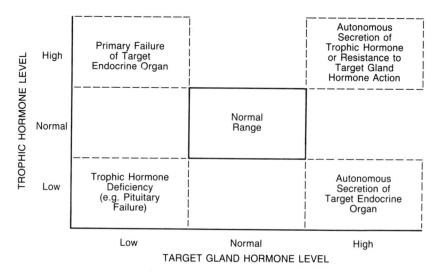

Fig. 5-3   Diagram depicting the possible interpretations of alterations in trophic and target gland hormone pairs (e.g., thyroxine and thyrotropin). (Redrawn with modification from Griffin JE: In: *Williams Textbook of Endocrinology*, 8th ed., JD Wilson and DW Foster, eds., Saunders, Philadelphia, p. 1664, 1992.)

to the pituitary trophic hormones and the hormones secreted by their target endocrine glands (Fig. 5-3), the concept is broader and applies to such hormone pairs as calcium and PTH, insulin and glucose, and plasma osmolality and vasopressin.

This paradigm is central to the assessment of endocrine status. For example, since the normal range of plasma thyroxine is broad, a low-normal T_4 coupled with an elevated TSH indicates early, compensated thyroid failure. Likewise, a high-normal PTH in a patient with a simultaneously elevated serum calcium has a completely different implication from the same PTH value in a patient with a low serum calcium. Basically, the measurement of hormone pairs allows assessment of the effects of a hormone upon its regulatory control mechanism. Other possible outcomes of measuring both members of a hormone pair are also indicated in Fig. 5-3. In the context of the pituitary and its target endocrine glands, finding low levels of both members of the hormone pair indicates the primary problem to be trophic hormone deficiency or pituitary failure (e.g., low TSH and T_4) (Fig. 5-3). High levels of target gland hormone coupled with low levels of trophic hormone suggest autonomous secretion of the target endocrine organ (i.e., typical hyperthyroidism results in suppression of TSH secretion). The finding of elevated levels of both members of a hormone pair is compatible with several different forms of deranged endocrine physiology (Fig. 5-3). Autonomous secretion of a trophic hormone can arise either at the normal site or at an ectopic location; for example, excess cortisol production may be due to the secretion of pituitary ACTH or from secretion of ACTH by lung tumors. Alternatively, the combined elevation of trophic and target endocrine gland hormones can result from resistance to the action of the target endocrine gland hormone (e.g., elevated LH and testosterone in androgen resistance). Insight into which of these two explanations for elevation of both members of a hormone pair is likely often derives from the clinical situation. Autonomous hypersecretion of the trophic hormone typically results in clinical evidence of target gland hormone excess, whereas target gland hormone resistance is usually associated with evidence of hormone deficiency. Knowledge of the relative frequency of autonomous trophic hormone secretion versus resistance to target gland hormone action is also helpful. Thus, since glucocorticoid resistance is rare, elevation of ACTH and cortisol usually signals the presence of autonomous ACTH secretion.

URINARY HORMONE EXCRETION

Measurement of urinary excretion of a hormone or hormone metabolite may reflect overall hormone secretion or mean plasma levels. Thus the 24-hour urinary 17-hydroxycorticoids and/or the urinary free cortisol may be a better estimate of glucocorticoid secretion throughout the day than an isolated plasma cortisol. The 17-hydroxycorticoids represent a summation of glucocorticoid metabolites, whereas the urinary free cortisol "sums" the plasma levels of cortisol above the saturation of binding proteins. Likewise, measurement of urinary metabolites of catecholamines may provide a useful clue to their episodic excess in the plasma.

Urinary hormone measurements have been used less since the development of reliable plasma immunoassays. Not only are reliable timed urine collections difficult to obtain, but urinary hormone measurements have some inherent limitations. In order to ensure completeness of urine collection, total creatinine should be measured on all specimens. Men excrete about 1.5 g of creatinine per day, and women excrete about 1.0 g. Some calculations of normal values for urinary hormone excretion try to account for variably complete collections by expressing the value per gram of urinary creatinine. Since many urinary hormone measurements involve determination of a metabolite rather than the hormone itself, drugs or disease states that alter hormone metabolism may thus alter the urinary measurement without an abnormality of hormone secretion. In addition, some urinary hormone metabolite measurements include hormones from more than one source and, thus, are of little value in assessing specific glandular secretion. Because urinary 17-ketosteroids are a measure of both adrenal and gonadal androgens, for example, they are a poor index of testicular function (see Chapter 10). Finally, since some hormones are not primarily excreted either intact or in metabolite form in the urine, measurement of the small amounts of these hormones present in the urine is obviously of little value. The thyroid hormones are examples of this latter category (see Chapter 13).

HORMONE PRODUCTION RATES

Measuring the actual secretion or production rate of a hormone may overcome most of the problems inherent in plasma hormone levels and urinary hormone excretion in assessing endocrine function. However, such measurements are technically difficult and usually involve the administration of radioisotopes. The production rate can be measured directly by the dilution of an infused radioisotope with endogenous nonradioactive hormone over a period of time. Alternatively, the metabolic clearance rate (MCR) of a hormone may be determined and together with mean plasma concentration (PC) used to estimate production rate (PR) by the formula PR = MCR × PC. In either case, the measurement of hormone production rates remains primarily a research procedure.

DYNAMIC TESTS OF ENDOCRINE FUNCTION

Dynamic endocrine tests provide information about endocrine function beyond that which is obtained from measurements of single hormones or even of hormone pairs. These tests are based on either the stimulation or the suppression of endogenous hormone production. The ultimate functional test of endocrine function is to demonstrate a normal response in target tissues to physiological or stressful stimuli *in vivo*. Thus, for example, if the urine can be maximally concentrated in response to water deprivation one can infer that the osmolality-sensing mechanisms in the hypothalamus, the secretion of vasopressin, the receptor for vasopressin, and postreceptor events in vasopressin action in the kidney are all normal (see Chapter 7). However, such *in vivo* tests of tissue effects are not available for most endocrine systems; and even when they are available, failure of the final expected response does not identify the level of the defect. Other functional tests of endocrine status are therefore necessary.

Stimulation Tests

Stimulation tests are usually used when hypofunction of an endocrine organ is suspected. They are designed to take advantage of known endogenous control mechanisms to assess reserve capacity to produce hormone. A trophic hormone can be administered to test the capacity of the target gland to increase hormone production. The trophic hormone can be a hypothalamic releasing factor such as thyrotropin-releasing hormone (TRH) or a substitute for a pituitary hormone (cortrosyn for ACTH or hCG for LH). The capacity of the target gland to respond is measured by the increase in the plasma level of the appropriate hormone (TSH, cortisol, and testosterone in these examples). Alternatively, a stimulatory test may be performed by producing an increase in the secretion of an endogenous trophic hormone and measuring the effect on a target gland hormone level. For instance, metyrapone is given to block a late step in cortisol synthesis, and the ability of the pituitary to respond is assessed by the increase in ACTH as well as its subsequent effect on a plasma cortisol precursor (see Chapter 14). Stimulation tests in which endogenous trophic hormone secretion is altered actually assess the overall capacity of the hypothalamic–pituitary–target organ axis to respond to challenge.

Some stimulation tests are useful in suspected hyperfunction of endocrine glands. For example, the presence of an exaggerated calcitonin response to pentagastrin or calcium administration may indicate a premalignant change called thyroid C-cell hyperplasia. Another stimulation test useful in deranged physiology associated with endocrine hyperfunction is the TRH test in suspected hyperthyroidism. In hyperthyroid states the response of TSH following TRH injection is blunted, so that in this case a failure of response is a positive test.

One stimulation test is useful in assessing potential hormone resistance. The increase in urinary cyclic AMP and phosphate following PTH administration is blunted or absent in patients with PTH resistance (pseudohypoparathyroidism) but normal in patients with hypoparathyroidism due to PTH deficiency. Tests to assess resistance to other hormones would be helpful in endocrine diagnosis.

Suppression Tests

Suppression tests are used when endocrine hyperfunction is suspected. They are designed to determine whether negative feedback control mechanisms are intact. A hormone or other regulatory substance is administered, and the inhibition of endogenous hormone secretion is assessed. For example, glucocorticoid (dexamethasone) is given to persons with suspected hypercortisolism to assess its capacity to inhibit ACTH secretion and thus cortisol production by the adrenal. In this type of test, failure to suppress indicates the presence of autonomous secretion either of the target endocrine gland hormone or of the trophic hormone. Other suppression tests use glucose in evaluation of suspected growth hormone excess and saline in assessment of excess aldosterone secretion.

Problems in Interpreting Dynamic Tests

A variety of factors influence the response to stimulation and suppression tests. Change in response to a test with age, the need for repeated stimulation to elicit a normal response, the inherent rhythmicity in the secretion of some hormones, and the response characteristics of deranged physiologic processes are important considerations in the interpretation of test results. For example, the TSH response to TRH stimulation is decreased in older men (but not in older women). In subjects with long-standing ACTH deficiency, repetitive stimulation may be required to bring out a normal glucocorticoid response to ACTH stimulation. Similarly, men presumed to have severe hypothalamic disease may show a subnormal LH response to an initial bolus of luteinizing hormone-releasing hormone (LHRH) but respond normally to acute stimulation following a week of daily infusions of LHRH. Apparently, the normal pituitary requires the priming effect of the repeated infusions of the hypothalamic hormone in order to demonstrate its ability to respond to a bolus. Such a protocol may help distinguish between hypothalamic and pituitary hypogonadism - (Fig. 5-4). Some usually reliable dynamic tests of endocrine function are occasionally misleading due to the unusual patterns of response of the specific endocrinopathy. An uncommon pattern of pituitary-dependent hypercortisolism is associated with periodic hormonogenesis resulting in seemingly paradoxical stimulation of cortisol levels in response to dexamethasone suppression testing.

Other disease states may pose difficulties in interpreting dynamic endocrine tests. For instance, hypothyroidism, hyperthyroidism, and hypercortisolism all may cause impaired growth hormone responsiveness to the usual stimuli. The response of growth hormone to insulin-induced hypoglycemia and other stimuli may also be blunted in the presence of obesity. In contrast, patients with severe malnutrition or

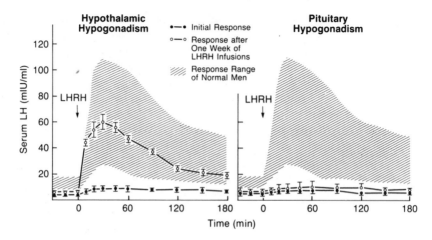

Fig. 5-4   Mean serum LH responses of 10 men with hypogonadotropic hypogonadism to a 250-μg intravenous bolus dose of lutenizing hormone-releasing hormone (LHRH) before and after daily infusions (500 μg over 4 hours) of LHRH for 1 week. Five men had presumed hypothalamic disease and five had presumed pituitary disease. (From Snyder PJ, et al: Repetitive infusion of gonadotropin-releasing hormone distinguishes hypothalamic from pituitary hypogonadism. J Clin Endocrinol Metab 48:864–868, 1979; © 1979, The Endocrine Society.)

chronic kidney or liver disease tend to have elevated basal levels of growth hormone with either lack of suppression or paradoxical increase following a glucose load. In addition, a number of psychiatric disorders are associated with abnormal dynamic endocrine tests in the absence of a specific endocrine disorder. Depression is the psychiatric disorder most often associated with aberrant dynamic endocrine tests. Quite commonly, patients with severe primary depression have an impaired suppression of plasma cortisol following dexamethasone administration, and the cortisol response may return to normal after treatment of the depression. In about one-fourth of all patients with acute psychiatric illness, the response of TSH to TRH is blunted or absent without thyroid disease.

Drugs may interfere with dynamic endocrine tests in several ways. Some drugs, such as high doses of glucocorticoids and L-dopa administration, directly impair pituitary function testing in regard to stimulation of secretion of growth hormone and TSH, respectively. A commonly used anticonvulsant medication, phenytoin, impairs testing of both the reserve of the pituitary–adrenal axis as described above with metyrapone and its suppressibility with dexamethasone by stimulating the hepatic metabolism of both test substances.

MEASUREMENT OF HORMONE RECEPTORS

With continued increase in our knowledge about the receptor mechanisms of hormone action (see Chapter 3), the measurement of receptor number and affinity is becoming more commonplace in the overall evaluation of endocrine function. The major form of endocrine pathophysiology in which receptor measurements are useful is hormone resistance. Diseases due to resistance to hormone action have now been described for most hormones. In general, the characteristic of such states is the lack or incompleteness of the expected hormone's effects in the presence of normal or increased circulating levels of the hormone. The types of tissues used for hormone–receptor measurements vary with the hormone and receptor of interest and include lymphocytes, red blood cells, tissue biopsies, and cultured skin fibroblasts. Examples of the types of receptor and postreceptor abnormalities detected by such studies include the reversible effects of obesity on insulin receptor number, the abnormalities of the GTP-binding protein in some subjects with PTH resistance, and impaired induction of the enzyme 25-hydroxyvitamin D 24-hydroxylase in one form of resistance to vitamin D.

Recently, mutations in cell surface receptors and their associated postreceptor proteins have been shown to result in hormone excess. Thus point mutations in TSH and LH receptors cause hyperthyroidism and Leydig cell hyperplasia, respectively. Activating mutations in the $G_s\alpha$ component of the GTP-binding protein have been associated with variable hyperfunction of multiple endocrine glands in the same individual.

Hormone–receptor measurements have proved informative in another way with regard to at least one form of hormone-responsive malignancy, cancer of the breast. Measurements in breast biopsies of the estrogen receptor and of a protein made in breast tissue in response to estrogen acting through its receptor, that is, the progesterone receptor, have been useful in predicting the response of such tumors to hormonal manipulation.

SUGGESTED READING

Geokas MC, Yalow RS, Straus EW, and Gold EM: Peptide radioimmunoassays in clinical medicine. Ann Intern Med 97:389–407, 1982.

Gorden P, and Weintraub BD: Radioreceptor and other functional hormone assays. In: *Williams Textbook of Endocrinology,* 8th ed., JD Wilson and DW Foster, eds., Saunders, Philadelphia, pp. 1647–1661, 1992.

Griffin JE: Dynamic tests of endocrine function. In: *Williams Textbook of Endocrinology,* 8th ed., JD Wilson and DW Foster, eds., Saunders, Philadelphia, pp. 1663–1670, 1992.

Klee GG, and Hay ID: Sensitive thyrotropin assays: Analytic and clinical performance criteria. Mayo Clin Proc 63:1123–1132, 1988.

Lefkowitz RJ: Clinical implications of basic research. N Engl J Med 332:186–187, 1995.

Pardridge WM: Transport of protein-bound hormones into tissues in vivo. Endocrine Rev 2:103–123, 1981.

Pardridge WM: Serum bioavailability of sex steroid hormones. Clin Endocrinol Metab 15:259–278, 1986.

Schall RF Jr, and Tenoso HJ: Alternatives to radioimmunoassay: Labels and methods. Clin Chem 27:1157–1164, 1981.

Spencer CA, LoPresti JS, Patel A, Guttler RB, Eigen A, Shen D, Gray D, and Nicoloff JT:

Applications of a new chemiluminometric thyrotropin assay to subnormal measurement. J Clin Endocrinol Metab 70:453–460, 1990.

Spiegel AM, Weinstein LS, and Shenker A: Abnormalities in G protein–coupled signal transduction pathways in human disease. J Clin Invest 92:1119–1125, 1993.

Verhoeven GFM, and Wilson JD: The syndromes of primary hormone resistance. Metabolism 28:253–289, 1979.

Walker WHC: An approach to immunoassay. Clin Chem 23:284–402, 1977.

6

The Anterior Pituitary and Hypothalamus

SAMUEL M. McCANN
SERGIO R. OJEDA

As indicated in Chapter 1, the core of the neuroendocrine system is represented by the hypothalamic–pituitary complex. The hypothalamus is composed of a diversity of neurosecretory cells arranged in groups, which secrete their products either into the portal blood system that connects the hypothalamus to the adenohypophysis (see below) or directly into the general circulation after storage in the neurohypophysis. The hypothalamic hormones secreted by the neurohypophysis are arginine vasopressin (AVP), also called antidiuretic hormone (ADH), and oxytocin (OT). The functions and regulatory mechanisms controlling the secretion of these hormones are discussed in Chapter 7.

The hypothalamic hormones delivered to the portal blood system are transported to the adenohypophysis where they stimulate or inhibit the synthesis and secretion of different trophic hormones. In turn, these hormones regulate gonadal, thyroid, and adrenal function, in addition to lactation, bodily growth, and somatic development. Because of the nature of their actions, the hypothalamic hormones are classified as releasing or inhibiting hormones. As will be seen later in this chapter, the adenohypophysis can be divided into three parts. The bulk of the adenohypophyseal hormones is produced in two of these parts, the pars distalis or anterior pituitary and the pars intermedia, also known as the intermediate lobe. The trophic hormones produced in the anterior pituitary include adrenocorticotropic hormone (ACTH), β-endorphin, thyroid-stimulating hormone (TSH), growth hormone (GH), prolactin (PRL), and the gonadotrophins, luteinizing hormone (LH) and follicle-stimulating hormone (FSH). The pars intermedia produces several hormones that are synthesized as part of a large-molecular-weight precursor known as proopiomelanocortin (POMC). Tissue-specific processing of this prohormone results in the formation of β-lipotropin, β-endorphin, α-melanocyte-stimulating hormone (α-MSH), and corticotropin-like intermediate peptide (CLIP), a product composition quite different from the predominant formation of ACTH, β-lipoprotein, γ-lipotropin, and β-endorphin in the anterior pituitary.

This chapter focuses on the hypothalamic control of adenohypophyseal function, with special reference to the anterior pituitary. No attempts are made to discuss the actions of the different pituitary trophic hormones on their target glands since this is considered in detail in Chapters 9 and 10 (gonadotropins and PRL), Chapter 13 (TSH), and Chapter 14 (ACTH). An exception to this is GH. Although Chapter 14 considers some aspects of the control and actions of GH, a broader discussion of its physiological actions will be presented here because GH is the only anterior pituitary hormone that does not have a clear-cut target gland.

ANATOMY

The pituitary gland (Fig. 6-1) has two parts: one portion is of neural origin, the neurohypophysis, which will be discussed in Chapter 7; the other portion, the adenohypophysis, is ectodermal in origin and is derived from the pharynx. In embryonic development, an evagination from the roof of the pharynx pushes dorsally to reach a ventrally directed evagination from the base of the diencephalon. The dorsally projecting evagination, known as Rathke's pouch, forms the adenohypophysis, whereas the ventrally directed evagination of neural tissue forms the neurohypophysis. The neurohypophysis has three parts: the median eminence, the infundibular stem, and the neural lobe itself. The median eminence represents the intrahypothalamic portion, and lies just ventral to the floor of the third ventricle protruding slightly in the midline. The main part of the neurohypophysis, the infundibular process (neural lobe), is connected to the median eminence by the infundibular stem. Tissue from the adenohypophysis spreads dorsally to surround the infundibular stem and median eminence and is known as the *pars tuberalis*. The infundibular stem and surrounding pars tuberalis constitute the pituitary stalk. The adenohypophysis can be divided into the *pars distalis*, which is the ventral portion, the *pars intermedia*, which is that portion adjacent to the neural lobe and separated from the pars distalis

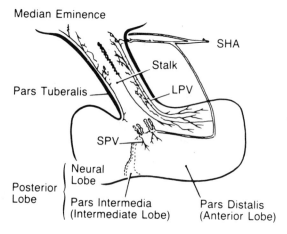

Fig. 6-1   Details of the anatomy of the pituitary gland showing the median eminence, pars tuberalis, stalk, infundibular stem and pars tuberalis, neural lobe, intermediate lobe, anterior lobe, posterior lobe, long portal vessels (LPV), superior hypophyseal arteries (SHA), and short portal vessels (SPV).

by the residual lumen of Rathke's pouch, and the pars tuberalis. The pars intermedia is rudimentary in humans. The pars distalis is frequently termed the *anterior lobe* because of its anterior position in most animals. The intermediate and neural lobes constitute the posterior lobe.

There is no significant functional innervation of the adenohypophysis by way of the hypothalamus or neural lobe. Instead, a humoral pathway is provided by means of the hypophyseal portal system of veins (Fig. 6-1). These take origin from capillary loops in the median eminence of the tuber cinereum and drain blood in parallel veins down the stalk. In the anterior lobe, the portal vessels break up into the sinusoids that provide its blood supply. Venous blood drains from the gland into the cavernous sinus. There may also be an arterial supply to the gland, but the portal vessels furnish the main blood supply to the anterior lobe. In addition to the vessels just described, which are called the *long portal vessels*, there are also short portal venous vascular connections that arise in the neural lobe and pass across the intermediate lobe to the anterior lobe. These are known as the *short portal vessels*.

The pituitary is a very vascular structure, with the various pituitary cell types lining the sinusoids of the gland. By light microscopy, the pituitary cells can be classed as acidophils, basophils, and chromophobes, depending on the affinity of their cytoplasm for either acidic or basic dyes. By electron microscopy, further subdivisions of these cells can be made (based on secretory granule size and number)

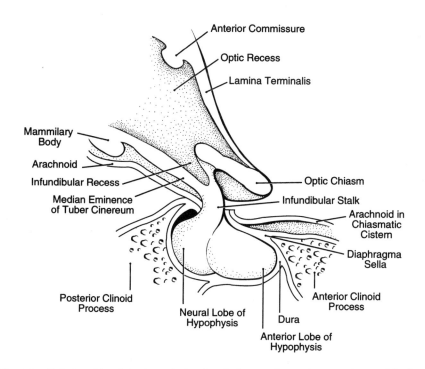

Fig. 6-2   Relationship of the hypothalamic–pituitary unit to other structures of the base of the brain. (Redrawn from Reichlin S: In: *Williams Textbook of Endocrinology*, 7th ed., JD Wilson and DW Foster, eds., Saunders, Philadelphia, p. 495, 1985.)

and cells secreting particular pituitary hormones can be characterized. For example, the somatotrophs secrete somatotropin (GH), the lactotrophs secrete PRL, and so forth. It was the long-held view that each cell type produced a single pituitary hormone. It is now clear that certain cells are able to secrete more than one hormone. An example of this type of cell is the gonadotroph. On the basis of immunocytochemical studies some of these cells secrete both FSH and LH. Other gonadotrophs may secrete one or the other of these hormones.

Uptake of hormones into and delivery of substances from the vascular system are favored by the presence of fenestrated capillaries within the gland. Since the vascular supply is primarily portal in nature, capillary pressure in the gland is very low and the permeability of these capillaries is high. The situation may be analogous to that in hepatic sinusoids with high permeability not only to pituitary protein hormones but also to the plasma proteins. Thus, a balance between hydrostatic pressure tending to produce ultrafiltration and colloid osmotic pressure tending to cause uptake of extracellular fluid is achieved at a very low sinusoidal pressure.

The *hypothalamus* is located at the base of the brain, ventral to the thalamus from which it is separated by the hypothalamic sulcus. Its anterior boundaries are provided by the optic chiasm; laterally it is separated from the temporal lobes by the hypothalamic sulci and its posterior limits are defined by the mammillary bodies. The base of the hypothalamus is the tuber cinereum, the central part of which is the median eminence. The anatomical relationships of the hypothalamic–pituitary unit to other structures of the base of the brain are shown in Fig. 6-2.

ANTERIOR PITUITARY HORMONES AND THEIR HYPOTHALAMIC CONTROL

Evidence for hypothalamic control of the anterior pituitary has come primarily from studies involving hypothalamic lesions or stimulation of animals such as the rat, guinea pig, and Rhesus monkey. The secretion of ACTH, GH, TSH, FSH, and LH can be increased by hypothalamic stimulation, whereas lesions that destroy the median eminence of the tuber cinereum result in decreased secretion of all of these hormones with the exception of PRL and α-MSH, which are increased. Thus, there is a net stimulatory influence of the hypothalamus on the secretion of most anterior pituitary hormones. Although these experiments clearly establish that the hypothalamus controls the release of pituitary hormones, they did not provide insight into the underlying mechanism.

The Neurohumoral Hypothesis of Control of Anterior Pituitary Hormone Secretion

From the very early days of neuroendocrinology, it was clear that the anterior lobe of the pituitary gland does not receive significant secretomotor innervation. The presence of the hypophyseal portal system of veins that drain blood from capillary loops in the median eminence down the hypophyseal stalk to the adenohypophyseal sinusoids suggested that the brain might exert a neurohumoral control over the pituitary (Fig. 6-3).

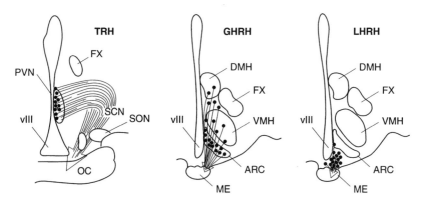

Fig. 6-3 Hypothalamic localization of the neurons that secrete thyrotropin-releasing hormone (TRH), growth hormone-releasing hormone (GHRH), and luteinizing hormone-releasing hormone (LHRH) based on human and animal studies. The neurons (solid dots) are shown in a coronal section through the plane of the densest cell bodies. The projection pathway of axons toward the median eminence is depicted by the solid lines forming an arrow. Arcuate nucleus, ARC; dorsomedial hypothalamic nucleus, DMH; fornix, FX; median eminence, ME; optic chiasm, OC; paraventricular nucleus, PVN; suprachiasmatic nucleus, SCN; supraoptic nucleus, SON; third ventricle, VIII; ventromedial hypothalamic nucleus, VMH. (From Riskind PN, and Martin JB: In: *Endocrinology*, 2nd ed., Vol. 1, LJ De Groot, ed., Saunders, Philadelphia, p. 101, 1989.)

Early evidence for the neurohumoral hypothesis was confusing, since experiments in which the pituitary stalk was sectioned gave variable results. This question was settled by placing an impervious plate between the cut ends of the stalk: a permanent derangement of pituitary function was the consequence. The earlier discrepancies had resulted from variable regeneration of hypophyseal portal vessels across the cut edges of the stalk. Further strong circumstantial evidence for the neurohumoral hypothesis came from ingenious grafting experiments in which the gland was removed from the hypophyseal capsule and then reimplanted under the median eminence so that it could be revascularized by hypophyseal portal vessels. Such grafts led to the return of normal anterior pituitary function, whereas grafts placed in other loci at a distance from the portal vessels so that they had to be revascularized by other arterial twigs led to a permanent dysfunction of the gland.

Identification of Hypothalamic Releasing and Inhibiting Hormones

The next logical step was to make extracts of the median eminence region, the area that is drained by the portal vessels, and to inject these into animals to determine if they could alter anterior pituitary function. It was soon shown that injection of extracts of the posterior pituitary released ACTH. Although synthetic vasopressin, a posterior pituitary peptide, could release ACTH, some workers were able to separate a substance in the posterior pituitary extracts, apparently different from vasopressin, that would also release ACTH. Unfortunately, this substance occurred in trace amounts and was not uniformly present in these extracts. The controversy cleared when it was shown that extracts of median eminence tissue contained more ACTH-releasing activity than could be accounted for by the amount of vasopressin present. This demonstrated that a corticotropin-releasing factor or hormone (CRH) existed in addition to vasopressin (Table 6-1).

The discovery of a CRH distinct from vasopressin provoked a search for similar factors that stimulate release of other anterior pituitary hormones. In rapid succession LH-, FSH-, GH-, and TSH-releasing activities, in addition to a PRL-inhibiting ac-

Table 6-1   Hypothalamic Releasing and Inhibiting Hormones[a]

Hypothalamic hormone	Purified	Synthesized	Stimulates	Inhibits	Cell bodies in
Vasopressin	Yes	Nonapeptide	ACTH	0	SO, PVN
Corticotropin-releasing hormone (CRH)	Yes	41 AA peptide	ACTH	0	PVN
LH-releasing hormone (LHRH)	Yes	Decapeptide	LH, FSH	0	POA, MBH
FSH-releasing factor (FSH-RF)	Yes	No	FSH	0	PVN?
Thyrotropin-releasing hormone (TRH)	Yes	Tripeptide	TSH, PRL	0	PVN
Growth hormone-releasing hormone (GHRH)	Yes	44 AA peptide	GH	0	ARC
Growth hormone-inhibiting hormone (GHIH) (somatostatin)	Yes	Tetradecapeptide	0	GH, PRL, TSH, gastrin, glucagon, insulin	APR
Prolactin (PRL)-inhibiting factors (PIFs),					
dopamine,	Yes	Yes	0	PRL	ARC
peptidergic PIF	No	No	0	PRL	?
PRL-releasing factors (PRFs)					
a. Oxytocin	Yes	Nonapeptide	PRL	0	SO, PVN
b. TRH, VIP, PHI, AII neurotensin substance P	Yes	—	PRL	0	PVN, SO
MSH-releasing factor (MRF)	Yes	?	MSH	0	
MSH-inhibiting factor (MIF)	Yes	?	0	MSH	
Pituitary adenylate cyclase-activating peptide (PACAP)	Yes	27 and 38 AA	ACTH, GH	0	SO, PVN

[a] AA, amino acid; ACTH, adenocorticotropic hormone; AAII, angiotensin II; APR, anterior periventricular region; ARC, arcuate nucleus; FSH, follicle-stimulating hormone; MBH, medial basal hypothalamus; MSH, melanocyte-stimulating hormone; PHI, peptide histidine-isoleucine; POA, preoptic area; PVN, paraventricular nucleus; SO, supraoptic nucleus; TSH, thyroid-stimulating hormone; VIP, vasoactive intestinal polypeptide.

tivity, were found in hypothalamic extracts. Evidence was also found for PRL- and MSH-releasing activities and a MSH-inhibiting activity.

An intensive effort was made to purify these new substances, to separate them from other active substances in hypothalamic extracts, and to determine their structure so that this could be confirmed by synthesis. Since only very small amounts of the releasing hormones are present in hypothalamic tissue, this proved to be a herculean task; isolation and determination of their structure required the processing of hundreds of thousands and even millions of hypothalamic fragments.

Hypothalamic Releasing and Inhibiting Hormones

Thyrotropin-Releasing Hormone (TRH)

TRH acts on the anterior pituitary gland to release TSH as well as PRL; it is a tripeptide (pyro-Glu-His-Pro amide; Table 6-2) derived from a large precursor that has a molecular weight of about 29,000 daltons. The TRH precursor contains five copies of the sequence Gln-His-Pro-Gly, which presumably originates the mature TRH molecule following glutamine cyclization and formation of the terminal amide. Within the hypothalamus, TRH is mainly synthesized in neurons of the parvocellular division of the paraventricular nucleus (Fig. 6-3 and Table 6-1) and also in some cells of the preoptic region. Axons project caudally from these regions and then project ventrally into the median eminence where the bulk of TRH is stored. TRH is also found outside the hypothalamus, even in the spinal cord, suggesting its involvement in other central nervous system functions.

Luteinizing Hormone-Releasing Hormone (LHRH)

LHRH, also known as gonadotropin hormone-releasing hormone (GnRH) because of its ability to stimulate the release of both LH and FSH, is a 10 amino acid peptide (Table 6-2) derived from a precursor of 92 amino acids, which in mammals is encoded by a single gene. In addition to LHRH, the mammalian gene encodes a 56 amino acid associated peptide termed gonadotropin-releasing hormone associated peptide (GAP). Although GAP has been shown to have PRL releasing activity, this effect does not appear to be consistent.

LHRH belongs to a family of decapeptides highly conserved through vertebrate evolution. Eight different variants of the decapeptide have been described in nonmammalian species. Recently, evidence was provided for the existence of a second LHRH gene in fish, which encodes two associated peptides instead of one. In most mammals examined, neurons producing LHRH are scattered through the anterior

Table 6-2 Structure of Three Hypothalamic Hormones: LHRH, TRH, and Somatostatin

TRH (P-Glu-His-Pro-NH$_2$)
    1   2   3
LHRH (P-Glu-His-Trp-Ser-Tyr-Gly-Leu-Arg-Pro-Gly-NH$_2$)
     1 2 3 4 5 6 7 8 9 10
Somatostatin(Ala-Gly-Cys-Lys-Asn-Phe-Phe-Trp-Lys-Thr-Phe-Thr-Ser-Cys)
      1 2 3 4 5 6 7 8 9 10 11 12 13 14

and medial part of the hypothalamus, and the preoptic septal areas. However, in primates—including humans—a substantial fraction of LHRH neurons is also localized to the medial basal hypothalamus and arcuate nucleus (Fig. 6-3 and Table 6-1). LHRH nerve terminals projecting to the portal vasculature are predominantly localized to the lateral portion of the median eminence, although in humans a medial distribution is also observed.

Corticotropin-Releasing Hormone (CRH)

CRH is a 41 amino acid peptide that stimulates the secretion of ACTH and β-endorphin from the adenohypophysis. CRH derives from a 196 amino acid precursor highly conserved among mammalian species. The CRH gene contains only two exons separated by a rather short intervening sequence; the mature peptide sequence is entirely encoded by the second exon. CRH synergizes with vasopressin to stimulate ACTH release. The hypothalamic neurons that produce CRH are mostly found in the medial parvocellular portion of the paraventricular nucleus, where they frequently colocalize with other peptides, particularly vasopressin (Fig. 6-3 and Table 6-1). As in the case of TRH, CRH-containing neurons are also found in several extrahypothalamic areas, suggesting the involvement of CRH in nonendocrine functions.

Somatostatin or Growth Hormone-Inhibiting Hormone (GHIH)

GHIH is a 14 amino acid peptide (Table 6-2) that acts on the pituitary gland to inhibit GH secretion. It can also inhibit TSH release. GHIH has a cyclic structure that results from the formation of two intramolecular disulfide bonds between its two cysteine residues. It is now clear that the original 14 amino acid somatostatin molecule belongs to a family of related peptides, which includes a 28 amino acid peptide, a 12 amino acid fragment derived from the amino terminus of somatostatin 28, and even larger forms having molecular weights of up to 16,000 daltons. The somatostatin gene contains two exons separated by a short intervening sequence; both somatostatin 14 and 28 are encoded by sequences present in the second exon. Since there appears to be only one somatostatin gene in mammals, the different somatostatin peptides are likely to derive from alternative messenger RNA processing and/or post-translational modifications of the propeptide.

Somatostatin is widely distributed in the organism, including several regions of the brain, the gastrointestinal tract, and pancreas. Within the hypothalamus, the principal source of somatostatin fibers projecting to the median eminence is neurons of the anterior periventricular area (Fig. 6-3 and Table 6-1).

Growth Hormone-Releasing Hormone (GHRH)

GHRH is a 44 amino acid peptide derived from a 108 amino acid precursor protein encoded by a single gene. The GHRH gene contains five exons that together with the corresponding intervening sequences span 10 kilobases of the human genome. Judging from the structure of the GHRH precursor protein, proteolytic processing may yield two peptides of unknown function, in addition to GHRH. GHRH stimulates GH secretion from the pituitary gland. Its location is the most restrictive of all releasing factors. The majority of the GHRH-producing neurons in the hypothalamus are found in and around the arcuate nucleus, close to the median eminence (Fig. 6-3 and Table 6-1).

Anterior Pituitary Hormones

All adenohypophyseal hormones are protein or polypeptide in nature. Their names, number of amino acids, and molecular weights are presented in Table 6-3.

Growth Hormone (GH)

As discussed in Chapter 12, there are several forms of GH, but it appears that the predominant species secreted under physiological conditions is a 191 amino acid form with a molecular weight of 22,650. Growth hormone is structurally related to human chorionic somatomammotropin, a peptide produced by the placenta. There are at least five GH-related genes and pseudogenes in the human genome. All of them, have five exons separated by four introns; alternative splicing of exon B leads to the synthesis of a 20 kDa GH variant, instead of the predominant 22 kDa form. This variant lacks amino acid residues 32 to 46, which are present in the larger form. Growth hormone exhibits species specificity that is undoubtedly due to differences in molecular structure among the species. The hormone is stored in specific secretory granules in cells specialized for its synthesis, known as *somatotrophs*, which are acidophils. The secretory granules in humans have a diameter of approximately 300

Table 6-3 Human Anterior Pituitary Hormones

	Amino acids	Carbohydrate (CHO)	Molecular weight
I. Corticotropin-related peptide hormones: Single small peptides derived from common precursor			
1. α-Melanocyte-stimulating hormone (α-MSH) (α-melanotropin)	13AA		1,823
2. Corticotropin (ACTH)	39AA		4,507
3. β-Lipotropin (β-LPH)	91AA		9,500
4. β-Endorphin, β-LPH (61–91)	31AA		3,100
II. Glycoprotein hormones: Composed of two dissimilar peptides. The α chain is similar in structure or identical. The β-chain differs with each hormone and confers specificity.			
1. Follicle-stimulating hormone (FSH) (follitropin)	α 89AA β 115AA	18% CHO 5% Sialic acid	32,000
2. Luteinizing hormone (LH) (lutropin)	α 89AA β 115AA	16% CHO 1% Sialic acid	32,000
3. Thyrotropin (TSH)	α 89AA β 112AA	16% CHO 1% Sialic acid	32,000
III. Somatomammotropin hormones: Single peptide chains with 2 or 3 SS[a] bonds; no carbohydrate			
1. Prolactin (PRL)	198AA		23,510
2. Growth hormone (GH) or somatotropin	191AA		22,650

[a] SS = disulfide.

μm. The somatotrophs are the most abundant cell type in the pituitary gland and they are filled with secretory granules. The store of GH in the pituitary is quite large and actually represents 4–8% of the gland by dry weight. The hormone is released from the somatotrophs by exocytosis. It is extruded into the extracellular space, particularly in the vicinity of the so-called vascular pole of the cell adjacent to a pituitary sinusoid.

The actions of GH and additional aspects of its control will be considered after the discussion of the hypothalamic hormones. Other details are provided in Chapter 12.

As indicated above, GH secretion is regulated by one releasing and one inhibiting hormone, GHRH and somatostatin, respectively. Under normal circumstances, the rate of GH release depends on a balance between the stimulatory effect of GHRH and the inhibitory effect of somatostatin. GHRH exerts its biological actions by binding to a membrane-anchored protein shown to be a guanine nucleotide-binding regulatory protein (G protein)–coupled receptor. As such, the GHRH receptor contains seven transmembrane domains. Its activation by GHRH results in cyclic AMP formation. The GHRH receptor gene appears to be expressed only in the anterior pituitary gland, in which two mRNA transcripts of 2.5 and 4 kilobases can be detected.

The somatostatin receptor also belongs to the superfamily of G protein–coupled receptors having seven transmembrane domains. Thus far, five different somatostatin receptors have been described, with an overall sequence homology of 40 to 60%. Despite these large divergencies in amino acid sequence, the somatostatin receptors recognize both somatostatin 14 and 28 with high affinity; all of them are negatively coupled to adenylate cyclase. Interestingly, their tissue distribution is also divergent; some have greater expression in the brain than in peripheral tissues and vice versa.

Prolactin (PRL)

Prolactin (Table 6-3) is a 23,000 molecular-weight protein hormone that bears considerable structural resemblance to GH. In fact, both hormones appear to have evolved from gene duplication of a common ancestral gene. Despite this, the PRL and GH genes are located on different chromosomes. While in humans there appears to exist a single PRL gene, in rodents there is evidence for at least three related genes, all of which are expressed in the placenta. As in the case of GH, there are several forms of PRL. The predominant form of 199 amino acids has a molecular weight of 23,000. The main function of PRL is to induce mammary gland growth and milk secretion in the properly prepared breast following delivery of the infant (see Chapter 9). In laboratory rodents, PRL affects gonadal function and helps maintain the structure and function of accessory sex organs. Whether PRL acts similarly in humans is not known. The fact that PRL receptors have been found in the ovary of primates, however, suggests a physiological role for the hormone in higher species. An important function of prolactin is to up-regulate immune function in animals and humans. It does so by stimulating lymphocyte proliferation and by providing together with GH the initial signals that prepare cells for proliferation and differentiation.

The human pituitary contains 50 times less PRL than GH. Prolactin is produced by lactotropic cells, which can constitute up to 25% of the normal population of anterior hypophyseal cells. Both the number and the size of the lactotrophs are increased by estrogen, and this becomes particularly noticeable during pregnancy. Serum PRL levels are higher in nonpregnant women than in men, reflecting the difference in estrogen production between the two sexes. However, not all changes in PRL secretion are related to sex steroids as shown by the increase in circulating PRL levels that occurs during sleep in both males and females.

Prolactin secretion is controlled by PRL-inhibiting and PRL-releasing factors. It was proposed that oxytocin, which is released by the suckling stimulus that also stimulates PRL release, is a PRL-releasing factor. It is now clear that terminals of oxytocinergic neurons end in juxtaposition to portal vessels, that there is a high concentration of oxytocin in portal blood, and that oxytocin can stimulate the secretion of PRL by pituitary tissue *in vitro*. Furthermore, antisera directed against oxytocin can partially inhibit the suckling-induced PRL release. Thus it appears that oxytocin is indeed a physiologically significant PRL-releasing factor. Similar evidence suggests that vasoactive intestinal polypeptide (VIP) and the related peptide, peptide histidine isoleucine (PHI), may also be physiologically significant PRL-releasing factors. Several other peptides such as angiotensin II, substance P, and neurotensin also stimulate PRL release. TRH, the first hypothalamic hormone to be synthesized, stimulates not only TSH but also PRL release.

Dopamine is a major PRL-inhibiting factor. Dopamine inhibits PRL secretion by binding to a dopamine receptor of the D_2 subtype. This receptor inhibits adenylate cyclase via coupling to a G protein of the G_i type. The D_2 receptor belongs to a family that comprises at least three other members. All of them contain seven transmembrane domains, a topology characteristic of all known G protein–linked receptors. Dopamine reaches the portal vessels after being released from terminals of the tuberoinfundibular dopaminergic tract that ends in juxtaposition to the hypophyseal portal capillaries (Fig. 6-3 and Table 6-1). Receptors for the catecholamine are found in the pituitary. Blockers of dopamine action can reverse the inhibition of PRL release normally exercised by the hypothalamus, and it is estimated that the concentration of dopamine in portal blood may be sufficient to hold PRL secretion in check, except when stimulated by stress or by the suckling stimulus. The elevation in PRL that occurs under these conditions is probably related not only to withdrawal of dopaminergic inhibition but also to secretion of the various PRL-releasing factors just described.

Adrenocorticotropin (ACTH)

Adrenocorticotropin, as its name implies, is the adenohypophyseal hormone that controls the function of the adrenal cortex. It is a relatively small molecular-weight polypeptide hormone consisting of a single chain of 39 amino acids. The complete structure is known and the hormone has been synthesized (Fig. 6-4, Table 6-3). Interestingly, the entire molecule is not needed for biological activity. The first 16 amino acids beginning with the N-terminal amino acid are all that is required for minimal biological activity. Progressive increases in activity occur as the length of the chain increases until full biological activity is present with a polypeptide consisting of the first 23 amino acids.

Recombinant DNA techniques have revealed that ACTH is part of the POMC precursor molecule. The POMC molecule and its constituents are illustrated in Fig. 6-5. This large peptide is produced in certain specific loci of the brain and in the anterior and intermediate lobes of the pituitary gland. Processing of the prohormone

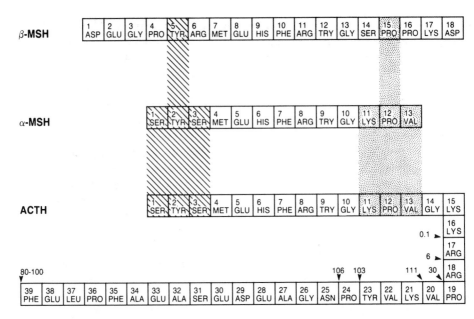

Fig. 6-4 The structure of human corticotropin (ACTH) and the melanocyte-stimulating hormones (MSHs). The shaded areas represent similarities in amino acid sequence. The figures with arrows on the ACTH molecule indicate the biological activity of peptide fragments beginning at the N-terminal amino acid to the point indicated by the arrow. (From Daughaday WH: The adenohypophysis. In: *Williams Textbook of Endocrinology,* 5th ed., RH Williams, ed., Saunders, Philadelphia, pp. 31–79, 1974.)

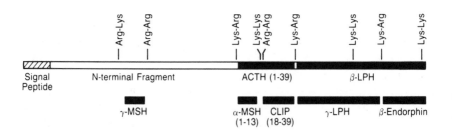

Fig. 6-5 Diagram of the structure of the proopiomelanocortin (POMC) molecule and the hormones derived from it by proteolysis. The precursor POMC protein contains a leader sequence (signal peptide) that is followed by a fragment that contains the sequence of α-MSH (amino acids 51–62), the ACTH molecule (1–39) that contains the sequences for α-MSH (ACTH 1–13) and corticotropin-like intermediate lobe peptide CLIP (ACTH 18–39), and the β-lipotropin (1–91) molecule that contains the sequences of γ-MSH (1–58) and β-endorphin (61–91). The latter also includes the sequence for met-enkephalin (first five amino acids of β-endorphin). Potential sites for proteolytic cleavage are indicated by the presence of the basic amino acids arginine-lysine (Arg-Lys). (From Reichlin S: In: *Williams Textbook of Endocrinology,* 7th ed., JD Wilson and DW Foster, eds., Saunders, Philadelphia, p. 502, 1985.)

varies depending on its cellular site. In the anterior lobe, it is processed to ACTH and β-endorphin, an endogenous opioid peptide of 31 amino acids. It has analgesic action but is not secreted in sufficient amounts to be physiologically significant in this regard. In the intermediate lobe of rodents, the processing of proopiomelanocortin is different, peptide bonds in the ACTH sequence being broken between the 13 and 18 position, resulting in the production of α-MSH (ACTH 1–13) and CLIP (ACTH 18–39). In the brain, it appears that the major products are ACTH, β-endorphin, and also α-MSH.

Evidence from light microscopic studies has suggested that in the adenohypophysis ACTH is secreted by basophils. With the electron microscope ACTH-producing cells appear irregularly shaped and replete with secretory granules. Radioimmunoassay of plasma ACTH is now the method of choice for evaluating ACTH secretion, although glucocorticoid levels in the blood are also measured as an index of ACTH secretion.

The secretion of ACTH is augmented by noxious stimuli of various sorts termed *stresses*, and this will be discussed in Chapter 14. In addition, ACTH secretion is also inhibited by glucocorticoids, an example of classic negative feedback regulation. As indicated above, the hypothalamus controls ACTH secretion via a CRH. The biological actions of CRH are initiated by the binding of the neuropeptide to a membrane protein structurally related to the calcitonin–VIP–GHRH family of G protein–coupled receptors. The CRH receptor is a 415 amino acid protein displaying seven putative transmembrane domains. It is encoded by a 2.7 kilobase mRNA and is positively coupled to adenylate cyclase. Interestingly enough, during attempts to isolate CRH, two additional brain peptides, substance P and neurotensin, were isolated from hypothalamic extracts.

Vasopressin has a physiological role in the control of ACTH release since there is a definite deficiency in ACTH release in animals with hereditary diabetes insipidus; such animals lack endogenously produced vasopressin. Another indication that vasopressin is a corticotropin-releasing factor is its direct action in stimulating ACTH release in a dose-related fashion from the pituitary incubated *in vitro*. Finally, vasopressin is present in high concentrations in portal blood collected from the pituitary stalk of monkeys and rats; these concentrations are in the range that should stimulate ACTH release, at least under stressful conditions, at a time when vasopressin is released in large amounts.

It is now apparent that vasopressin and CRH cooperate in the control of ACTH release. Vasopressin can potentiate the action of CRH at both hypothalamic and pituitary levels. Antisera against CRH given intravenously largely block the stress response, which can also be inhibited partially by vasopressin antisera. Furthermore, it appears that the vasopressin that reaches the pituitary comes predominantly from vasopressinergic neurons that extend to the median eminence to end in juxtaposition to portal vessels. Many of the vasopressinergic neurons apparently also contain CRH, so that there may be corelease of the two peptides into portal blood.

Melanocyte-Stimulating Hormone (MSH)

In lower forms such as reptiles and amphibia, MSH is clearly formed in and secreted by the intermediate lobe of the pituitary gland. Since this structure is rudimentary in humans and yet MSH is present, the hormone may also be formed in the human anterior lobe. In fact, the immunohistochemical analysis of adult human pituitary glands has shown that α-MSH cells are more frequently present in the anterior lobe than in the rudimentary pars intermedia. In both the scattered intermediate-type cells that form the rudimentary pars intermedia and the anterior lobe, α-MSH colocalizes with ACTH. It is appropriate to discuss MSH at this point since its chemical structure is very closely related to that of ACTH. There are two forms of MSH in nearly all species. The first is known as α-MSH and contains 13 amino acids (Fig. 6-4, Table 6-3). These 13 are identical with the first 13 amino acids of ACTH. β-MSH is a somewhat larger molecule and exhibits considerable species variation in both size and amino acid sequence. It consists of 22 amino acids. Amino acids 11–17 of β-MSH and ACTH are identical. In view of the overlap in structure between the MSHs and ACTH, it is not surprising that ACTH has MSH-like activity. The converse, however, is not true and MSH has no ACTH-like activity, except possibly during embryonic development.

Melanocyte-stimulating hormone, as its name implies, acts on the melanocytes in lower forms such as amphibians and reptiles to stimulate the dispersion of melanin granules within these cells, which causes darkening of the skin and permits the animal to blend with its environment. Even in humans, continued administration of MSH for a period of days will produce skin darkening. α-MSH and related peptides have been shown to delay the extinction of learned avoidance behaviors and food-motivated behaviors in a variety of experimental models. α-MSH administered intraventricularly has also been shown to have antipyretic and anti-inflammatory effects. It also inhibits the release of CRH and LHRH at low concentrations (10^{-13}M) in the rat. Although for a long time these effects were considered as pharmacological, the recent cloning of a family of receptors for the melanocortin peptides in both humans and rodents and the identification of their sites of expression in the brain have provided compelling evidence for a physiological role of MSH in the regulation of mammalian brain function. To date, four types of melanocortin receptors (MC-1R to MC-4R) have been molecularly characterized, and three of them were found to bind MSH (only MC-3R binds ACTH selectively). The secretion of MSH has been postulated to be controlled by both MSH-releasing and MSH-inhibiting factors. However, the physiologic significance of these factors is not clear and it appears that the major hypothalamic inhibitory control over MSH secretion is exercised by dopaminergic nerves in the intermediate lobe.

Pituitary Glycoprotein Hormones

The pituitary hormones discussed above are relatively simple proteins. The adenohypophysis secretes other more complex hormones that have a carbohydrate moiety attached to the protein component that in turn is composed of two interconnected amino acid chains. These hormones are called *glycoproteins* because of their carbohydrate moiety. There are three pituitary glycoprotein hormones, namely, thyrotropin (thyroid-stimulating hormone, TSH), follicle-stimulating hormone (FSH), and luteinizing hormone (LH), each with a molecular weight of about 30,000. They consist of two subunits, an α-subunit that is identical in the three and three different β-subunits that confer biological specificity to each hormone (Table 6-2). The α- and β-chains are synthesized separately and they are combined before joining the carbohydrate moieties. At the time of secretion, an excess of α-chains is secreted so that one can detect α-chain in the plasma by radioimmunoassay. When the secretion rate is elevated, a small amount of free β-chain may also appear in the plasma.

Thyroid-Stimulating Hormone (TSH)

As discussed in Chapter 13, this hormone stimulates secretion of the hormones thyroxine and triiodothyronine by the thyroid gland. As indicated above, the α-chain is identical to that of the gonadotropins and the biological specificity is confered by the β-chain, which is unique for TSH. There is a high degree of conservation in the amino acid sequence of the TSH-β subunit among different mammalian species. In all cases, the subunit appears to be encoded by a single gene, the expression of which is subjected to regulatory control by thyroid hormones. Like other glycoprotein hormones in the pituitary, TSH is secreted by basophilic cells, termed thyrotrophs, and these cells are characterized by the presence of abundant, relatively small secretory granules. The thyrotrophs make up to 15% of the cells of the adenohypophysis. TSH secretion is under negative feedback control by thyroid hormones (see Chapter 13) and is stimulated by the hypothalamic peptide, TRH. Thyrotropin-releasing hormone initiates its biological actions by binding to a membrane-anchored receptor that, like the GHRH, somatostatin, and CRH receptors, belongs to a family of G protein–coupled receptors containing seven highly conserved transmembrane domains. The TRH receptor is a protein of 393–412 amino acids and appears to be the product of a single gene. Binding of TRH to the receptor results in phospholipase C activation, calcium mobilization, and cyclic AMP formation (see below).

Follicle-Stimulating Hormone (FSH)

FSH is the gonadotropin that stimulates the ovarian follicles to grow. When given to a hypophysectomized animal, its only action is to stimulate growth of the follicles beyond the early antrum phase. As indicated in Chapter 9, primordial follicles are surrounded by only one layer of granulosa cells. As they grow into graafian follicles, a small, fluid-filled lumen, known as the *antrum*, eventually forms between the layers of proliferating granulosa cells. This process requires the presence of FSH. In the presence of LH, FSH increases estrogen secretion by the ovary. In the male, FSH acts to promote spermatogenesis. Further details of these actions are given in Chapters 9 and 10.

Like other glycoprotein hormones, FSH shows up in basophils under light microscopy. Electron microscopy and immunocytochemical methods reveal that it is secreted by cells called *gonadotrophs*. Some of these secrete only FSH, but others secrete both FSH and LH. FSH is produced in cells that have very small secretory granules relative to those of other pituitary cells.

Luteinizing Hormone (LH)

Pituitary control of the gonads requires still another glycoprotein hormone, LH, also known as *interstitial cell-stimulating hormone* (ICSH) because it stimulates the interstitial cells of both ovary and testes. The function of this hormone is to induce ovulation and the formation of the corpus luteum in the female, and to stimulate steroid secretion in both the ovary and the testes (see Chapters 9 and 10). The hormone is structurally similar to FSH and again possesses the common α-chain. Its β-chain is different from that of FSH and TSH and confers specific biological activity to the molecule. Once more, LH is found in basophils. Apparently, some gonadotrophs secrete only LH, but others secrete both FSH and LH.

The secretion of LH is stimulated by LHRH. The peptide releases not only LH but, to a lesser extent, FSH as well, which has led some to postulate that LHRH is sufficient to account for the hypothalamic stimulation of both FSH and LH release. Since dissociation of FSH and LH release can occur following hypothalamic lesions and stimulations and also in a variety of physiological states, an FSH-releasing factor may exist; however, this point remains controversial. Several groups have reported partial purifications of the putative FSH-releasing factor.

The LHRH receptor was recently cloned and found to be a 327 amino acid protein with a molecular weight of 37,000 daltons. The nucleotide sequence of the receptor indicates that it, as in the case of the GHRH, somatostatin, CRH, and TRH receptors, contains seven transmembrane domains characteristic of G protein–coupled receptors. Surprisingly, the LHRH receptor lacks the cytoplasmic C-terminus region, which appears to be important for the biological activity of other G protein–coupled receptors. Very recently the 5'-flanking region of the mouse LHRH receptor gene was isolated and found to contain the regulatory elements for tissue-specific expression and for LHRH-dependent regulation.

With the elucidation of the structures of these releasing and inhibiting hormones, it has been possible to prepare many analogs of each of the factors. In the case of LHRH, the initial aim was to obtain inhibitory analogs that might suppress fertility. However, it was accidentally found that analogs more active than the natural compound could easily be prepared if unnatural amino acids were substituted at positions likely to be subject to proteolysis. The agonist analogs have already proven to be important clinically since the so-called super LHRH analogs are capable of inducing ovulation in infertile women.

Multiple Actions of Releasing and Inhibiting Hormones

We have already alluded to the fact that TRH stimulates not only TSH but also PRL release. Whether the latter action has physiologic significance remains uncertain. Similarly, LHRH has FSH-releasing activity, which has led some to change the name LHRH for the name *gonadotropin-releasing hormone* (GnRH). Somatostatin inhibits not only the secretion of GH but also the secretion of most other pituitary hormones if given in sufficient dosage. It also suppresses the functions of many other systems, probably because it is distributed widely throughout the body (see below). Injected somatostatin has such an evanescent action in suppressing GH release because of enzymatic degradation that it has never been shown to suppress growth. Thus the name *somatostatin* is a misnomer. Recently, more stable agonist analogs of the compound have been reported to suppress growth. Since the inhibitory actions of the peptide are so pervasive, it has been suggested the name be changed from *somatostatin* to *panhibin*.

Distribution of the Mammalian Peptides in Lower Forms

The initial clue that these hypothalamic peptides might be found in lower forms came from work on CRH, since the amphibian skin peptide sauvagine has CRH activity when tested on the pituitaries of rats. Similarly, urophysin I, a peptide obtained from the urophysis, a neurohypophysis-like organ in the caudal spinal cord of fish, has CRH activity. Subsequently it was shown that yeast mating factor had structural homology with the mammalian LHRH and released LH from rat pituitaries, albeit with one ten-thousandth the potency of the mammalian hormone. It had previously been demonstrated that LHRH would induce mating behavior in higher forms, including mammals. Although all the evidence is not yet in, it appears that these peptides exist throughout the phylogenetic scale down to unicellular organisms. They probably have a similar function in unicellular organisms as in higher organisms. As differentiation increases up the phylogenetic scale, their localization is restricted to certain areas, but their basic functions do not change. Thus LHRH is a reproductive peptide, whereas CRH and vasopressin are stress peptides.

Mechanism of Action of Releasing and Inhibiting Hormones

Peptide hormones act by combining with highly specific receptors on the cell membrane of target cells. The peptide–receptor interaction then activates one or more of the several possible pathways for stimulation of secretion (see Chapter 3). One pathway involves activation of adenylate cyclase with the generation of cyclic AMP. The cyclic AMP formed activates a protein kinase that may mediate most of the effects of the cyclic nucleotide on cellular function. In the case of the pituitary, it has been suggested that activation of a protein kinase by cyclic AMP results in phosphorylation of certain membrane constituents that promote exocytosis, the mechanism for secretion. The secretory granules migrate to the cell surface and fuse with the cell membrane, and the granular core containing the hormone is extruded into the extracellular space, particularly at the vascular pole of the cell. Of the various pituitary cells, it would appear that the somatotrophs utilize the cyclic AMP pathway based on the ability of cyclic AMP to activate secretion and of its augmentation by inhibition of the breakdown of cyclic AMP by phosphodiesterase inhibitors. Similarly, the stimulatory effect of CRH on ACTH release appears to be mediated by cyclic AMP. In the gonadotrophs, the cyclic AMP system seems to play a minor role, possibly only in enhancing synthesis of LH.

Calcium plays an important part in the releasing process. Extracellular calcium is required for optimal release of all pituitary hormones. It may also be mobilized from intracellular stores in the case of some of the hormones as well. For those releasing factors that operate via the cyclic AMP mechanism, a cyclic AMP–dependent protein kinase may be involved in opening calcium channels allowing entrance of the ion and possibly also may mobilize it from intracellular stores.

The products of arachidonic acid metabolism can promote exocytosis, and their relative importance varies with the pituitary cell type, just as the relative importance of cyclic AMP seems to vary with the cell type. Prostaglandins appear to be important in the release of growth hormone and ACTH, and they can activate the production of cyclic AMP. In other cells such as the gonadotrophs, it appears that metabolites of arachidonic acid different from prostaglandins, such as the leukotrienes and epoxides, may also be involved in the exocytosis process.

Arachidonic acid release from the cell membrane may result from a receptor-mediated activation of phospholipase C, a membrane-bound enzyme that catalyzes the hydrolysis of polyphosphoinositides into inositol triphosphate and diacyglycerol. The latter not only contains arachidonate, which is then made available to phospholipase A for subsequent metabolism, but also interacts with protein kinase C, a calcium-dependent, phospholipid-activated enzyme. Protein kinase C phosphorylates several proteins that then participate in the process of hormone release. Inositol triphosphate, the other product of phospholipase C activity, mobilizes intracellular calcium from the endoplasmic reticulum.

The releasing process is not completely understood, but each cell type utilizes these various pathways to varying degrees. The role of guanylate cyclase and its product, cyclic GMP, is not yet clear.

The inhibiting hormones such as somatostatin and dopamine act in part by inhibition of adenylate cyclase and in part also by other mechanisms such as decreasing the availability of calcium to the cell.

If the cell membrane is depolarized by placing pituitary tissue in a high potassium medium, release is also induced. It is not known whether depolarization of the cell membrane actually accompanies the releasing process of normally secreting pituitary cells, but when it is artificially induced by a high potassium concentration, it probably increases calcium uptake by the cells, which then mediates the releasing process.

The means by which the secretory granules migrate to the cell surface remain an enigma. It has been postulated that this might involve microtubules or microfilaments; however, colchicine, which disrupts microtubules, not only fails to block the release of several pituitary hormones but actually augments it, suggesting that the tubules might hold the granules in the interior of the cell, and that after dissolution of the tubules, the granules then migrate spontaneously to the surface, possibly because of electrostatic forces between the granule and the inside of the cell membrane.

The releasing hormones act on the cell very rapidly (within less than a minute) to either promote or inhibit release of particular pituitary hormones; however, their precise effect on the biosynthesis of these hormones has not been elucidated. With sufficient stimulation of release, synthesis is promoted, probably via an increase in the mRNA required for the synthesis of the pituitary hormone in question. Whether this is secondary to the release process or represents another primary action of the releasing hormone on the biosynthetic process itself remains to be determined.

In general, very little is known about the intracellular signaling mechanisms that mediate the stimulatory–inhibitory effects of releasing–inhibiting hormones on pituitary hormone secretion. Surprisingly, releasing hormones such as TRH and LHRH have been found to utilize the same intracellular signal transduction pathways to exert their stimulatory effects on TSH–PRL and LH–FSH release. Both receptors are coupled to the same G protein; activation by their respective ligands results in the stimulation of phospholipase C, enhanced phosphoinositide turnover, calcium mobilization, cAMP formation, and arachidonic acid release. This situation is not unique to releasing hormones as it has also been observed in the case of growth factors. It is possible that the specificity in these seemingly common signaling pathways is provided by a family of newly discovered ''adaptor'' proteins that appear to connect the different components of the signaling pathway.

Factors Affecting Responsiveness of the Adenohypophysis to Releasing and Inhibiting Hormones

The hormonal state is very important in determining the responsiveness of the adenohypophysis to the various releasing hormones. In the case of TRH, pituitary responsiveness is enhanced by removal of the thyroid and consequent loss of negative feedback by thyroxine and triiodothyronine. Conversely, responsiveness is suppressed by the administration of thyroid hormones. In fact, the negative feedback of thyroid hormone seems to take place predominantly in the pituitary gland to modulate its responsiveness to TRH (Fig. 6-6, for other details see Chapter 13).

Negative feedback results in decreased transcription of the pro-TRH gene. At the pituitary level, thyroid hormone excess results in a decreased number of TRH receptors on thyrotrophs and decreased transcription of TSH-α and -β genes. TRH release is under inhibitory control via circulating thyroid hormones.

In the case of CRH, there is also good evidence that adrenal steroids feed back directly at the pituitary to inhibit the response to CRH (Fig. 6-6); however, this feedback also takes place at the hypothalamic level to inhibit release of CRH and vasopressin.

The interplay between gonadal steroids and the hypothalamic–pituitary axis is particularly complex (see Chapters 9 and 10). Following removal of the gonads and the elimination of feedback by the gonadal steroids, predominantly estrogen in the female and testosterone in the male, levels of both FSH and LH are elevated. The release of these pituitary hormones is pulsatile and occurs rhythmically. The timing of the discharge varies among species and even among individual animals within a species. In the human, these pulses occur hourly and are called the *circhoral rhythm* of LH release. This rhythm is brought about by pulsatile release of LHRH and possibly of FSH-releasing factor as well (if such a discrete factor actually exists). Following removal of negative feedback, the enhanced LHRH release increases not only the discharge of gonadotropins but also their synthesis, so that the quantities stored in the gland increase. This augmented storage is associated with an increase in responsiveness to the neurohormone. Small doses of estrogen or androgen can suppress the responsiveness to LHRH, and this occurs quickly. At least in the case of estrogen administration, it can take place within 1 hour. With long-term therapy, there is suppression not only of release, but also of its synthesis, and the pituitary content of gonadotropin consequently falls. This is associated with a further decline in responsiveness to LHRH, which may be caused in part by down-regulation of LHRH receptors.

In the female there are complex endocrine relationships during the menstrual

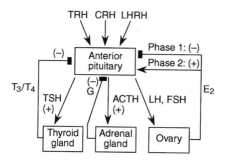

Fig. 6-6 Modulation of anterior pituitary responsiveness to hypothalamic releasing hormones. (−) Inhibits the response to the releasing hormone; (+) facilitates the response. G, glucocorticoids; E$_2$, estradiol; T$_3$, triiodothyronine; T$_4$, thyroxine; Phase 1, early phase; Phase 2, later phase, only in females.

cycle. Responsiveness to LHRH during the early follicular phase is minimal. Responsiveness increases as the follicular phase progresses and reaches its height at the time of the preovulatory discharge of LH. Treatment of women with estrogen has a biphasic effect on pituitary responsiveness to LHRH; the initial suppression is followed by an augmented responsiveness (Fig. 6-6). Thus, estrogen secreted by the preovulatory follicles is probably responsible for the enhanced responsiveness to LHRH that occurs in the late follicular phase.

In addition, the characteristics of the preovulatory discharge of LH in response to LHRH change, the response becoming much more rapid and pulse-like. This further change in responsiveness probably emanates from LHRH itself, since it can be induced by a priming injection of LHRH in the late follicular phase of the cycle. Responsiveness then declines after the ovulatory discharge.

The mechanism by which estrogen augments responsiveness to LHRH is not known. It might induce the formation of additional receptors for the neurohormone, or, alternatively, it might alter the synthesis of LH and provide a larger pool of releaseable LH. Similarly, the mechanism for the self-priming action (positive feedback) of LHRH remains to be elucidated, but this could be explained by an effect on synthesis of a releaseable pool of LH or, alternatively, by an effect of LHRH to induce new LHRH receptors. The ability of LHRH to up-regulate its own receptors has been demonstrated.

This remarkable increase in responsiveness to LHRH as a function of the steroid milieu undoubtedly accounts in part for the preovulatory discharge of LH, but it is believed that an enhanced release of LHRH is also involved. This is induced once again by estrogen from the preovulatory follicle (see Chapter 9). This increased release of LHRH is probably responsible for bringing on the self-priming action of LHRH that characterizes the late preovulatory phase. Evidence for the increased release of LHRH at the preovulatory surge of LH includes the detection of increased levels of the hormone in peripheral blood in some animals and humans and the observation of very high levels of LHRH in portal blood collected from the transected pituitary stalk at this stage of the cycle in animals.

It would appear, then, that the preovulatory discharge of LH is brought about by enhanced release of LHRH coupled with a marked increase in responsiveness to the neurohormone. The result is a discharge of LH far greater than that necessary to induce ovulation. Perhaps this is part of a fail-safe mechanism to ensure full ovulation even when follicular development is not optimal; it might serve to prolong the reproductive life of the individual.

Putative Synaptic Transmitters Involved in Controlling the Release of Releasing Hormones

The neurons that produce releasing factors are in synaptic contact with a host of putative neurotransmitters. The most abundant neurotransmitter in the hypothalamus is the excitatory amino acid glutamate. The hypothalamus is also supplied with monoaminergic nerve fibers. There is a heavy input of noradrenergic fibers from neurons whose cell bodies lie in the brain stem. The distribution of these neurons has been mapped by fluorescent histochemistry. There are also terminals from epinephrine-containing neurons that end in the hypothalamus. These also appear to originate from neurons whose cell bodies are located in the brain stem.

Axons of serotonin-containing neurons whose cell bodies lie in the medial raphe nuclei project to the suprachiasmatic, anterior hypothalamic, and median eminence regions. Assays for choline acetyltransferase indicate the widespread distribution of cholinergic fibers within the hypothalamus as well. There is also an abundance of histamine that appears to be located in synaptosome-like structures, and this amine is concentrated particularly in the median eminence region where it may also serve as a synaptic transmitter. There is evidence as well for a γ-aminobutyric acid-containing (or GABAergic) system that is localized partially in the infundibular nucleus with projections to the median eminence. High concentrations of GABA have been found in portal blood and GABA has a direct inhibitory action on the secretion of prolactin, raising the possibility that it may be a prolactin-inhibiting factor. Other GABAergic neurons are small interneurons localized throughout the hypothalamus where they may impinge on dendrites or somata of releasing hormone neurons to inhibit their secretory activity.

Great effort has been put into experiments with rodents and monkeys to determine the role of these possible transmitters in controlling the release of the various releasing hormones. The most extensive studies have been done on gonadotropins. As discussed in Chapter 9, it appears that among the important neurotransmitter systems involved in the control of LHRH secretion are those that employ excitatory amino acids to stimulate LHRH release and GABA to inhibit LHRH neuronal activity. In addition there are excitatory noradrenergic synapses that may contribute to inducing not only the preovulatory release of LHRH but also the increased LHRH release that follows castration. In the case of the preovulatory release of LHRH, the synapses may be in the preoptic-anterior hypothalamic region, whereas in the case of the increased release in the castrate, the synapses may be on the other population of LHRH neurons, located in the arcuate nucleus. The stimulatory effect of the noradrenergic system on LHRH release appears to be complemented by intrahypothalamic neurons that produce neuropeptide Y. This peptide has been shown to not only stimulate LHRH release at the level of the median eminence but also to synergize with LHRH in the pituitary gland following its release into the portal blood.

Dopamine has been postulated to both stimulate and inhibit LH release. The evidence is confusing, but the view held at the present time is that this catecholamine may have only a minor role.

Serotonin, when injected into the third ventricle in castrates, can inhibit LH release; this indicates that it is an inhibitory transmitter, but other evidence suggests that it may facilitate preovulatory LH release. Histamine can stimulate LH release following its intraventricular injection in large doses. It is still not clear if this has physiological significance.

There is considerable evidence for a cholinergic link in gonadotropin release, since atropine can block gonadotropin release when it is administered systemically, microinjected into the third ventricle, or implanted within the hypothalamus.

In the case of ACTH, there is considerable evidence that CRH is under the stimulatory control of the noradrenergic system; there is some evidence for cholinergic and serotoninergic stimulatory components as well. TRH has not been investigated extensively, but may be under adrenergic control. Growth hormone release also appears to be under adrenergic control with a stimulatory α-receptor and an inhibitory β-receptor component, but the relative importance of dopamine and norepinephrine has yet to be clearly established.

Prolactin is definitely under inhibitory control via dopamine. There is a tuberoinfundibular dopaminergic tract, the neurons of which have cell bodies lying in the arcuate nucleus and axons projecting to the external layer of the median eminence. Here they terminate in juxtaposition to hypophyseal portal capillaries. Dopamine agonists, such as bromocriptine, can lower plasma prolactin in hyperprolactinemic patients.

In addition to these classical low-molecular-weight transmitters, the hypothalamus is a repository for a veritable host of neuropeptides. Not only are there the releasing and inhibiting factors just discussed, and vasopressin and oxytocin, but brain opioid peptide systems are also present. Within the hypothalamus, infundibular neurons produce POMC. Their axons project to other parts of the hypothalamus and other brain regions and apparently establish synaptic contacts with other cells, secreting β-endorphin, ACTH, and possibly α-MSH into the synaptic cleft. Studies using opiate receptor blockers, such as naloxone, have shown that the opioid peptides are involved in stress-induced PRL release. The peptide involved appears to be β-endorphin based on studies with specific antibodies against β-endorphin that block stress-induced PRL release. β-Endorphin also appears to have a physiologically significant inhibitory role in the control of gonadotropin secretion.

Enkephalin and dynorphin neurons and even other classes of opioid peptides are also localized in neurons within the hypothalamus. The functional significance of these other opioid peptides remains to be clarified.

In addition to these brain peptides, many peptides that were originally thought to be localized exclusively in other organs have now been found in the brain. Among these are angiotensin II, which was first thought to be formed only in the circulation after release of renin from the juxtaglomerular apparatus of the kidney. There are angiotensin II-producing neurons within the hypothalamus, and they may play a stimulatory role in controlling ACTH and PRL secretion. The recently discovered atrial natriuretic peptides, whose main locus is the atria of the heart, have now been found within neurons in the hypothalamus and may suppress ACTH as well as vasopressin secretion (see Chapter 7). Of particular interest is the recently identified peptide pituitary adenylate cyclase activating peptide (PACAP; Table 6-1). It is apparently released into the portal vessels and acts in the pituitary to induce cyclic AMP formation. Despite its potency in this regard, it has not been found to be an effective stimulus for hormone release. Recent experiments have shown that it may regulate POMC gene transcription.

Many gastrointestinal peptides originally thought to be localized only in the gut have been found in brain neuronal systems. An example is vasoactive intestinal polypeptide, which appears to have important roles in controlling pituitary hormone secretion by both hypothalamic and pituitary actions. As indicated above, it may be a physiologically significant PRL-releasing factor. Cholecystokinin is also present in the brain and appears to have significance in the control of pituitary hormone secretion, as do neurotensin and substance P. Antagonists against these various peptides and antisera directed against them have been used to determine their physiological significance.

Short Loop Feedback of Pituitary Hormones to Alter Their Own Release

A variety of pituitary hormones exert short loop negative feedback actions to suppress their own release. In this case, the hormone inhibits its own release without reaching the general circulation (Fig. 6-7). This is in contrast to the so-called long loop feedback of pituitary target gland hormones such as gonadal steroids. It was originally thought that the short loop feedback of pituitary hormones might be mediated in the hypothalamus by reverse flow in portal vessels, which would deliver the pituitary hormones to the hypothalamus. There is some evidence for reverse flow under certain circumstances, but it appears that most of the pituitary hormones may actually be produced in the brain. In the case of ACTH, it is now known that the POMC molecule is synthesized in neurons in cells of the medial basal hypothalamus (arcuate nucleus). ACTH, a fragment of this molecule, may be secreted from cells producing POMC. There is also evidence that PRL-secreting and LH-secreting neurons exist within the brain. Short loop negative feedback has been clearly established for PRL and GH. In the latter case, it may be mediated not only by GH itself but also by IGF-I (see below), either delivered via the peripheral circulation or made directly in the brain. There is also considerable evidence of short loop feedback for LH and FSH as well as ACTH. Little work has been done on TSH, but in all probability the same mechanism applies to this hormone.

Ultrashort Loop Feedback of Releasing and Inhibiting Hormones to Modify Their Own Release

Ultrashort loop feedback is an alteration in the release of hypothalamic peptide induced by the peptide itself acting within the brain. It may occur as direct recurrent inhibition or via interaction with an interneuron that in turn alters the discharge rate of the peptidergic neuron (Fig. 6-7).

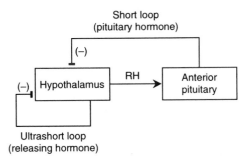

Fig. 6-7 Short and ultrashort loop feedback systems in the hypothalamic–pituitary unit. RH, releasing hormone.

There is evidence from animal experiments for ultrashort loop negative feedback of the releasing hormone neurons to inhibit their own release in the case of somatostatin, GHRH, and LHRH.

Extrapituitary Actions of Releasing Hormones

The distribution of releasing hormones in brain regions outside the hypothalamus (e.g., brain stem, cortex) has stimulated a search for their extrapituitary actions. As indicated earlier, TRH is found in other brain regions and even in the spinal cord; somatostatin is distributed widely throughout the nervous system and also has been found in the delta cells of the pancreatic islets of Langerhans. Since somatostatin can inhibit the release of both insulin and glucagon, it probably acts locally to control the release of these hormones from the islets.

The clearest behavioral effect of releasing hormones is the induction of mating behavior by relatively low doses of LHRH in animals. The occurrence of LHRH in the preoptic area, the region of the brain that is known to be involved in mating behavior, plus the onset of this behavior shortly after the preovulatory discharge of LHRH, suggested this effect of LHRH. Indeed, LHRH induces mating behavior in many species including primates and possibly humans. This is not caused by the gonadotropins released, since these hormones have no effect on mating behavior and since induction of the behavior is seen in the hypophysectomized rat. Further studies have shown that the effect can be obtained by microinjecting LHRH into the preoptic-anterior hypothalamic and arcuate median eminence regions, whereas similar injections into lateral hypothalamus or cortex are ineffective. There is a latency between the injection of LHRH either into the brain or systemically and the onset of mating behavior, suggesting that some intervening steps may be involved.

Thyrotropin-releasing hormone has effects that indicate an arousal action of the hormone, that is, it shortens the duration of pentobarbital anesthesia. The doses required are very large; however, if the hormone is present at synaptic sites in the brain, it is conceivable that these responses to high doses could be physiological. Somatostatin, on the other hand, has been shown to depress animals. Thus, the concept is emerging that the releasing factors may have important extrapituitary actions. One could even envision the possibility that they may serve as peptidic neurotransmitters and that this might be a role as important as that of governing the release of anterior pituitary hormones. There is evidence that the action of releasing factors in the CNS may integrate complex hormonal and neural mechanisms, for example, blood pressure and fluid control (vasopressin) or reproductive processes (LHRH).

PITUITARY GROWTH HORMONE, ITS ACTION AND ITS CONTROL

Now that the overall hypothalamic control of anterior pituitary hormone secretion has been considered, the actions of GH and control of its secretion will be discussed in more detail.

Actions of Growth Hormone

Growth hormone has powerful effects on growth and metabolism. Its absence is associated with dwarfism. Its excessive secretion leads to giantism or acromegaly (see below). To study the precise action of GH, it is best to utilize hypophysectomized animals or hypopituitary patients. The hormone can then be administered and its metabolic effects noted. Administration of GH produces the following effects:

1. Decrease in blood amino acid concentrations.
2. Decrease in blood urea nitrogen.
3. Positive nitrogen balance, defined as the difference between daily nitrogen intake in food and excretion in urine and feces as nitrogenous wastes.
4. Increased DNA, RNA, and protein synthesis, that is, a protein anabolic effect.
5. Elevated blood glucose via decreased utilization of carbohydrates and decreased sensitivity to the plasma glucose-lowering action of insulin.
6. Increased oxidation of fat leading to a decrease in the respiratory quotient (R_Q = CO_2 output divided by O_2 intake). The respiratory quotient for the metabolism of carbohydrates to CO_2 and water is 1.0, whereas that for the metabolism of fats is 0.7. Protein metabolism leads to an R_Q of about 0.8. The increased utilization of fats is often accompanied by an elevation of the free fatty acids in blood because of lipid mobilization from fat stores and by an increase in ketone body formation (β-hydroxybutyric acid, acetoacetic acid, acetone) associated with the increased fatty acid oxidation (see Chapter 16).
7. Growth.
8. Stimulation of the growth and calcification of cartilage.
9. High doses of GH given over a long period of time can induce diabetes mellitus in dogs and cats. The action is via an exhaustion of the insulin-secreting beta cells of the islets of Langerhans, which are overstimulated by the elevated plasma glucose as a result of GH secretion. The syndrome is called *metahypophyseal diabetes*. Whether this is the cause of the impaired glucose tolerance observed in acromegalic patients has not been determined.

Mechanism of Action of Growth Hormone

The complex metabolic effects of GH can be viewed as a means of increasing protein synthesis and turning to fat as the main source of calories in order to spare amino acids for protein synthesis. At the same time, utilization of glucose is suppressed.

The mechanism by which growth hormone accomplishes these myriad effects is by no means clear. The following facts are known:

1. Growth hormone stimulates the uptake of amino acids by muscle and liver cells, as demonstrated by its clear-cut enhancement of the uptake of the nonutilizable amino acid, α-aminoisobutyric acid. Administering GH leads to the accumulation of this amino acid within cells. The effect on amino acid uptake may figure in the anabolic action of GH.

2. Although some synthesis of protein can be accomplished under the influence of GH even if RNA synthesis is blocked, it is apparent that later on increased synthesis of RNA is stimulated by GH. Whether this is a primary action of the hormone remains to be determined. Ultimately, not only RNA, but also DNA synthesis is increased and finally cell division occurs.

3. The mechanism of action of GH on fat and carbohydrate metabolism is less well understood but the hormone appears to block the phosphorylation of glucose after carbohydrate has penetrated cells.

4. It is now clear that GH requires an intermediary in order to influence cartilage and bone metabolism. This material was initially called *sulfation factor* because of its ability to stimulate the incorporation of [^{35}S]sulfate into cartilage, but it is now called *somatomedin C* or *insulin-like growth factor I* (IGF-I). It is a small peptide produced by the liver and other tissues (see Chapter 12).

The initial evidence for the existence of IGF-I was that GH added to cartilage *in vitro* did not stimulate the incorporation of labeled sulfate, whereas plasma from animals treated with GH did so. As discussed in Chapter 12, the active peptide has been isolated and its structure determined.

Long bones grow by proliferation of cartilage cells in the epiphyseal plate, which is followed by growth due to ossification occurring in diaphyseal and epiphyseal ossification centers. In the absence of GH in hypophysectomized animals, the epiphyseal ossification center atrophies and there is a very thin epiphyseal plate of cartilage. Administration of GH causes a thickening of this cartilage plate and this has served as a bioassay for the hormone.

IGF-I may be involved in mediating some or even all of the other actions of GH on metabolism.

Control of Growth Hormone Secretion

From radioimmunoassay studies it is apparent that plasma GH values are very high in the first few days of life. Then levels decline to those seen in adulthood; however, the response to various stimuli is exaggerated in children, which leads to a net increase in GH secretion throughout the 24-hour period as compared to that of adults. This, coupled with a high tissue sensitivity to the hormone and to IGF-I, presumably accounts for growth.

Two types of stimuli appear to enhance secretion of GH from the somatotrophs. The first type of stimulus is metabolic. If insulin is injected to produce hypoglycemia, an immediate release of GH ensues. The response is caused by the hypoglycemia rather than the insulin per se, since no release of GH occurs if glucose is infused to maintain normal blood sugar levels. Since one of the metabolic effects of GH is to inhibit carbohydrate utilization, this would seem to be a useful physiological response. Another metabolic signal for releasing GH is an increase in the plasma levels of certain amino acids. In particular, oral or intravenous injection of arginine will result in a rise of GH secretion. Again, the response would seem to have a physiological purpose in that GH stimulates protein anabolism by promoting uptake of amino acids by cells.

The second type of stimulus that provokes GH release is nonspecific and noxious such as cold exposure, fright, and even muscular exercise. One could argue that this "stress response" is useful; for instance, it would change the metabolic mixture during exercise by shifting metabolism from carbohydrates to fat.

GH release also occurs during onset of deep sleep, stages III and IV, and is inhibited during rapid eye movement (REM) sleep, which is associated with dreaming. In all instances, the release of the hormone occurs in bursts, that is, it is pulsatile. Clearly, GH release is not constant but highly variable and responsive to changes in the external and internal environment of the organism.

Another important regulator of GH secretion would appear to be the concen-

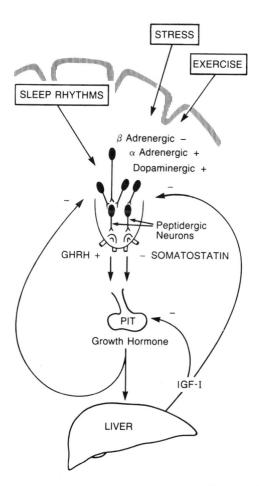

Fig. 6-8   Schematic diagram of the hypothalamic control of growth hormone secretion. For details see text. (Modified from Reichlin S: In: *Williams Textbook of Endocrinology*, 7th ed., JD Wilson and DW Foster, eds., Saunders, Philadelphia, p. 529, 1985.)

tration of GH itself in plasma. Administering GH to rats can cause a reduction in pituitary weight and GH content; in humans it will block its own release in response to hypoglycemia. As indicated above, this feedback of a pituitary hormone to reduce its own secretion is known as *short loop feedback*.

The regulation of GH secretion is accomplished by means of the hypothalamus (Fig. 6-8). Placement of hypothalamic lesions, particularly in the ventromedial nucleus of the hypothalamus, impairs GH secretion, and conversely, stimulation in this same portion of the hypothalamus can augment the secretion of the hormone. Presumably, the various metabolic and environmental stimuli that alter the hormone's secretion rate act via the hypothalamus. The short loop feedback action of GH may be exerted at both a hypothalamic and a pituitary level. The feedback of GH may be mediated not only by GH itself but by IGF-I, either reaching the hypothalamus via the general circulation or made locally within the hypothalamus.

In addition to the short loop feedback of GH just described, it is now quite clear that there are also ultrashort loop negative feedbacks of somatostatin and GHRH operating within the hypothalamus. Intraventricular injection of somatostatin in rats causes a paradoxical elevation of GH mediated by a decrease in somatostatin release. There may also be an attendant increase in GHRH release. Similarly, microinjection of GHRH into the third ventricle can suppress rather than elevate GH. Since these actions are opposite to the direct effect on the pituitary, they must occur within the brain. *In vitro* studies have shown that GHRH produces a dose-related stimulation of somatostatin release, so that it is apparent that this ultrashort loop feedback action is mediated, at least in part, by release of somatostatin. The physiologic significance of these ultrashort loop feedbacks has not yet been proved.

CLINICAL ASSESSMENT OF ANTERIOR PITUITARY FUNCTION

Clinical assessment of anterior pituitary function to determine whether it is normal includes the history, physical examination, and usually some form of laboratory testing. Since physiological regulation of anterior pituitary hormone secretion needs to be considered in assessing pituitary function, the evaluation for possible deficiency or excess of each of the clinically relevant anterior pituitary hormones is briefly discussed in this section (see also Chapter 5).

Growth Hormone

A random serum GH measurement is not useful for assessing adequacy of secretion since levels may be low in normal individuals. Provocative tests are necessary. Screening tests include the measurement of GH following exercise or 90 minutes after the onset of sleep. If screening tests are abnormal, more definitive tests are required. The two most commonly used definitive tests are the measurement of GH following insulin-induced hypoglycemia or the administration of L-dopa. Serum ACTH, cortisol, glucagon, and epinephrine are also increased following hypoglycemia.

As with possible GH deficiency, a random serum GH may not be reliable in assessing growth hormone excess. Since GH is secreted episodically, if the random sample is drawn at the height of a secretory episode, seemingly elevated levels may be obtained in normal individuals. Growth hormone levels are normally suppressed by the rise in glucose that occurs 30 to 90 minutes following an oral glucose load. Patients with GH excess do not suppress and often have a paradoxical increase in GH after glucose. IGF-I levels are also measured to assess the possibility of GH excess (see Chapter 12).

Prolactin

Although PRL secretion can be stimulated by a number of agents (including commercially available TRH) and inhibited by L-dopa, it is not necessary to perform such maneuvers for routine clinical assessment. A single basal serum PRL determination is usually sufficient to determine the presence of PRL excess, and PRL deficiency is not usually a clinical concern. Since PRL is the only anterior pituitary hormone with predominant negative control by the hypothalamus, it is often elevated by lesions that interfere with the hypothalamic–pituitary portal blood flow. Such elevation is less pronounced than that seen with primary overproduction of PRL by adenomas of the pituitary (see below).

Thyroid-Stimulating Hormone

TSH deficiency may be assessed by a single serum measurement. As with all anterior pituitary hormones that have target endocrine glands, however, the interpretation of the adequacy of TSH secretion requires measurement of the other member of the hormone pair, thyroxine. Deficiency of TSH secretion is not usually manifested by a lower than normal serum TSH value, but instead the TSH is inappropriately normal in the presence of a low serum thyroxine (see also Chapter 5).

Likewise, TSH excess may be assessed by a single serum measurement. Uncommonly, thyroid hormone excess is associated with inappropriately normal or increased levels of TSH. This is in contrast to the usual form of thyroid hormone overproduction, in which the TSH level is below normal or undetectable (see Chapter 13). Thus TSH levels must be interpreted in light of the other member of the hormone pair, thyroxine, because excess TSH may only be manifested as an inappropriately normal level.

Gonadotropins (LH, FSH)

In a woman, the presence of normal spontaneous cyclic menses indicates that gonadotropin secretion is normal (see Chapter 9). In women who do not have evidence of normal ovarian estrogen secretion, a single serum sample for measuring LH and FSH is sufficient to determine whether primary ovarian failure is present (elevated gonadotropins) or whether hypothalamic–pituitary abnormalities exist (low or normal gonadotropins). Similarly, in a man, gonadotropins must be interpreted in light of testicular function (see Chapter 10). Since LH is secreted in pulses with resultant pulses of testosterone secretion, aliquots from three serum samples obtained at 20-minute intervals should be pooled for a more accurate estimate of basal levels of LH and testosterone. Again, elevated levels of gonadotropins with low testosterone indicate primary testicular failure, whereas low or normal gonadotropin levels signal hypothalamic or pituitary disease. LHRH testing may help distinguish hypothalamic from pituitary causes of low gonadotropins (see Chapter 10).

Gonadotropin excess unrelated to gonadal failure may occur in women or men as a result of a pituitary adenoma and may be assessed by basal plasma levels. The α-subunit common to all the glycoprotein hormones is often secreted in excess by such gonadotroph cell adenomas.

Adrenocorticotropin

The clinical assessment of ACTH secretion usually involves the measurement of a product of the target gland, e.g., cortisol. Thus ACTH deficiency is assessed by screening for sufficiency of the pituitary–adrenal axis. This is done with the short ACTH stimulation test by measuring the serum cortisol 1 hour after ACTH injection. Basal cortisol levels alone are not sufficient for diagnostic screening. If the response to the short ACTH stimulation test is abnormal, either hypothalamic–pituitary or adrenal abnormalities may be present. Prolonged ACTH administration may be stimulated by insulin-induced hypoglycemia or the drug metyrapone to identify a pituitary problem (see Chapter 14).

ACTH excess as a possible cause of excess cortisol secretion is assessed by suppression tests since basal cortisol levels are not reliable indicators of glucocorticoid excess and plasma ACTH may not be elevated in patients with pituitary-dependent cortisol excess. Dexamethasone, a synthetic potent glucocorticoid, is given at 11:00 P.M., and the serum cortisol is measured at 8:00 A.M. the next day. If the individual is normal, ACTH and consequently cortisol will be low the morning after dexamethasone is given. Failure of the cortisol to suppress below 5 μg/dl indicates the need for definitive testing with a prolonged suppression test to determine the cause of the adrenal hyperfunction (see Chapter 14).

DERANGED HYPOTHALAMIC–PITUITARY FUNCTION

Hypothalamic Disease

A variety of hypothalamic disorders including tumors, trauma, infections, congenital malformations, genetic defects, and vascular alterations can affect the secretion of different releasing hormones either selectively or collectively, depending on the location and extent of the lesion. In general, lesions of the median eminence–tuber cinereum region decrease pituitary function. An exception to this is the secretion of PRL, which increases after such lesions because of the loss of its inhibitory hypothalamic tone.

A deficit of individual hypothalamic hormones results in isolated pituitary hormone deficiency. Prominent examples of this condition are the syndromes of hypothalamic hypothyroidism caused by TRH deficiency, olfactory-genital dysplasia (Kallmann's syndrome or isolated gonadotropin deficiency) characterized by a deficit in LHRH secretion associated with hyposmia, and idiopathic dwarfism caused by GHRH deficiency. Kallmann's syndrome is particularly interesting because the deficit in LHRH secretion is related to the failure of embryonic LHRH neurons to migrate from their site of origin (the olfactory placode) to their final destination in the hypothalamus due to a defect in a neuroadhesion molecule.

Hypopituitarism

If hypopituitarism develops in a child, the defect in GH secretion leads to the development of a pituitary dwarf. In this condition there is a proportional reduction in body size, in contrast to the conspicuous shortening of the long bones in the so-called achondroplastic dwarf. If the deficiency extends to other anterior pituitary hormones, other symptoms and signs will be present. These can include weakness, fatigue, and headaches, but they are of little diagnostic significance. The development of hypopituitarism in adulthood may also produce weakness, fatigue, and headaches, plus signs and symptoms associated with decreased secretion of gonadotropins such as loss of axillary and pubic hair and breast atrophy. The reduction in GH is probably at least in part responsible for the pallor and fine wrinkling of the facial skin seen in adult hypopituitarism. In women, menstrual cycles cease and there is genital atrophy. In men, secondary testicular failure can cause loss of libido and potency, muscle strength, beard growth, and testicular size, the latter because of the diminished gonadotropin secretion, the former because of the loss of testosterone. Hypopituitarism may be caused by hypothalamic disease or by disorders that directly affect the pituitary gland, such as tumors, infarct, and damage by radiation therapy.

Acromegaly and Gigantism

An excess of GH release leads to the development of gigantism if the hypersecretion has been present during early life, which is a rare condition, and to acromegaly if hypersecretion occurs after body growth has stopped. In the case of gigantism, a rather symmetrical enlargement of the body results in a true giant with overgrowth of long bones, connective tissue, and visceral organs. In the case of acromegaly, the epiphysial and diaphysial centers of the long bones (i.e., femur, tibia, etc.) fuse and elongation of these bones is no longer possible. In this condition there is an overgrowth of cancellous bones, resulting in a protruding jaw, termed prognathism, thickening of the phalanges, overgrowth of soft tissue that thickens the skin, and overgrowth of the visceral organs. Acromegaly and gigantism are caused by eosinophilic adenomas consisting of somatotrophs, leading to the excessive secretion of GH.

Hyperprolactinemia

This is the most common example of pituitary hyperfunction and is brought about by microadenomas of the lactotrophs. Hyperprolactinemia may also result from treatment with the dopamine receptor blockers that are commonly prescribed for psychiatric illnesses. In women, hyperprolactinemia frequently leads to amenorrhea and galactorrhea. The amenorrhea is caused by the disturbance in gonadotropin secretion arising from prolactin excess. Galactorrhea is caused by the direct effect of PRL on the breast. In men, galactorrhea is not common; however, hyperprolactinemia can cause loss of libido, impotence, and decreased sperm density usually associated with a decrease in the plasma levels of LH and testosterone. The diagnosis can be made by measuring plasma PRL and then examining for the presence of micro- or macroadenomas of the pituitary by computed tomographic scanning or magnetic resonance imaging. Hyperprolactinemia can be treated with a dopamine agonist, such as bromocriptine, or by surgical removal of the adenoma.

SUGGESTED READING

Albarracin CT, Kaiser UB, and Chin WW: Isolation and characterization of the 5'-flanking region of the mouse gonadotropin-releasing hormone receptor gene. Endocrinology 135:2300–2306, 1994.

Bell GI, and Reisine T: Molecular biology of somatostatin receptors. Trends Neurosci 16:34–38, 1993.

Brazeau P, Vale W, Burgus R, Ling N, Butcher M, Rivier J, and Guillemin R: Hypothalamic polypeptide that inhibits the secretion of immunoreactive pituitary growth hormone. Science 179:77–79, 1973.

Burgus R, Butcher M, Amoss M, Ling N, Monahan M, Rivier J, Fellows R, Blackwell R, Vale W, and Guillemin R: Primary structure of the ovine hypothalamic luteinizing hormone-releasing factor (LRF). Proc Natl Acad Sci USA 69:278–282, 1972.

Burgus R, Dunn TF, Desiderio D, Ward DN, Vale W, and Guillemin R: Characterization of ovine hypothalamic hypophysiotropic TSH-releasing factor. Nature 226:321–325, 1970.

Chen R, Lewis KA, Perrin MH, and Vale WW: Expression cloning of a human corticotropin-releasing-factor receptor. Proc Natl Acad Sci USA 90:8967–8971, 1993.

Conn PM, Janovick JA, Stanislaus D, and Kuphal D: Molecular and cellular basis of gonadotropin releasing hormone action in the pituitary and central nervous system. In: *Vitamins and Hormones,* Vol. 50, G Litwack, ed., Academic Press, New York, 1995 (in press).

Daughaday WH: The anterior pituitary. Nature (London) 301:568–613, 1983.

Gershengorn MC: Mechanism of thyrotropin releasing hormone stimulation of pituitary hormone secretion. Annu Rev Physiol 48:515–526, 1986.

Gharib SD, Wierman ME, Shupnik MA, and Chin WW: Molecular biology of the pituitary gonadotropins. Endocr Rev 11:177–199, 1990.

Goodman RH, Jacobs JW, Dee PC, and Habener JF: Somatostatin-28 encoded in a cloned cDNA obtained from a rat medullary thyrocarcinoma. J Biol Chem 257:1156–1159, 1982.

Grossman A: Brain opiates and neuroendocrine function. Clin Endocrinol Metab 12:725–746, 1983.

Guillemin R, Brazeau P, Böhlen P, Esch F, Ling N, and Wehrenberg WB: Growth hormone-releasing factor from a human pancreatic tumor that caused acromegaly. Science 218:585–587, 1982.

Hoffman AR, and Crowley WF Jr: Induction of puberty in men by long-term pulsatile administration of low-dose gonadotropin-releasing hormone. N Engl J Med 307:1237–1241, 1982.

Hökfelt T: Aminergic and peptidergic pathways in the nervous system with special reference to hypothalamus. In: *The Hypothalamus,* Vol. 56, S Reichlin, RJ Baldessarini, and JB Martin, eds., Raven Press, New York, pp. 69–135, 1978.

Hsueh AJW, and Jones PBC: Extrapituitary actions of gonadotropin-releasing hormone. Endocr Rev 2:437–461, 1981.

Jackson IMD: Thyrotropin-releasing hormone. N Engl J Med 306:145–155, 1982.

Kaiser UB, Katzenellenbogen RA, and Conn PM: Evidence that signalling pathways by which thyrotropin-releasing hormone and gonadotropin-releasing hormone act are both common and distinct. Mol Endocrinol 8:1038–1048, 1994.

Kourides IA, Gurr JA, and Wolf O: The regulation and organization of thyroid-stimulating hormone genes. Recent Prog Horm Res 40:79–120, 1984.

Krieger DT, Liotta AS, and Brownstein MJ: ACTH, β-lipotropin and related peptides in brain, pituitary and blood. Recent Prog Horm Res 36:272–344, 1980.

Lechan RM, Wu P, Jackson IMD, Wold H, Cooperman S, Mandel G, and Goodman RH: Thyrotropin-releasing hormone precursor: Characterization in the rat brain. Science 231:159–161, 1986.

Lin C, Lin S-C, Chang C-P, and Rosenfeld MG: Pit-1-dependent expression of the receptor for growth hormone releasing factor mediates pituitary cell growth. Nature 360:765–768, 1992.

Mayo KE: Molecular cloning and expression of a pituitary-specific receptor for growth hormone-releasing hormone. Mol Endocrinol 6:1734–1744, 1992.

Mayo KE, Vale W, Rivier J, Rosenfeld MG, and Evans RM: Expression-cloning and sequence of a cDNA encoding human growth hormone-releasing factor. Nature 306:86–88, 1983.

McCann SM, and Krulich L: Role of transmitters in control of anterior pituitary hormone release. In: Endocrinology, 2nd ed., Vol. 1, LJ DeGroot, ed., WB Saunders, Philadelphia, pp. 117–130, 1989.

McCann SM, Mizunuma H, and Samson WK: Differential hypothalamic control of FSH secretion: A review. Psychoneuroendocrinology 8:299–308, 1983.

Panetta R, Greenwood MT, Warszynska A, Demchyshyn LL, Day R, Niznik HB, Srikant CB, and Patel YC: Molecular cloning, functional characterization, and chromosomal localization of a human somatostatin receptor (somatostatin receptor type 5) with preferential affinity for somatostatin-28. Mol Pharmacol 45:417–427, 1994.

Reichlin S: Somatostatin. N Engl J Med 309:1495–1501, 1556–1563, 1983.

Reichlin S: Neuroendocrinology. In: Williams Textbook of Endocrinology, 8th ed., JD Wilson and DW Foster, eds., Saunders, Philadelphia, pp. 135–219, 1992.

Riskind PN, and Martin JB: Functional anatomy of the hypothalamic-anterior pituitary complex. In: Endocrinology, 2nd ed., Vol. 1, LJ DeGroot, ed., WB Saunders, Philadelphia, pp. 97–107, 1989.

Rivier J, Spiess J, Thorner M, and Vale W: Characterization of a growth hormone releasing factor from a human pancreatic islet tumor. Nature 300:276–278, 1982.

Rivier J, Spiess J, and Vale W: Characterisation of rat hypothalamic corticotropin-releasing factor. Proc Natl Acad Sci USA 80:4851–4855, 1983.

Schally AV, Arimura A, and Coy DH: Recent approaches to fertility control based on derivatives of LHRH. Vit Horm 38:257–323, 1980.

Seeburg PH, and Adelman JP: Characterization of cDNA for precursor of human luteinizing hormone releasing hormone. Nature 311:666–668, 1984.

Shibahara S, Morimoto Y, Fururanti Y, Notake M, Takahash N, Shimizu S, Horikawa S, and Numa S: Isolation and sequence analysis of the human corticotropin-releasing factor precursor gene. EMBO 2:775–779, 1983.

Tepperman J, and Tepperman HM: Metabolic and Endocrine Physiology, 5th ed., Yearbook Medical Publishers, Chicago, 1987.

Tsutsumi M, Zhou W, Millar RP, Mellon PL, Roberts JL, Flanagan CA, Dong K, Gillo B, and Sealfon SC: Cloning and functional expression of a mouse gonadotropin-releasing hormone receptor. Mol Endocrinol 6:1163–1169, 1992.

Vale W, Rivier C, and Brown MR: Chemical and biological characterization of corticotropin releasing factor. Recent Prog Horm Res 39:245–290, 1983.

Zimmerman EA, and Nilavar G: The organization of neurosecretory pathways to the hypophysial portal system. In: Pituitary Hyperfunction: Physiopathology and Clinical Aspects, F Camanni and EE Muller, eds., Raven Press, New York, pp. 1–25, 1984.

7

The Posterior Pituitary and Water Metabolism

WILLIS K. SAMSON

STRUCTURE OF THE POSTERIOR PITUITARY GLAND

The neurohypophysis, also called the *posterior pituitary* or *neural lobe*, is the ventral extension of hypothalamic tissue derived from a developmental downgrowth of the neuroectoderm forming the floor of the third cerebroventricle. It weighs approximately 0.10–0.15 g in humans and is well developed at birth, having been present since the fifth month of intrauterine life. In addition to containing glial elements called *pituicytes*, the posterior pituitary is composed of unmyelinated nerve fibers and axon terminals of neurons whose cell bodies reside primarily in the supraoptic and paraventricular hypothalamic nuclei (Fig. 7-1). These hypothalamo-neurohypophyseal fibers deliver the two primary posterior pituitary hormones, oxytocin (OT) and arginine vasopressin (AVP), to the neural lobe in association with specific proteins, the neurophysins, once thought to be carrier proteins but now known to be portions of the OT and AVP precursor molecules.

The neurons produce either OT or AVP, never both, and recent studies indicate that in addition to one of these two hormones, other neuropeptides, such as corticotropin-releasing hormone (CRH), and neurotransmitters are also produced in OT- or AVP-containing cells. The phenomenon of colocalization of neuromodulatory agents has aroused a great deal of clinical interest in the role of neuropeptides such as OT and AVP in brain function. OT- and AVP-containing nerve fibers originating in the supraoptic and paraventricular nuclei also project to a variety of other brain structures that are thought to be the sites of their observed central nervous system actions, and to the vicinity of the hypophyseal portal vessels in the median eminence. Release from these fibers of both OT and AVP explains the high levels of these hormones in portal blood and provides the framework for the actions of OT and AVP as modulators of anterior pituitary function.

The arterial blood supply of the posterior pituitary is via the inferior (and to some degree the superior) hypophyseal arteries, which originate from the cavernous and postclinoid portions of the internal carotid artery. The venous drainage is comprised of efferent vessels joining the intercavernous sinuses and eventually the internal jugular vein.

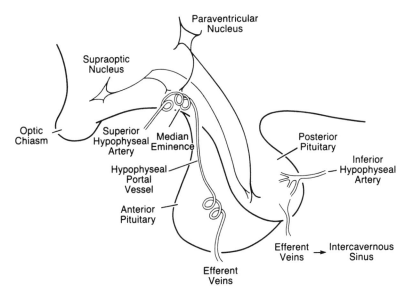

Fig. 7-1 Blood supply and innervation of the posterior pituitary by oxytocin- and vaso-pressin-containing neurons.

POSTERIOR PITUITARY HORMONES

Both OT and AVP (also called *antidiuretic hormone*, ADH) are nonapeptides containing internal disulfide bonds linking cystine residues at positions 1 and 6, giving them a ring structure that is necessary for some of the peptides biological activity (Fig. 7-2). Both are assembled on the ribosomes as parts of large precursor molecules that consist of the respective hormone and its associated carrier protein (a neuro-

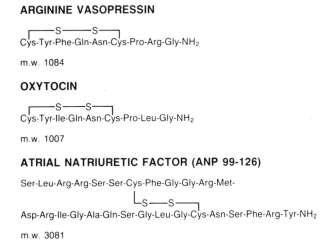

Fig. 7-2 Structures of vasopressin, oxytocin, and atrial natriuretic factor.

physin, approximately 10,000 molecular weight). The AVP-associated neurophysin, nicotine-stimulated neurophysin, is a polypeptide with a C-terminal extention that contains the 39 amino acid, AVP-associated C-terminal glycoprotein. Estrogen-stimulated neurophysin, associated with OT, is a polypeptide which lacks the C-terminal glycoprotein extension. After ribosomal assembly the precursor molecules are transferred to the Golgi complex and packaged into neurosecretory granules that move along the axon at a flow rate of 8 mm/hour (faster than normal axoplasmic flow). The precursor and the associated neurophysins have not been demonstrated conclusively to possess biologic activity; however, during axoplasmic transport the neurophysin carrier is proteolytically cleaved from the associated neurohormone and both the neurophysin and neurohormones (OT or AVP) are released into the capillary spaces in the posterior pituitary. These hypothalamo-neurohypophyseal cells are neurons that can generate and propagate action potentials that upon arrival at the nerve terminals, cause depolarization and exocytosis of the secretory granule contents (stimulus–secretion coupling). Both OT and AVP circulate in the blood unbound to other proteins and are rapidly removed (mainly by the kidney, but also by the liver and brain) from the circulation. The older literature reflects a plasma half-life ($t_{1/2}$) for both hormones of about 5 minutes; however, more recent studies employing sensitive radioimmunoassays for each peptide and calculations based on more physiological circumstances reveal a two component disappearance curve with a much faster $t_{1/2}$ (less than 1 minute) and a second component of approximately 2–3 minutes.

VASOPRESSIN (AVP, ADH)

Primary Action: Antidiuresis

AVP, which circulates in basal levels of around 1 pg/ml plasma (10^{-12} M), binds to a specific membrane receptor, the V_2 receptor, on the peritubular (serosal) surface of cells of the distal convoluted tubules and medullary collecting ducts inducing adenylyl cyclase activity. The cyclic AMP (cAMP) formed is responsible for activation of a protein kinase, which initiates a phosphorylation cascade resulting in the insertion of the protein, aquaporin, in the luminal membrane and thereby enhances permeability of the cell to water. This AVP-dependent increase in membrane permeability to water permits back diffusion of solute-free water remaining in the urine after proximal tubule handling down the osmotic gradient from hypotonic urine to the hypertonic interstitium of the renal medulla, resulting in an increase in urine osmolality (relative to glomerular filtrate or plasma). The net result is an increase in urine osmolality and a decrease in urine flow.

Control of Secretion

Plasma Osmolality

Water deprivation results in increased plasma osmolality, which is thought to be sensed by specialized brain cells, called *osmoreceptors*, located in two highly vascularized regions (where the blood–brain barrier is absent) of the central nervous system: the vascular organ of the lamina terminalis and the subfornical organ. The increased osmolality of plasma results in a loss of intracellular water from the osmoreceptors and a stimulation of AVP release. The osmoreceptor cells themselves do not release AVP, instead, as revealed in animal studies, they are thought to communicate with the AVP neurons using acetylcholine as their neurotransmitter. The sensitivity of the osmoreceptor to changes in plasma osmolality is increased by angiotensin II, which potentiates AVP release in response to osmotic stimuli. As can be seen in Fig. 7-3, the AVP response to changes in plasma osmolality is extremely sensitive: indeed increases as small as 1% result in enhanced AVP secretion.

Nonosmotic Factors

A fall in blood volume of greater than 8% (hemorrhage), quiet standing (orthostatic hypotension), and positive pressure breathing, all of which reduce cardiac output and therefore central blood volume, are potent stimuli for AVP release (Fig. 7-3), whereas maneuvers that increase total blood volume (isotonic saline or blood infusion, cold water immersion) suppress AVP release. Low (left atrial) and high (carotid and aortic) pressure baroreceptors sense small alterations in blood volume and communicate with central nervous system structures via medullary afferents of the ninth and tenth cranial nerves. These pressure receptors normally exert a tonic inhibition on AVP release via medullary efferents to the hypothalamus; thus decreases in blood volume *unload* the baroreceptors (decrease afferent flow), resulting in *less* inhibition (α-adrenergic) of AVP release and *increased* circulating AVP. Hypovolemia also results in renin release from the juxtaglomerular apparatus of the kidney (see Chapter 14) and the formation of angiotensin II, both of which sensitize the osmoreceptor cells of the hypothalamus leading to enhanced AVP release. Therefore, altering blood volume can "reset" the central osmotic threshold and possibly the sensitivity for AVP release (Fig. 7-3), such that the AVP response to any given increase in plasma osmolality is more accentuated as the level of volume depletion increases. A variety of other stimuli can affect AVP release. Increased P_aCO_2 or reduced P_aO_2, pain, stress, increased temperature, β-adrenergic agents, estrogens, progesterone, opiates, barbiturates, nicotine, and prostaglandins have been demonstrated to stimulate AVP release. A decrease in temperature, α-adrenergic agents, ethanol, and cardiac hormones can exert inhibitory actions on AVP release.

Secondary Actions

It has been demonstrated that AVP acts within the central nervous system to lower body temperature and to facilitate memory consolidation and retrieval. Indeed, studies in aged humans revealed the improvement of short-term memory after AVP administration. Vasopressin-containing fibers that project to the hypophyseal portal plexus in the median eminence deliver AVP to the portal blood and anterior pituitary where AVP acts via a subtype of vasopressin receptor, which differs from that in the kidney and vasculature, to potentiate the release of adrenocorticotropin in response to CRH. This action of AVP seems coordinated to potentiate the adrenocorticotropin (ACTH) response to stress.

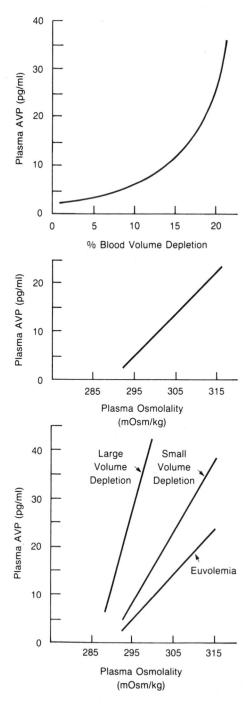

Fig. 7-3 Effects of volume depletion (top) and osmolality (center) on vasopressin release. Interaction of the two variables (bottom) is also illustrated. (Idealized for humans from rat data reported by Dunn FL, et al: J Clin Invest 52:3212–3219, 1973.)

In addition to its effects in the kidney, via V_2 receptors, AVP exerts potent vascular actions as well. Interaction of AVP with specific (V_1) receptors on vascular smooth muscle results in profound contraction, shunting of blood away from the periphery, and elevated central venous pressures. Total peripheral resistance increases in a linear fashion with increases in circulating AVP. However, the rise in total peripheral resistance seen in response to AVP is buffered by baroreceptor-mediated falls in cardiac output and general sympathetic tone so that other than having a prolonged effect on regional distribution of blood flow, the overall increase in arterial pressure is minimal.

Disorders of AVP Secretion

Deficient Secretion

Usually caused by destruction or dysfunction of the hypothalamo-neurohypophyseal complex, diabetes insipidus is characterized by the inability to produce a concentrated urine, frequent urination (with low specific gravity and osmolality), and often excessive thirst. Treatment of this disorder involves use of a vasopressin analog, desmopressin, which fails at therapeutic doses to react with vascular (V_1) AVP receptors (therefore is a nonpressor) yet is recognized by renal tubular (V_2) receptors; it has the additional advantage of a prolonged duration due to a longer ($t_{1/2}$) in plasma, and enhanced antidiuretic potency.

Patients with diabetes insipidus cannot reduce urine flow during standard water deprivation testing because of their inability to concentrate their urine. Diabetes insipidus can result from central nervous system (CNS) lesions (failure to secrete AVP) or from a renal disorder (failure to respond to AVP). The differential diagnosis can be made by AVP administration. Patients with diabetes insipidus of central origin respond whereas those with nephrogenic diabetes insipidus do not. The renal form of diabetes insipidus is primarily a local lesion characterized by unresponsiveness of the tubule to appropriate levels of vasopressin and is not associated with lesions of the hypothalamus or posterior pituitary gland.

Inappropriate Production

AVP release exceeding that predicted by plasma volume or tonicity can result from CNS disease or trauma, drug interactions, or ectopic production by tumors resulting in water retention and concentration of urine in excess of plasma. Paradoxically, sodium excretion into the urine increases despite the low serum sodium (probably due to elevated levels of atrial natriuretic factors caused by the expanded plasma volume, see below). Symptoms include altered mental status, headache, drowsiness, nausea, and often coma. Fluid restriction is appropriate treatment; however, acute administration of hypertonic saline is often considered to raise plasma osmolality. In the longer term, treatment with naloxone (to decrease central AVP release) or demeclocycline (to block AVP action on the kidney) is often employed. This so-called Syndrome of Inappropriate AntiDiuretic Hormone (SIADH) secretion is characterized by normal renal and adrenal function, yet by

1. Hyponatremia (serum sodium concentrations of 100–115 mEq/liter)
2. Continued renal sodium excretion

3. Absence of clinical evidence of volume depletion or edema
4. Inappropriately high urine osmolality

OXYTOCIN (OT)

Primary Action

Oxytocin stimulates milk ejection by contracting the myoepithelial cells surrounding the alveoli and ducts in the mammary gland. Additionally, it stimulates rhythmic myometrial contractions in the uterus, aiding in expulsion of the fetus. Although not necessary for its initiation, labor proceeds more slowly in the absence of OT. Therapeutically, oxytocin (also called pitocin) is employed postpartum to sustain contractions and to decrease bleeding.

Control of Release

Oxytocin is normally present in barely detectable levels (1–10 pg/ml plasma) but is present in readily detectable quantities during ovulation, parturition, and lactation, as well as in males and females during certain mild stresses. Vaginal stimulation during intercourse or delivery and stimulation of touch receptors in the nipples result in afferent neural input (via the spinothalamic tract and a variety of brain stem relays) to OT-producing cells in the paraventricular and supraoptic nuclei. The stimulatory neurotransmitter involved is thought to be acetylcholine or dopamine. Extremely severe pain, increased temperature, and loud noise can inhibit OT release (probably via an opioid mechanism), whereas hemorrhage and psychogenic stress (restraint, novel environment, mild apprehension and fear) stimulate oxytocin secretion. The OT release seen during the immediate preovulatory period is possibly due to rising estrogen levels present at that time (remember the OT-associated carrier is also known as the *estrogen-stimulated neurophysin*). Indeed administration of exogenous estrogens can stimulate OT release.

Secondary Actions

In addition to possessing some renal and vascular actions that mimic AVP, probably due at least in part to an interaction with the vasopressin receptor, OT can potentiate CRH-induced adrenocorticotropin release, again by interacting with the AVP receptor. Central nervous system actions of OT include amnestic effects and a stimulation of maternal behavior. Recently OT has been shown to act within the CNS to curb salt appetite. Increases in oxytocin levels in plasma often parallel those of prolactin (during mild stress, ovulation, lactation) and recent evidence indicates that OT can act directly at the lactotroph to stimulate prolactin release, suggesting a physiological role for OT as a prolactin-releasing factor.

Disorders of Oxytocin Secretion

Oxytocin excess has not been clearly demonstrated. A deficiency in OT secretion normally results in difficulty in nursing due to inadequate milk ejection.

HORMONES OF CARDIAC/BRAIN ORIGIN

Identity of Atrial Natriuretic Factors (ANF)

In addition to neuronal control of cardiovascular and renal function, several hormonal systems interact at a variety of sites to assure the maintenance of fluid and electrolyte homeostasis. As discussed above, AVP plays a central role in this scheme and, as will be discussed in detail in Chapter 14, the renin–angiotensin system is another major factor. Recently, a third hormonal system, the atrial natriuretic factors, was identified that exerts profound influences on renal, vascular, and adrenal function, and thereby on the control of fluid homeostasis.

It has been known for quite some time that atrial distension resulted in profound *diuresis* (increased urine flow), and the proposed mechanisms for this effect included not only neuronal reflex activation of CNS structures (vagal afferents) but also the possible release of a volume regulatory substance from the heart. Infusions of extracts of mammalian cardiac atrial tissue were found to stimulate significant increases in urine volume and urinary sodium excretion (*natriuresis*). Additionally, these extracts were shown to relax precontracted vascular and gastrointestinal smooth muscle strips *in vitro* (a *spasmolytic* action). Several laboratories then using a combination of biochemical and molecular techniques identified the natriuretic, diuretic, and spasmolytic substances in atrial extracts as a family of peptides derived from the same precursor molecule (151 amino acids in length). The active fragments of this precursor, called *preatriopeptigin* (identified by cDNA cloning of the human gene), reside in the carboxy terminus and are small peptides 24 to 28 amino acids in length. These peptides share in common the 17 member ring structure formed by an internal disulfide linkage and vary in the extent of their N- and C-terminal extension; however, the major circulating form in humans is 28 amino acids in length (Fig. 7-2) and possesses the full C-terminal extension of the prohormone. Significant amounts of the N-terminally shortened 25 and 24 amino acid forms are present in blood. These forms possess bioactivity equal to that of the 28 amino acid form. These peptides have been given various names (atriopeptins, cardionatrin, auriculin), but for simplicity, the entire family of peptides is called *atrial natriuretic factors* (ANF) or *peptides* (ANP). Normal plasma levels of the major form (α human ANP 1–28) vary between 50 and 150 pg/ml. Highest levels are present in the aorta (vena cava to aorta concentration gradient 76–177 pg/ml) and a plasma half life of about 1–3 minutes has been reported. In addition to being filtered into the urine, major degradation sites include the kidney, liver, and brain.

Recently, two additional natriuretic peptides, both encoded by unique genes, have been described and characterized. B-type natriuretic peptide (BNP) in its final post-translationally modified state is comprised of 32 amino acids and shares significant sequence homology with ANP. BNP is produced also in the heart with a significant increase in gene transcription and hormone secretion in volume and pressure overload states such as congestive heart failure. It is thought to exert a spectrum of biological activities similar to those of ANP. The third member of the natriuretic peptide family is C-type natriuretic peptide (CNP). This peptide has been reported to be more abundant in brain than ANP or BNP and, while little CNP gene product has been detected in the heart, the endothelium is a rich source of CNP. It differs from ANP in that this 22 amino acid peptide lacks the C-terminal extension downstream from the internal disulfide ring and differs at five sites within the ring.

Control of Release

Cardiac myocytes (which produce ANF) are located in a position to detect pressure changes accompanying increases in venous return or afterload. Indeed, stretch associated with increases in venous return is the major stimulus for ANF release since interventions (head down tilt, passive leg elevation, cold water immersion, saline or blood loading) that result in increased right atrial pressure and an increase in right atrial dimensions result in concomitant increases in ANF release and diuresis; a significant, positive correlation between right atrial pressure and ANF levels has been demonstrated in humans. Therefore the heart responds to an increase in venous return (Fig. 7-4) by releasing hormones that act at the kidney (and other sites, see below), resulting in a lowering of fluid volume (via diuresis) and a decrease in venous return. Recent results indicate that increases in plasma osmolality also release ANF, perhaps reflecting the natriuretic action as well. Increasing sodium intake in normal volunteers from 10 to 200 mmol sodium per day resulted in significant, 2-fold elevations in plasma ANF. Furthermore these subjects released even more ANF in response to interventions that elevated venous return.

Natriuretic Peptide Receptors

Three subtypes of natriuretic peptide receptors have been cloned and sequenced. Two have extracellular binding domains linked via a transmembrane spanning segment with an intracellular, particulate guanylyl cyclase. The third lacks the intracellular extension, and although it has been designated the nonbiologic, clearance receptor (originally thought to function only as a biologic sponge which removed the natriuretic peptides from the circulation), it is now apparent that this subtype transduces important antimitogenic signals of the natriuretic peptides and is linked to activation of inositol phosphate hydrolysis and inhibition of adenylyl cyclase. Table 7-1 summarizes the three receptor subtypes.

Renal Actions

The hallmark renal actions of ANF are natriuresis and diuresis. Infusion of ANF into healthy volunteers results in increases in urine volume, creatinine clearance, free water clearance, and the excretion rates of sodium and chloride, and, to a lesser degree, potassium. These effects are at least partially due to ANF's ability to increase glomerular filtration rate, a reflection of its action to stimulate postglomerular vasoconstriction. Additionally, ANF exerts a direct tubular action to inhibit solute transport in the distal nephron and antagonizes the effect of AVP in the tubule and duct. A final renal action of ANF is exerted directly at the level of the juxtaglomerular apparatus to inhibit renin secretion, thus reducing angiotensin II formation.

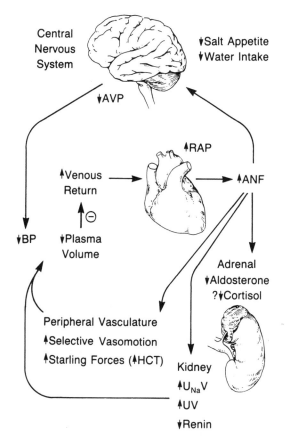

Fig. 7-4 Atrial natriuretic factor (ANF) is released in response to increased venous return and the associated increase in right atrial pressure (RAP). ANF acts in the brain, adrenal, kidney, and peripheral vasculature to result in a drop in blood pressure (BP) and venous return, thus eliminating the signal for its release. AVP, vasopressin; HCT, hematocrit; $U_{Na}V$, urine sodium excretion; UV, urine volume.

Table 7-1 The Natriuretic Peptide Receptor Family

Receptor subtype	Ligand preference	Signalling mechanism
NPR$_A$ receptor (guanylyl cyclase A, GC-A)	ANP>BNP>>CNP	particulate guanylyl cyclase
NPR$_B$ receptor (guanylyl cyclase B, GC-B)	CNP>ANP>BNP	particulate guanylyl cyclase
NPR$_C$ receptor (clearance receptor)	ANP=BNP=CNP	inositol phosphase turnover, adenylyl cyclase

Adrenal Actions

In addition to preventing angiotensin II formation by inhibiting renin release, ANF blocks the action of angiotensin II in the adrenal. Indeed, both basal and stimulated (angiotensin II and ACTH) aldosterone secretion are inhibited by the cardiac hormones. Although the renal effects of ANF (natriuresis) are too rapid to be explained by alterations in aldosterone levels, in chronic volume expansion ANF may play a part in the suppression of aldosterone levels.

Vascular Actions

Intravenous infusion of ANF in humans results in decreased mean arterial blood pressure and a decreased cardiac output. This decrease in mean arterial pressure could be due solely to an *in vivo* vasodilatory effect of ANF that matches its *in vitro* spasmolytic action. Increases in forearm blood flow have been observed; however, in certain vascular regions, vasoconstrictive effects are observed (efferent renal artery). Alternatively, the drop in mean arterial pressure could be secondary to the observed decrease in cardiac output caused by decreased venous return. Strong evidence in favor of this exists. Detailed experiments have revealed that ANF reduces circulatory capacitance as a result of decreased blood volume (and vascular recoil) and active vasoconstriction, thereby decreasing venous return. In addition to urinary fluid loss, the observed increase in hematocrit during ANF infusion may be due to transcapillary fluid shifts initiated by increases in postcapillary resistance. No direct chronotropic or inotropic effects have been observed *in vitro* in isolated heart preparations.

The abundant production of CNP in vascular endothelial cells reflects a paracrine action of the peptide to control vascular tone in concert with several other vasorelaxant and vasoconstrictor factors produced locally. Circulating endothelin-3 or locally produced endothelin-1, members of a family of powerful vasoconstrictor peptides, stimulate the release from the endothelium of both CNP and nitric oxide (NO). These factors are released abluminally, away from the vascular lumen toward the surrounding smooth muscle, and induce vasodilation, by unique but potentially convergent mechanisms. While CNP activates particulate guanylyl cyclase via the NPR_B receptor, NO diffuses into the vascular smooth muscle cell and activates soluble guanylyl cyclase, both resulting in elevated cytosolic cGMP. Thus there exists at the endothelial interface with the vascular smooth muscle a balance of redundant vasodilatory substances including CNP and nitric oxide, and additional endothelial-derived factors such as prostacyclins and the recently described peptide adrenomedullin, with local vasoconstrictors such as endothelin-1. An interplay of these agents, together with local neural influences, determines vascular tone and therefore, blood pressure.

Central Nervous System Actions

In addition to its renal, adrenal, and vascular actions, ANF exerts a profound influence on central nervous system-mediated events that relate to fluid and electrolyte homeostasis. Specific ANF receptors have been localized to discrete brain regions and the production of ANF in neuronal elements is well established. ANF-containing neurons terminate in regions known to be important in the neuroendocrine and neuronal control of renal and cardiovascular function. Indeed, ANF exerts a powerful inhibitory action on AVP release (Fig. 7-4) and significantly impairs fluid consumption by inhibiting water intake and salt appetite. These central actions seem well coordinated to match the peripheral effects of the hormone such that in addition to exerting an opposite action (diuresis) in the kidney, ANF inhibits the release of AVP (the antidiuretic hormone). Also, the renal diuretic and natriuretic actions are suitably paired with ANF's central nervous system actions to inhibit water and salt intake.

Disorders of ANF Release

Whereas derangements in ANF secretion have been observed (i.e., elevation during congestive heart failure, supraventricular tachycardia, chronic renal failure, and primary hyperaldosteronism), these conditions of elevated plasma ANF usually reflect exaggerated drive for the release of the hormone. The extreme elevation of plasma ANF in the face of continued water reabsorption during congestive heart failure suggests a derangement of ANF receptor mechanisms and not hormone levels per se. Just the same, exaggerated ANF release might be an important factor in SIADH (see above) and aldosterone escape (see Chapter 14).

INTEGRATED CONTROL OF WATER AND SODIUM METABOLISM

The integrated control of body fluid homeostasis requires a balance of renal, vascular, behavioral, neural, and endocrine influences. Most importantly, these all converge on the kidney as the final site of regulation. Central nervous system involvement in fluid balance involves the actions of neuropeptides in ingestive behavior. Angiotensin II of either central or peripheral origin acts as a stimulator of water intake and atrial natriuretic factor can inhibit the action of angiotensin II on water intake and can limit the hunger for salt. By acting centrally to inhibit vasopressin secretion, ANF exerts yet another powerful effect on water metabolism.

These opposing central actions are matched peripherally as well. Vasopressin (the antidiuretic hormone) acts in the kidney to increase tubular permeability, resulting in a back diffusion of solute-free water, an increase in urine osmolality, and a decrease in urine volume. ANF can reverse the action of AVP and, by its action on glomerular filtration rate (increases), present a greater filtered load to the nephron, with the final effect being increased urine volume (diuresis) and sodium excretion (natriuresis). ANF also counteracts the action of aldosterone in the kidney by opposing its sodium reabsorptive action. In the adrenal gland, ANF blocks angiotensin II-stimulated aldosterone secretion directly and indirectly because of its inhibitory action on renin release and therefore angiotensin II formation.

It appears, then, that in addition to hemodynamic and direct, neuronal influences on the kidney, a well-developed hormonal system of checks and balances operates to maintain a stable fluid and electrolyte milieu. Hormones (cardiac) designed to void the system of an excess volume or electrolytes not only counteract those that act to recapture fluid and solute but also act to inhibit either the release (AVP, aldosterone, renin) or the generation (angiotensin II) of such hormones. Thus excessive influence of the volume conservatory hormones results in volume overload and ANF release. ANF, in return, acts (1) intrarenally to correct the volume expansion, (2) in the adrenal to inhibit aldosterone release, (3) within the vasculature to decrease venous return, and (4) centrally to inhibit AVP release and salt and water intake. Clearly any alteration in the production, release, or bioactivity (including response to) of any one of these hormones can result in imbalanced fluid and electrolyte control.

SUGGESTED READING

Baylis PH, and Thompson CJ: Osmoregulation of vasopressin secretion and thirst in health and disease. Clin Endocrinol 29:549–576, 1988.

Dawood MY, Khan-Dawood FS, Wahi RS, and Fuchs F: Oxytocin release and anterior pituitary and gonadal hormones in women during lactation. J Clin Endocrinol Metab 52:678–682, 1981.

Dunn FL, Brennan TJ, Nelson AE, and Robertson GL: The role of blood osmolality and volume in regulating vasopressin secretion in the rat. J Clin Invest 52:3212–3219, 1973.

Ganten D, and Pfaff D: *Neurobiology of Vasopressin,* Springer-Verlag, Berlin, 1985.

Ganten D, and Pfaff D: *Neurobiology of Oxytocin,* Springer-Verlag, Berlin, 1986.

Ganten D, and Pfaff D: *Behavioral Aspects of Neuroendocrinology,* Springer-Verlag, Berlin, 1990.

Gavras H: Pressor systems in hypertension and congestive heart failure. Role of vasopressin. Hypertension 16:587–593, 1990.

Reichlin S: *The Neurohypophysis, Physiological and Clinical Aspects,* Plenum, New York, 1984.

Rubanyi GM, and Polokoff MA: Endothelins: molecular biology, biochemistry, pharmacology, physiology, and pathophysiology. Pharmacol Rev 46:325–415, 1994.

Samson WK, and Quirion R: *Atrial Natriuretic Peptides,* CRC Press, Boca Raton, FL, 1990.

Samson WK: Natriuretic peptides: a family of hormones. Trends Endocrinol Metab 3:86–90, 1992.

Schrier RW: *Vasopressin,* Raven Press, New York, 1985.

Wilkins RM, Nunez DJ, and Wharton J: The natriuretic peptide family: turning hormones into drugs. J Endocrinol 137:347–359, 1993.

8

Sexual Differentiation

FREDRICK W. GEORGE

Sexual differentiation is a sequential process beginning with the establishment of chromosomal sex at fertilization, followed by the development of gonadal sex, and culminating in the development of secondary sexual characteristics, collectively termed the male and female phenotypes (Fig. 8-1).

This paradigm, which has become the central dogma of sexual differentiation, was formulated by Alfred Jost in the late 1940s to explain the results of fetal castration experiments carried out in rabbits. He found that removal of the gonads (ovaries or testes) before sexual differentiation invariably resulted in the castrated embryos developing as phenotypic females. The presence of the Y chromosome in the male somehow dictates the development of a testis; the hormonal secretions of the testis then impose male development on the phenotypically indifferent fetus. In the absence of a Y chromosome a testis fails to develop and a female phenotype results. Thus a central concept in mammalian sexual differentiation is that the male is the induced phenotype, whereas the female develops passively in the absence of male determinants. Under normal circumstances the chromosomal sex of the embryo corresponds to the phenotypic sex of the embryo. Occasionally, however, the process of sexual differentiation is incomplete, and phenotypic sexual development is ambiguous. Clinically recognized aberrations of sexual development occur at different levels and range from common defects involving the terminal stages of male development (testicular descent and growth of the penis) to more fundamental abnormalities involving hormone formation and action that result in various degrees of ambiguity of phenotypic sex. Although most of these abnormalities impair the reproductive capabilities of the affected individuals, they are not usually life threatening. Consequently, animals and humans with naturally occurring defects in sexual differentiation survive and come to the attention of physicians and scientists. Many of these abnormalities in sexual differentiation occur as the result of single gene mutations; the systematic study of families harboring such mutations has provided much insight into this developmental process.

This chapter reviews the sequence of anatomic events in normal sexual differentiation and describes several disorders of the process in humans that have helped to define this hormonally mediated process.

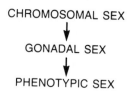

Fig. 8-1   The Jost paradigm for sexual differentiation.

ANATOMIC EVENTS IN SEXUAL DIFFERENTIATION

Development of Chromosomal Sex

Normal human somatic cells have a complement of 44 autosomes and two sex chromosomes; the genotype of the female is 46,XX, whereas that of the male is 46,XY. By a process of meiosis (reduction division), the chromosomal number is reduced by half in the gametes so that all oocytes have a 23,X complement and spermatozoa have either a 23,X or a 23,Y complement. At fertilization when the haploid egg fuses with the haploid sperm, the diploid complement of the soma is re-established. Depending on the genotype of the fertilizing spermatozoa, the zygote will be either 46,XX (female) or 46,XY (male). The chromosomal sex of the fertilized zygote is therefore dependent on the genotype of the fertilizing sperm cell.

The Y chromosome is one of the smallest of the human chromosomes. Although it is thought to carry only a few functional genes, the undisputed function of the mammalian Y chromosome is to serve as the repository for the primary gene that controls testis differentiation and male development. The presence of a single Y chromosome in cells of an individual ensures testicular differentiation and the development of a predominantly male phenotype, even in situations in which the karyotype includes multiple X chromosomes. The region of the Y chromosome that encodes the primary testis determinant, Sex-determining Region of the Y (SRY), has been localized to the short arm (Yp) of the Y chromosome between the centromere and the pseudoautosomal region. The *SRY* gene is expressed in the urogenital ridge of the male embryo at the time of testicular differentiation and encodes a DNA-binding protein with homology to other known transcriptional regulators. Although *SRY* is thought to be the primary determinant of testis differentiation, complete development of both the male and the female phenotypes requires the expression of genes located on the X chromosome and autosomes. Consequently, the overall process of sexual differentiation is more correctly viewed as a cascade of differentiative events initiated by the *SRY* gene and involving genetic determinants of the X and Y chromosomes and autosomes.

Development of Gonadal Sex

The gonads are first recognizable during the fourth week of human embryogenesis as a proliferation of the coelomic epithelium and condensation of the underlying mesenchyme on each side of the midline between the primitive kidney (mesoneph-ros) and the dorsal mesentery. These primitive "gonadal ridges" are initially devoid of germ cells. During the fifth week of human gestation, the germ cells proliferate and migrate from their site of origin in the primitive gut to the gonadal ridges (Fig. 8-2). After germ cell migration is complete about 1 week later, the indifferent gonads are composed of three principal cell types: (1) germ cells, (2) supporting cells derived from the coelomic epithelium that differentiate into either the Sertoli cells of the testis or the granulosa cells of the ovary, and (3) stromal (interstitial) cells derived from the mesenchyme of the gonadal ridge. During the seventh week of gestation, the fetal testis begins to differentiate. First, the differentiating Sertoli cells become grouped into primitive "seminiferous" tubules. The formation of these primitive tubules is accompanied by the secretion of müllerian-inhibiting hormone (MIH) by the fetal Sertoli cells which causes regression of the müllerian ducts of the male embryo. Since the onset of müllerian duct regression is one of the first

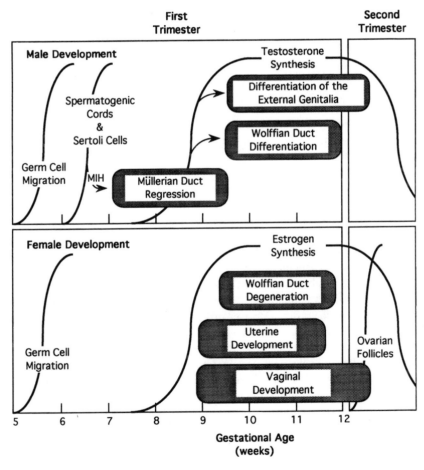

Fig. 8-2 Gonadal differentiation and the timing of the events of phenotypic sexual differentiation.

recognizable events in male differentiation, it is possible that activation of the *MIH* gene is under direct control of the *SRY* gene transcript. At approximately 8 weeks of development, the interstitial cells of the fetal testis differentiate into steroid-synthesizing Leydig cells and begin to secrete testosterone, which virilizes the wolffian ducts and external genitalia (Fig. 8-2). During this period, the fetal ovaries grow but do not undergo identifiable cellular changes from that seen in the indifferent fetal gonad. Cells corresponding to those of the testis that undergo organization into primitive seminiferous tubules multiply in the ovary but do not organize into follicles. It is not until the 15th week of gestation that primordial follicles begin to form, some 8 weeks after the initiation of testicular differentiation (Fig. 8-2). Although the fetal ovary begins to synthesize estrogen at the same time that the fetal testis begins to synthesize testosterone (Fig. 8-2), the meaning of this ovarian endocrine differentiation is not clear since female phenotypic differentiation occurs in the absence of gonads (see below).

Development of Phenotypic Sex

The time sequence of events in the development of the male and female phenotypes is depicted in Fig. 8-2.

Indifferent Phase

Before approximately 8 weeks of gestational age, the sexual phenotype of the embryo cannot be recognized. This is termed the *indifferent phase* of development. During this indifferent phase the embryo has acquired a dual genital duct system that will be the precursor for the male and the female internal genitalia (Fig. 8-3, top). This dual duct system develops within the substance of the mesonephric (primitive) kidney. The first duct to form is the wolffian (or mesonephric) duct, which is the excretory duct for the mesonephric kidney. Caudally, as the wolffian duct joins the primitive urogenital sinus, an outgrowth of the dorsomedial wall of the wolffian duct (the ureteric bud) gives rise to the excretory ducts and collecting tubules of the metanephric (permanent) kidney.

The paramesonephric (müllerian) ducts begin to form in 6-week-old embryos of both sexes as an evagination of the coelomic epithelium into the mesonephros just lateral to the wolffian duct. The müllerian ducts cannot form if the wolffian ducts fail to develop, although the precise nature of the müllerian–wolffian interaction that occurs near the urogenital sinus is uncertain. The male internal accessory organs of reproduction (the seminal vesicle, vas deferens, and epididymis) develop from the wolffian ducts, whereas the female reproductive tract (fallopian tubes, uterus, and upper vagina) develop from the müllerian ducts. Thus, the internal genitalia of the male and the female develop from different embryonic anlage.

The external genitalia of the female and the male, on the other hand, develop from a common anlage (Fig. 8-3, bottom). Before the sixth week of development, the structures that participate in the formation of the external genitalia (the genital tubercle, the urethral folds and groove, and the labioscrotal swellings) of both sexes are identical.

FEMALE   INDIFFERENT STAGE   MALE

INTERNAL GENITALIA

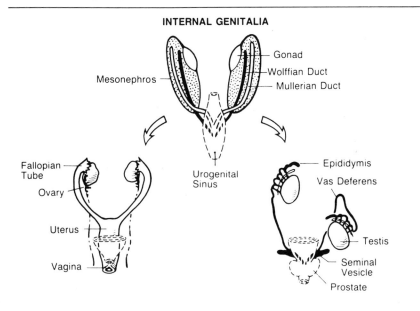

EXTERNAL GENITALIA

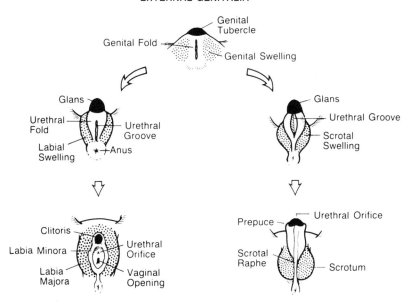

Fig. 8-3   Differentiation of the internal and external genitalia of the human fetus.

Male Development

The initial event in the virilization of the male urogenital tract is the onset of müllerian duct regression, which coincides with the development of the spermatogenic cords in the fetal testis at approximately 8 weeks of gestation.

The transformation of the wolffian ducts into the male genital tract begins after the onset of müllerian duct regression (Fig. 8-2). The mesonephric tubules adjacent to the testis (the epigenital tubules) lose their primitive glomeruli and establish contact with the developing rete and spermatogenic tubules of the testis to form the efferent ductules of the testis. The portion of the wolffian duct immediately caudal to the efferent ductules becomes elongated and convoluted to form the epididymis, and the middle portion of the duct develops thick muscular walls to become the vas deferens (Fig. 8-3, top). At about the thirteenth week of gestation, the seminal vesicles begin to develop from the lower portions of the wolffian ducts near the urogenital sinus. The terminal portions of the ducts between the developing seminal vesicles and the urethra become the ejaculatory ducts and the ampullae of the vas deferens.

The prostatic and membranous portions of the male urethra develop from the pelvic portion of the urogenital sinus. At about 10 weeks of gestation prostatic buds begin to form in the mesenchyme surrounding the pelvic urethra. The budding is most extensive in the area surrounding the entry of the wolffian ducts (the ejaculatory ducts) into the male urethra. Although differentiation of the prostate occurs early in embryogenesis, growth and development of the gland continue into postnatal life.

Beginning in the ninth week of gestation, and continuing through the twelfth week, development of the external genitalia of the male and female diverges (Fig. 8-3, bottom). In the male, the genital swellings enlarge and migrate posteriorly to form the scrotum. Subsequently, the genital folds fuse over the urethral groove to form the penile urethra. The line of closure remains marked by the penile raphe. Incomplete fusion of the genital folds over the urethral groove results in a condition known as *hypospadias*. When the formation of the penile urethra is almost complete, the prepuce starts to develop and by 15 weeks of gestation completely covers the glans penis.

Thus, formation of the male genital tract is largely accomplished between 7 and 13 weeks of gestation. Two additional aspects of male development take place during the latter two-thirds of gestation—the completion of descent of the testes and the growth of the external genitalia.

Testicular descent is a complex and poorly understood process. The overall process involves movement of the testis from the dorsal abdominal wall to the internal inguinal ring, formation of the processus vaginalis, and movement of the testis through the inguinal canal into the scrotum. The final phase of this process occurs around the time of birth.

When formation of the male urethra is largely completed at approximately 11 weeks of gestation, the size of the phallus in the male does not differ substantially from that of the female. During the last two trimesters, however, the male external genitalia, prostate, and structures of the wolffian duct grow progressively so that the external genitalia of the male are much larger than those of the female at the time of birth.

Female Development

The internal reproductive tract of the female is formed from the müllerian ducts, and the wolffian ducts, for the most part, degenerate. The cephalic ends of the müllerian ducts (the portions derived from coelomic epithelium) are the anlagen of the fallopian tubes, and the caudal portions fuse to form the uterus (Fig. 8-3, top). The cervix of the uterus can be recognized by 9 weeks of development, and the formation of the muscular walls of the uterus (myometrium) from the mesenchyme that surrounds the müllerian ducts is completed by approximately 17 weeks of development.

Development of the vagina begins at approximately 9 weeks of gestation with the formation of a solid mass of cells (the uterovaginal plate) between the caudal buds of the müllerian ducts and the dorsal wall of the urogenital sinus. The cells of the uterovaginal plate subsequently proliferate, thus increasing the distance between the uterus and urogenital sinus. At 11 weeks of development a lumen begins to form in the caudal end of the vaginal plate, and by 20 weeks gestation the vagina is completely canalized. It is currently believed that the upper one-third of the vagina is derived from the müllerian ducts and the remainder is derived from the urogenital sinus.

After 10 weeks of gestation the genital tubercle of the female begins to bend caudally, the lateral portions of the genital swellings enlarge to form the labia majora, and the posterior portions of the genital swellings fuse to form the posterior fourchette. The urethral folds flanking the urogenital orifice do not fuse but persist as the labia minora (Fig. 8-3, bottom). Thus, most of the urogenital sinus of the female remains exposed on the surface as a cleft into which the vagina and urethra open.

Breast Development

At 5 weeks of development, paired lines of epidermal thickening extend from the forelimb to the hindlimb on the ventral surface of the embryo. Between 6 and 8 weeks of development these ''mammary lines'' largely disappear, except for a small portion on each side of the thoracic region that condenses and penetrates the underlying mesenchyme. This single pair of mammary buds undergoes little change until the fifth month of embryonic development when secondary epithelial buds appear and the nipples begin to form. Throughout the remainder of gestation these secondary buds (15–25 in number) grow into the underlying mesenchyme and begin to form buds of their own. Although in some species sexual dimorphism in breast development is apparent during embryogenesis, such dimorphism has never been documented in humans, and the development of the breast in boys and girls is identical before the onset of puberty.

ENDOCRINE CONTROL OF PHENOTYPIC DIFFERENTIATION

Fetal castration experiments in rabbits convincingly demonstrated that the fetal testis is necessary for male phenotypic development. Development of the female phenotype, on the other hand, apparently does not require secretions from the fetal ovaries, that is, female phenotypic development occurs in the absence of gonads. Two substances from the fetal testis are known to be essential for male development: a polypeptide hormone (MIH) causes regression of the müllerian ducts in the male, and testosterone virilizes the wolffian ducts, urogenital sinus, and genital tubercle.

Onset of Endocrine Function of Testes and Ovaries

Differentiation of fetal ovaries and testes can be first detected histologically with the development of the primitive spermatogenic tubules in the fetal testis at about 7 weeks of gestational development; the formation of MIH is associated with the appearance of the spermatogenic tubules in the testis. Thus, MIH is the primordial hormone of the fetal testis, and regression of the müllerian ducts is the first event in male development. Shortly thereafter, at approximately 8–9 weeks of gestation, steroidogenic (Leydig) cells begin to differentiate in the interstitium of the testis, and these cells begin to synthesize testosterone (Fig. 8-2). Whether or not the onset of testosterone synthesis in the human fetal testis and the early aspects of phenotypic virilization are regulated by fetal pituitary or placental gonadotropins is open to debate, but virilization that takes place during the latter two-thirds of gestation (such as growth of the penis) is probably gonadotropin dependent (see Chapter 11). The onset of estrogen formation by the human fetal ovaries occurs at approximately the same time that fetal testes begin to synthesize testosterone (Fig. 8-2), despite the fact that development of ovarian follicles does not commence until the fifth month of development (Fig. 8-2). Thus, in the human embryo endocrine differentiation in terms of steroid hormone formation occurs at the same time in fetal ovaries and testes.

Role of Testicular Hormones in Male Development

Müllerian Duct Regression
Regression of the müllerian ducts begins in male fetuses at 8–9 weeks of gestational age and is mediated by a large (140,000 molecular weight) dimeric glycoprotein, MIH, that is synthesized and secreted by the fetal Sertoli cells. The structure of the MIH gene and protein are conserved across species, and analysis of the C-terminus of MIH indicates that it is a member of the transforming growth factor-β "family" of growth inhibitors. The concept that müllerian duct regression is an active process in male development is supported by studies of individuals with the persistent müllerian duct syndrome (see below). In the female, it is essential that MIH *not* be expressed during fetal development of the urogenital tract, otherwise the reproductive tract would not develop. Therefore, the gene encoding MIH is likely to be under strict transcriptional control.

Virilization
Virilization of the male fetus results from the action of androgen on the wolffian ducts, urogenital sinus, and external genitalia. Although testosterone is the principal androgen secreted by the fetal and adult testis, many of the differentiating, growth-promoting, and functional actions attributed to androgens are mediated by the 5α-reduced metabolite of testosterone, dihydrotestosterone (Fig. 8-4).

The current view of how androgens act within target cells is depicted in Fig.

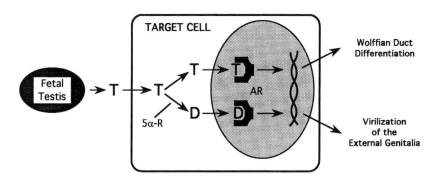

Fig. 8-4 Androgen action. T, testosterone; D, dihydrotestosterone; 5α-R, 5α-reductase; AR, androgen receptor.

8-4. Testosterone, the principal androgen secreted by the fetal testis, enters cells passively by diffusion. Once inside the target cell, testosterone either binds directly to specific receptor proteins located in the nucleus or, if the tissue expresses 5α-reductase activity, is converted to the potent androgen dihydrotestosterone before binding to the androgen receptor. The androgen receptor belongs to a "family" of structurally related hormone receptor proteins, including the steroid, thyroid, and retinoid receptors that function as transcriptional regulators (see Chapter 3). After binding, the receptor–androgen hormone complex undergoes a poorly understood transformation process that promotes binding of the complex to specific sites, termed *androgen response elements* (AREs) within the promoter region of androgen-responsive genes. As a consequence, proteins are synthesized in the cell cytoplasm.

Although genetic studies clearly indicate that the same receptor protein mediates the action of both testosterone and dihydrotestosterone, the two androgen–receptor complexes perform distinct physiologic roles. Testosterone–receptor complexes mediate the differentiation of the wolffian ducts, whereas dihydrotestosterone–receptor complexes are responsible for virilization of the external genitalia and formation of the prostate gland (Fig. 8-4). This deduction was originally based on the observation that anlage of the external genitalia (the urogenital sinus and genital tubercle) metabolize testosterone to dihydrotestosterone before the time of phenotypic differentiation. Fetal wolffian ducts, in contrast, are unable to form dihydrotestosterone until after virilization is advanced. This concept of different roles for the two hormones is supported by the study of patients with an inherited defect in dihydrotestosterone formation (5α-reductase deficiency) who have normally differentiated male wolffian-derived structures but failure of external virilization (see p. 160).

Female embryos have the same androgen receptor system and the same ability to respond to androgens as male embryos. Thus, differences in anatomical development between males and females depend on differences in the hormonal signals themselves and not on differences in the receptors for hormones.

In summary, three hormones are involved in male phenotypic differentiation. The regression of the müllerian duct is mediated by MIH secreted by the Sertoli cells of the fetal testis. Testosterone, the androgen secreted by the fetal testis, promotes virilization of the wolffian ducts, whereas dihydrotestosterone, which is formed from testosterone in peripheral, androgen-dependent tissues, is responsible for virilization of the external genitalia and the male urethra.

DISORDERS OF SEXUAL DIFFERENTIATION

Many different disorders of sexual differentiation have been identified in humans that substantiate the paradigm that chromosomal sex dictates gonadal sex and that gonadal sex, in turn, dictates phenotypic sex. In addition, many of these disorders have provided insight into the androgen-mediated events in sexual differentiation.

Disorders of Chromosomal Sex

Two disorders of chromosomal sex that result from nondisjunction of the sex chromosomes during meiosis—Klinefelter's syndrome (47,XXY) and gonadal dysgenesis (45,X)—illustrate the role of the sex chromosomes in controlling gonadal differentiation.

Klinefelter's Syndrome

Individuals with the Klinefelter syndrome typically have a 47,XXY karyotype and a predominantly male phenotype. Before puberty, patients have small testes with decreased numbers of spermatogonia, but otherwise they have a normal male phenotype. Most persons with the disorder come to the attention of physicians after the time of expected puberty because of infertility, breast enlargement (gynecomastia), or incomplete virilization. Typical histologic changes in the testes include hyalinization of the spermatogenic tubules, absence of spermatogenesis, and normal Leydig cells. Bilateral, painless, gynecomastia usually appears during adolescence and may eventually become disfiguring. Most patients have a male psychosexual orientation and function physically and socially as men.

Endocrine findings include low to normal plasma testosterone and elevated levels of circulating estradiol. The precise source of the elevated plasma estradiol is not known but may be due to an increased testicular secretion, a decreased metabolic clearance rate, or an increased rate of testosterone conversion to estradiol in peripheral tissues. The net result of this imbalance in the androgen to estrogen ratio is varying degrees of insufficient virilization and enhanced feminization.

Circulating gonadotropins (especially follicle-stimulating hormone) are elevated as a consequence of the damage to the seminiferous tubule resulting in decreased feedback on the hypothalamic–pituitary system (see Chapter 10). Progressive sclerosis of the seminiferous tubules eventually leads to diminished testicular blood flow, diminished testosterone secretion, and enhanced levels of luteinizing hormone.

Other individuals with a phenotype similar to that of the Klinefelter syndrome have a single Y chromosome and more than two X chromosomes. Thus, a single Y chromosome is sufficient for testicular differentiation and the development of a male phenotype despite the presence of multiple X chromosomes. Most of these individuals have absence of sperm in the ejaculate (azoospermia). Therefore, although having more than one X chromosome appears to be detrimental to spermatogenesis and normal testicular function later in life, the presence of a solitary Y chromosome is sufficient to induce testicular differentiation and the development of the male phenotype at the time of sexual differentiation.

Gonadal Dysgenesis (Turner's Syndrome)

Gonadal dysgenesis is a disorder in phenotypic females characterized by primary amenorrhea, lack of secondary sexual characteristics, short stature, multiple congenital anomalies, and bilateral streak gonads. The common chromosomal karyotype associated with this disorder is 45,X. Other affected individuals have mosaicism of this chromosomal complement with a 46,XX-containing cell line or structural abnormalities of one of the X chromosomes. The diagnosis is usually made in infancy or at the onset of puberty, when amenorrhea and failure of feminization are noted in conjunction with the other somatic abnormalities. The external genitalia are unambiguously female but remain immature. There is no spontaneous breast development. The internal genitalia consist of small but otherwise normal fallopian tubes and uterus. Although primordial germ cells migrate to the gonads and are present transiently in embryogenesis, they disappear before birth. After the age of puberty the gonads are identifiable only as fibrous streaks in the broad ligament. Although individuals with gonadal dysgenesis differentiate as phenotypic females, the secondary sex characteristics do not develop properly due to lack of female sex hormones. Thus, it is apparent that normal ovarian development requires the presence of two functionally intact X chromosomes.

Disorders of Gonadal Sex

Disorders of gonadal sex occur when chromosomal sex is normal, but for one reason or another, differentiation of the gonads is abnormal. Depending on the time during embryogenesis when the gonadal defect is manifest, the chromosomal sex may or may not correspond to the phenotypic sex of the individual.

Pure Gonadal Dysgenesis

Pure gonadal dysgenesis denotes a syndrome in which bilateral streak gonads are associated with an immature female phenotype. In contrast to patients with 45,X gonadal dysgenesis, these patients are of normal height, do not have associated somatic defects, and, most importantly, have a normal 46,XX or 46,XY chromosomal complement. Although some individuals with a 46,XY karyotype have mutations of the *SRY* gene, the cause of the gonadal dysgenesis in most individuals is unknown. Nevertheless, since gonadal development is arrested before phenotypic differentiation, the sexual phenotype is female. Although estrogen deficiency is variable, it is usually as profound as that seen in typical 45,X gonadal dysgensis.

The Vanishing Testis Syndrome

A spectrum of phenotypes has been described in 46,XY males with absent or rudimentary testes in whom endocrine function of the testis was present at some time during embryonic differentiation of the male phenotype. Thus, this disorder is quite distinct from that of pure gonadal dysgenesis in which no evidence can be inferred for gonadal function during embryonic development.

The disorder varies in its manifestations from complete failure of virilization, through varying degrees of incomplete virilization of the external genitalia, to otherwise normal males with anorchia. The most severely affected individuals are 46,XY phenotypic females who lack testes and accessory male reproductive organs and are sexually infantile. The disorder in these individuals differs from the 46,XY form of pure gonadal dysgenesis in that no müllerian duct derivatives are present. Thus, in these individuals, testicular failure must have occurred during the interval between the onset of formation of MIH and the secretion of testosterone, that is, after development of the seminiferous tubules but before the onset of Leydig cell function. In other patients, the clinical features indicate that the testicular failure occurred sometime later in gestation, and at the other extreme are phenotypic men with anorchia who lack müllerian derivatives and are completely virilized except for the absence of the epididymis.

Disorders of Phenotypic Sex

Disorders of phenotypic sex are classified as those disorders in which the phenotypic sex is ambiguous or is completely in disagreement with chromosomal and gonadal sex.

Female Pseudohermaphroditism

Female embryos that are exposed to androgen during the time of sexual differentiation can be profoundly virilized (female pseudohermaphroditism). Congenital adrenal hyperplasia, associated with an inherited deficiency of the 21-hydroxylase enzyme, is the most common cause of virilization in newborn females. This enzymatic defect impairs the synthesis of cortisol in the adrenal gland (Fig. 8-5), and the normal feedback mechanism in which circulating cortisol regulates pituitary adrenocorticotropin (ACTH) secretion is disrupted. Consequently, there is hypersecretion of ACTH. ACTH then overstimulates the adrenal gland to synthesize steroids proximal to the enzymatic defect, such as progesterone and 17-hydroxyprogesterone. 17-Hydroxyprogesterone is the precursor for the adrenal androgens, androstenedione and, to a lesser extent, testosterone. Androstenedione is converted to testosterone in many tissues of the body. The net result is an increase in circulating androgens that act to virilize the female.

The genitalia of females with virilizing congenital adrenal hyperplasia exhibit a spectrum of masculinization from simple clitoral enlargement to complete labioscrotal fusion and a penile urethra. The internal female structures and ovaries are unaltered, and male wolffian duct derivatives are not present. These facts indicate that either the onset of adrenal androgen secretion occurs late in embryogenesis after wolffian regression has occurred or that the wolffian ducts are not capable of responding to the adrenal androgens that are formed. The labial folds are rugated and resemble an empty scrotum. Thus, the external appearance of affected females is similar to that of males with bilateral cryptorchidism and hypospadias. In a few

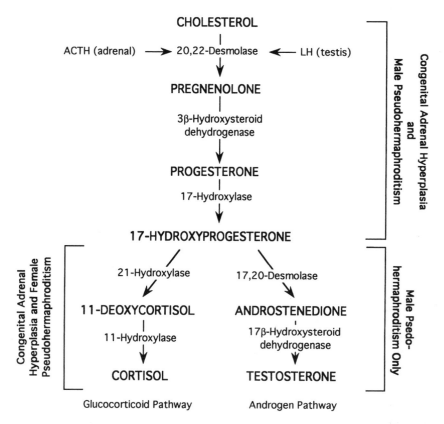

Fig. 8-5 Pathways for the synthesis of glucocorticoid and androgenic hormones from cholesterol in adrenal and testis.

cases, when virilization is so severe that a complete penile urethra is formed, errors in sex assignment are made at birth.

Male Pseudohermaphroditism

Defective virilization of the 46,XY male embryo with testes (male pseudohermaphroditism) can result from defects in androgen synthesis in the testis, defects in androgen action in the target cell, or defects in müllerian duct regression.

Testosterone Synthesis Defects. In addition to depicting the pathway by which glucocorticoids are synthesized in the adrenal, Fig. 8-5 summarizes the steroidogenic pathway by which androgen (testosterone) is synthesized from cholesterol in the testis. Three enzymes (20,22-desmolase, 3β-hydroxysteroid dehydrogenase, and 17α-hydroxylase) are necessary for the formation of both glucocorticoids and androgens, and enzymatic defects that occur in this part of the steroidogenic pathway result in adrenal insufficiency as well as male pseudohermaphroditism. Defects in the androgen synthetic pathway distal to the formation of 17-hydroxyprogesterone (17,20-desmolase or 17β-hydroxysteroid dehydrogenase deficiency) result in only male pseudohermaphroditism. However, since testosterone synthesis is normal in more than 80% of male pseudohermaphrodites, most disorders of male phenotypic development are assumed to result from abnormalities of androgen action. In these disorders, testosterone synthesis and müllerian duct regression are normal; however, male development is incomplete due to impairments in androgen action in the target cells. Two general categories of androgen resistance are recognized: (1) 5α-reductase deficiency that results in the inability to form the active androgen, dihydrotestosterone, in target tissues, and (2) androgen receptor defects that result in the inability of the androgen target tissue to respond to the hormone.

5α-Reductase Deficiency. Deficiency of the 5α-reductase enzyme that is responsible for the conversion of testosterone to the potent androgen dihydrotestosterone in peripheral tissues is a rare cause of androgen resistance in humans. Individuals with this autosomal recessive disorder have a generally female habitus. The urethral opening is commonly at the base of a small phallus. A blind-ending vaginal pouch is present that usually opens into the urethra near the urethral orifice in the urogenital sinus. Female internal genitalia are not present, but the male internal genitalia (epididymides, vasa deferentia, and seminal vesicles) are well developed. The fact that the defect in embryonic virilization is limited to the urogenital sinus and the external genitalia suggests that testosterone is sufficient for virilization of the wolffian ducts but that dihydrotestosterone is required for differentiation of the prostate and external genitalia. The testes are endocrinologically normal, and plasma testosterone levels are in the normal male range. At the time of puberty, some degree of masculinization occurs.

There are two genes that code for different isoenzymes of steroid 5α-reductase that have distinct biochemical and pharmacological properties. One of these genes is located on chromosome 2 and encodes the type 2 5α-reductase, and mutations in this gene cause 5α-reductase deficiency in humans.

Androgen Receptor Defects. The most severe form of androgen resistance, testicular feminization, is caused by profound defects in the function of the androgen receptor. Patients with testicular feminization usually come to the attention of physicians after the time of expected puberty when they are evaluated for primary amenorrhea. The karyotype is 46,XY, but the general habitus is female in character. Breast development at the time of expected puberty is that of a normal female. Axillary, facial, and pubic hair are absent or scanty. The external genitalia are unambiguously female, and the vagina is short and blind ending. All internal genitalia are absent except for testes which may be located in the abdomen, along the course of the inguinal canal, or in the labia majora. Plasma testosterone and plasma luteinizing hormone levels are elevated, and 5α-reduction of testosterone to dihydrotestosterone in peripheral tissues is normal. Nevertheless, patients with this disorder are completely resistant to androgen action, and the androgen receptor is usually functionally missing in these patients. The internal genitalia, as well as the external genitalia, fail to virilize indicating that testosterone and dihydrotestosterone act via the same receptor. Furthermore, since müllerian duct derivatives are not present in affected individuals, the müllerian-inhibiting function of the fetal testis is intact.

Mutations of the androgen receptor can also result in less severe defects in androgen receptor function and relatively minor degrees of undervirilization and/or infertility. Most of the mutations associated with these clinical syndromes are point mutations of the hormone-binding domain of the receptor leading to decreased or qualitatively abnormal androgen binding. Less commonly, point mutations in the DNA-binding domain impair transactivation of the receptor with normal hormone binding.

Persistent Müllerian Duct Syndrome. The persistence of the müllerian ducts (uterus and fallopian tubes) in a genotypic, and otherwise normal phenotypic male is a rare condition. Subjects are commonly discovered when it is noticed that inguinal hernias contain müllerian derivates (fallopian tubes and/or uterus). Penile development is normal and the patients masculinize normally at puberty. Thus, androgen secretion by the fetal and postnatal testis is presumably normal. However, since müllerian regression does not occur properly, it is assumed that either the fetal testis fails to elaborate the hormone at the appropriate time in embryogenesis or that the müllerian ducts fail to respond to the hormone. Persistence of the müllerian ducts is usually accompanied by failure of testicular descent, and it is conceivable that MIH plays a role in this process, possibly by influencing the cranial anchoring of the testis to the peritoneal fold.

CONCLUSION

Figure 8-6 summarizes our current understanding of the process of sexual differentiation and indicates sites of defects leading to abnormalities of sexual differentiation. The chromosomal (genotypic) sex of the embryo is established at fertilization. If the chromosomal sex of the embryo is male, the *SRY* gene on the Y chromosome causes the indifferent gonad to develop into a testis. In the absence of the Y chromosome, the indifferent gonad develops into an ovary, and the phenotype of the embryo will be female. If, on the other hand, a testis develops, two testicular hormones, MIH and testosterone, act to transform the indifferent urogenital tract into one that is characteristic of the male. MIH, secreted by the fetal Sertoli cells of the testis, causes regression of the müllerian ducts; testosterone, secreted by the Leydig cells of the fetal testis, is responsible for the remainder of male development. Thus, in simplistic terms, the default sexual phenotype is female, and only when the action of hormones elaborated by the fetal testis is imposed on the indifferent embryo does the phenotype of the male emerge. Anything that impairs the synthesis or function of either of these two testicular hormones causes aberrant or incomplete male phenotypic development. The individual phenotypes can vary considerably, depending on the severity of the defect. Virilization of female embryos can occur if they are exposed to androgens at the appropriate time in embryogenesis, as happens in virilizing congenital adrenal hyperplasia. Male phenotypic sexual differentiation, then, is determined by the presence (in males) or the absence (in females) of specific hormonal signals during embryogenesis.

Although we now understand the hormonal and genetic factors that are necessary for mammalian sexual differentiation in considerable detail, the most fundamental issues in the embryogenesis of the urogenital tract remain poorly understood.

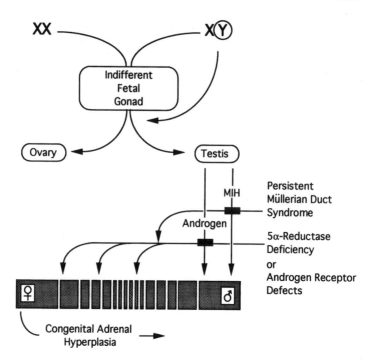

Fig. 8-6 Schematic diagram of the process of sexual differentiation indicating sites at which defects leading to abnormal sexual development have been identified. MIH, müllerian-inhibiting hormone.

Issues such as the nature and timing of chromosomal signals and cellular changes that cause gonadal differentiation and the mechanisms by which the same hormonal signal is translated into different physiological effects in different tissues will have to be clarified before we understand the entire program by which the myriad of genetic determinants and hormones interact in the development of phenotypic sex.

SUGGESTED READING

Bogan JS, and Page DC: Ovary? Testis?-A mammalian dilemma. Cell 76:603–607, 1994.

Brook CDG: Persistent müllerian duct syndrome. Pediatr Adolescent Endocrinol 8:100–104, 1981.

Donohoue PA, Parker K, and Migeon CJ: Congenital adrenal hyperplasia. In: *The Metabolic and Molecular Bases of Inherited Disease*, 7th ed., CR Scriver, AL Beaudet, WS Sly, D Valle, eds., McGraw-Hill, New York, pp. 2929–2966, 1995.

George FW, and Wilson JD: Sex Determination and Differentiation. In: *The Physiology of Reproduction*, 2nd ed., E Knobil and JD Neill, eds., Raven Press, New York, pp. 3–26, 1994.

Goodfellow PN, and Lovell-Badge R: *SRY* and sex determination in mammals. Annu Rev Genet 27:71–92, 1993.

Griffin JE, McPhaul MJ, Russell DW, and Wilson JD: The androgen resistance syndromes:

5α-reductase deficiency, testicular feminization, and related disorders. In: *The Metabolic and Molecular Bases of Inherited Disease,* 7th ed., CR Scriver, AL Beaudet, WS Sly, and D Valle, eds., McGraw-Hill, New York, pp. 2967–2998, 1995.

Josso N, Cate RL, Picard J-Y, et al: Anti-müllerian hormone: The Jost factor. Recent Prog Horm Res 48:1–59, 1993.

Jost A: A new look at the mechanisms controlling sexual differentiation in mammals. Johns Hopkins Med J 130:38–53, 1972.

9

Female Reproductive Function

SERGIO R. OJEDA

The production of germ cells is essential for the continuation of a species. This function, in the female, is accomplished by the ovaries. In addition, the ovaries secrete steroidal and nonsteroidal hormones that not only regulate the secretion of anterior pituitary hormones but also act on various target organs including the ovaries themselves, the uterus, fallopian tubes, vagina, mammary gland, and bone.

STRUCTURAL ORGANIZATION OF THE OVARY

Morphologically, the ovary has three regions: an outer cortex that contains the oocytes and represents most of the mass of the ovary, the *inner medulla*, formed by stromal cells and cells with steroid-producing characteristics, and the *hilum*, which in addition to serving as the point of entry of the nerves and blood vessels, represents the attachment region of the gland to the mesovarium (Fig. 9-1).

The cortex, which is enveloped by the germinal epithelium, contains the functional units of the ovary—the follicles—that are present in different states of development or degeneration (atresia), each enclosing an oocyte. In addition to the oocyte, ovarian follicles have two other cellular components: granulosa cells that surround the oocyte, and thecal cells that are separated from the granulosa cells by a basal membrane and are arranged in concentric layers around this membrane. The follicles are embedded in the stroma, which is composed of supportive connective cells similar to that of other tissues, interstitial secretory cells, and neurovascular elements.

The medulla has a heterogeneous population of cells, some of which are morphologically similar to the Leydig cells in the testes. These cells predominate in the ovarian hilum; their neoplastic transformation results in an excessive production of androgens that induce virilization.

OVARIAN HORMONES

The ovary produces both steroidal and peptidergic hormones. Whereas the steroids are synthesized in both interstitial and follicular cells, peptidergic hormones are

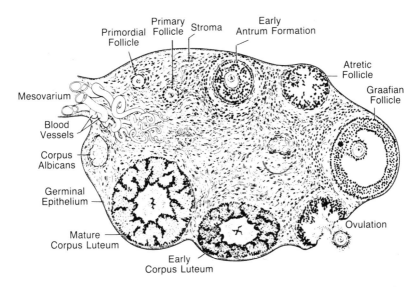

Fig. 9-1 Microscopic view of the human ovary. (Redrawn from Ham AW, and Li TS: *Histology*, 4th ed., Lippincott, Philadelphia, p. 847–899, 1961, with permission.)

primarily produced in follicular cells and, after ovulation, by cells of the corpus luteum.

Biosynthesis, Transport, and Metabolism of Steroid Hormones

Biosynthesis

The initial precursor for steroid biosynthesis is cholesterol (Fig. 9-2), which derives from animal fats of the diet or from local synthesis. The cholesterol that arrives at the ovary via the blood stream is mostly transported in association with low-density lipoproteins (LDL). LDL bind to specific receptors located in the plasma membrane of ovarian cells. Thereafter, the LDL–receptor complexes are internalized and hydrolyzed within lysosomes that release cholesterol for steroid biosynthesis. Excess cholesterol is esterified and stored in lipid droplets for later use. Under normal conditions cholesterol utilized in steroid biosynthesis comes from the plasma LDL-associated pool and intracellular lipid droplets. A less important source originates from *de novo* local synthesis.

The main steroids produced by the ovary are progesterone and estradiol (Fig. 9-2). Whereas androgens, particularly androstenedione and testosterone, are also secreted into the bloodstream, a significant portion of them is converted to estradiol through the action of an aromatase enzyme complex. In addition to these steroids, the ovary also secretes estrone, 17α-hydroxyprogesterone, 20α-hydroxyprogesterone, and 5α-reduced androgens such as 5α-dihydrotestosterone and 3α-androstanediol.

The first step in the synthesis of ovarian steroids is the conversion of cholesterol, which has 27 carbons, to the C-21 compound pregnenolone in a reaction catalyzed by an enzyme complex known as *cholesterol side-chain cleavage enzyme*. The re-

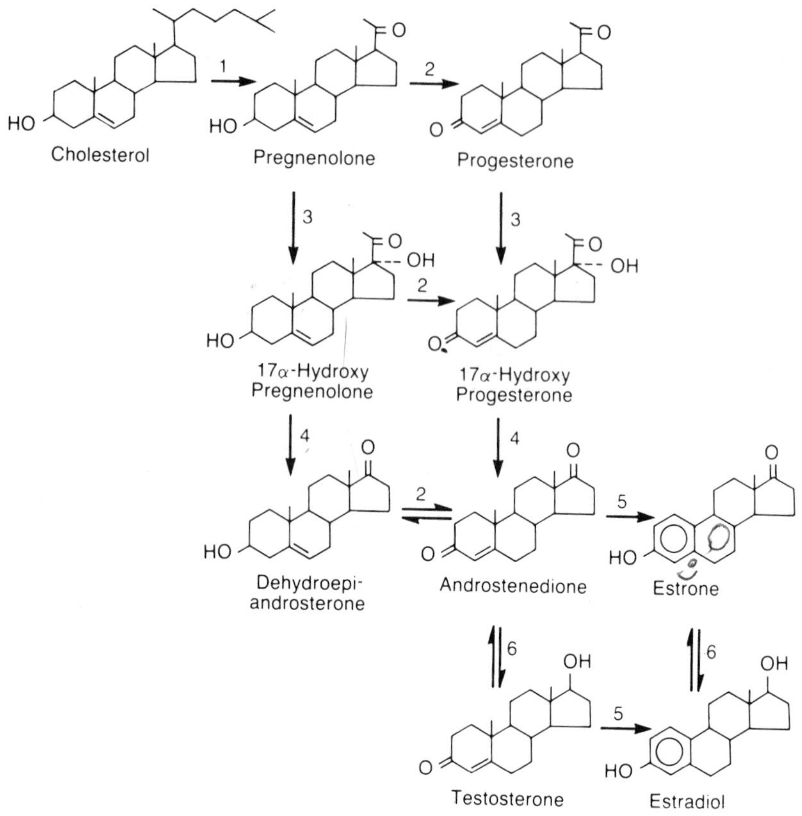

Fig. 9-2   Biosynthesis of steroids in the ovary. 1, Cholesterol side-chain cleavage enzyme complex; 2, 3β-hydroxysteroid dehydrogenase; 3, 17α-hydroxylase; 4, 17,20-lyase; 5, aromatase; 6, 17β-hydroxysteroid dehydrogenase.

action occurs in the mitochondria and is a rate-limiting step in the steroid biosynthetic pathway. Pregnenolone can then be converted into progesterone by the action of a 3β-hydroxysteroid dehydrogenase enzyme, or into 17α-hydroxypregnenolone by a 17α-hydroxylase. Both progesterone and 17α-hydroxypregnenolone can, in turn, be metabolized to 17α-hydroxyprogesterone (Fig. 9-2).

In contrast to cholesterol, progestogen compounds have only 21 carbons. Further metabolism to androgens and estrogens results in further reduction in the number of carbons to 19 (androgens) and 18 (estrogens).

17α-Hydroxyprogesterone is the substrate for a 17,20-lyase enzyme that, by cleaving the C-20,21 side chain, yields androstenedione. This androgen is either directly metabolized to estrone or, preferentially, converted into testosterone in a reaction catalyzed by the enzyme 17β-hydroxysteroid dehydrogenase. Both androstenedione and testosterone are substrates for estrogen production. Whereas androstenedione is converted into estrone, testosterone metabolism results in estradiol formation. In both cases the reaction is catalyzed by an aromatase enzyme that generates the estrogens through a complex series of reactions involving hydroxylations, oxidations, removal of the C-19 carbon, and aromatization of the A ring of the androgen. All of the reactions involved in the metabolism of progesterone to estradiol, including the initial formation of progesterone from pregnenolone, occur in the endoplasmic reticulum.

With the exception of the 3β- and 17β-hydroxysteroid dehydrogenases, the enzymes involved in ovarian steroid biosynthesis belong to a group called *mixed function oxidases* that utilize different forms of cytochrome *P*-450 and require molecular oxygen and nicotinamide adenine dinucleotide phosphate (NADPH) for their activity. They consist of the cytochrome *P*-450, which binds to the substrate, and an electron transport system. In the case of the cholesterol side-chain cleavage enzyme, the electron transport system consists of an iron–sulfur protein and a flavoprotein, NADPH-adrenodoxin reductase. In the case of the 17α-hydroxylase, 17,20-lyase, and aromatase enzyme systems, the electron transport system is provided by a flavoprotein NADPH–cytochrome *P*-450 reductase.

The bulk of androgen production occurs in interstitial cells and the cells of the follicular theca (see p. 169 and Fig. 9-3). Estradiol is mainly produced by granulosa cells of antral follicles. Progesterone, on the other hand, is secreted by all steroidogenic cells of the ovary regardless of their localization.

For those students interested in the subject, a detailed description of steroid chemistry, pathways, and biosynthetic enzymes can be found in the review by Kellie and the book *The Physiology of Reproduction* (Chapter 11), listed at the end of this chapter.

Transport and Metabolism of Steroid Hormones

Both 17β-estradiol and estrone are present in the bloodstream. Whereas estradiol is secreted directly by the ovary, estrone derives largely from peripheral conversion of estradiol or androstenedione. Estrone is further metabolized to estriol, a reaction that occurs predominantly in the liver. Most circulating estrogens (>70%) are bound to proteins, preferentially albumin, for which they have a low affinity, or to a steroid-binding protein known as *testosterone-binding globulin* (TeBG), for which they have a higher affinity. Estrogens have a lower affinity for TeBG than androgens; this results in a greater availability of estrogens to the tissues because only free steroids can be transported into their target cells (see Chapter 5).

Metabolism of estrogens via oxidation or conversion to glucuronide and sulfate conjugates occurs predominantly in the liver. The resulting metabolites are excreted in the bile and reabsorbed into the bloodstream via the enterohepatic circulation. Their final excretion route, however, is the urine.

Androgens in women are produced by both the ovaries and the adrenals. There are three main circulating androgens in women: dehydroepiandrosterone produced by the adrenal, androstenedione derived from both the ovary and the adrenal, and testosterone, which in addition to being produced by the ovary and adrenal is formed in peripheral tissues, mainly from androstenedione and to some extent from dehydroepiandrosterone and Δ5-androstenediol. Indeed, testosterone is the primary physiologically active androgen in women. It may act on target tissues directly or after its conversion to 5α-dihydrotestosterone. As in men (see Chapter 10), 97–99% of the total circulating testosterone in women is bound to TeBG. As with estrogens, androgens are excreted mostly in the urine.

Progesterone also appears to circulate bound to plasma proteins, but the characteristics of this association are not known. Progesterone is converted in the liver to pregnanediol, which, following conjugation to glucoronic acid, is excreted in the urine.

Peptide Hormones of Ovarian Origin

One of the first peptide hormones to be recognized as a product of the ovary was relaxin. It is mainly produced by the corpus luteum during pregnancy but has also been found in decidual tissue and human seminal plasma. The secretion of relaxin from the corpus luteum is stimulated by human chorionic gonadotropin (hCG). Although several peptides showing relaxin-like activity have been reported, they do not appear to be the products of related genes, but rather the result of limited proteolysis of relaxin during the process of isolation. In humans, however, there is good evidence for the existence of two relaxin-related genes. Human relaxin consists of an α- and a β-peptide chain covalently linked by disulfide bonds. Interestingly, both the position of the bonds and the processing of the prohormone to yield mature relaxin are identical to those of insulin, suggesting that both proteins, as well as the insulin-like growth factors (IGFs; see Chapter 12), derive from the duplication of a common ancestral gene. The main effect of relaxin is to induce relaxation of the pelvic bones and ligaments, inhibit myometrial motility, and soften the cervix. In addition, relaxin has been shown to induce uterine growth. It is clear, therefore, that the hormone plays an important role in both maintaining uterine quiescence and favoring the growth and softening of the reproductive tract during pregnancy.

For many years it was suspected that the ovary produces a peptide hormone that exerts selective inhibitory control over the secretion of follicle-stimulating hormone (FSH). This protein, originally described in testicular extracts and termed *inhibin*, has now been isolated from follicular fluid and found to be a heterodimeric protein consisting of an α- and a β-polypeptide chain connected by disulfide bonds. Two forms of inhibin (A and B) exists, each with a molecular weight of 32,000. The α-subunits are identical and the β-subunits are different. Combination of the β-subunits of inhibin among themselves via disulfide bond linkage results in the formation of another family of proteins with completely different functions. The first of these functions to be recognized was the selective stimulation of FSH secretion from the adenohypophysis. Due to this activity, this inhibin-associated protein was termed *FSH-releasing protein* (FRP) or *activin*. It is now clear that there are three forms of activins resulting from the differential dimerization of β-subunits, which can form βA/βA, βA/βB, and βB/βB dimers, respectively.

It now appears that the functions of activin are much more diverse than originally anticipated. The messenger RNA encoding the βA subunit of activin is expressed in the brain, pituitary gland, germ cells of the testis, and erythropoietic tissue, suggesting a role for activin in cell differentiation. Such a role is supported by the results of recent investigations that have shown that the differentiation of erythrocytes is promoted by a factor encoded by the same messenger RNA that encodes the βA subunit of activin. These nonreproductive functions, though initially surprising, are consistent with the finding that inhibins and activins are members of a large family of peptides that includes transforming growth factor-β (TGF-β), müllerian-inhibiting hormone (MIH, see Chapter 8), and many other peptides involved in the control of growth and differentiation.

In addition to activins and inhibins, follicular cells produce another group of proteins with molecular weights ranging from 32 to 43 kDA, which inhibit FSH secretion. These proteins, termed *follistatins*, bind activins with high affinity and thus are able to neutralize their biological activity.

Quite unexpectedly, the ovary has been found to produce peptides of the proopiomelanocorticotropin (POMC) family including β-endorphin, adrenocorticotropin (ACTH), and α-melanocyte-stimulating hormone (α-MSH). The ovary also produces vasopressin and oxytocin. These two peptides are synthesized in luteal cells. In the case of oxytocin there is clear evidence that the hormone is secreted into the general circulation. The physiological role that these peptides may play in the ovary is uncertain. However, they appear to exert paracrine effects.

Pituitary Control of Ovarian Hormone Formation

Steroids

The production of ovarian steroids is under the control of two main hormones secreted by the anterior hypophysis, luteinizing hormone (LH) and FSH. In rodents, prolactin and growth hormone (GH) have also been shown to act directly on the ovary. Experiments in rats demonstrated that depending on its level prolactin may either inhibit or support the stimulatory effect of LH on ovarian steroidogenesis, through modulation of the number of LH receptors. GH, on the other hand, has been found to facilitate FSH actions in rats by a mechanism involving local production of IGF-I.

Although the secretion of progesterone can be stimulated by both LH and FSH, the secretion of androgens is enhanced only by LH. The formation of estradiol, on the other hand, depends on both LH and FSH. LH stimulates production of androgens that are utilized in estrogen production. FSH directly activates the aromatase enzyme complex that catalyzes the conversion of androgens to estrogens.

LH receptors are found on steroidogenic cells of the stroma and thecal cells of the follicles, both of which produce progesterone and androgens. When the follicles become antral the granulosa cells acquire LH receptors. Since these cells have a limited 17,20-lyase activity, their exposure to LH results primarily in increased production of progesterone and little, if any, androgens. FSH receptors, on the other hand, are located exclusively on granulosa cells that are the only ovarian cells with substantial aromatase activity.

These interactions have formed the basis for the formulation of a "two cell–two gonadotropin" hypothesis to explain the gonadotropic control of ovarian steroidogenesis. According to this hypothesis the follicular theca, under the influence of LH, produces androgens that upon diffusion to the granulosa cell compartment of the follicle are converted to estrogens via an FSH-supported aromatization reaction (Fig. 9-3).

Peptide Hormones

Production of inhibin is stimulated by FSH. The role of LH in the control of inhibin production is more complex. Inhibin production in preovulatory follicles decreases

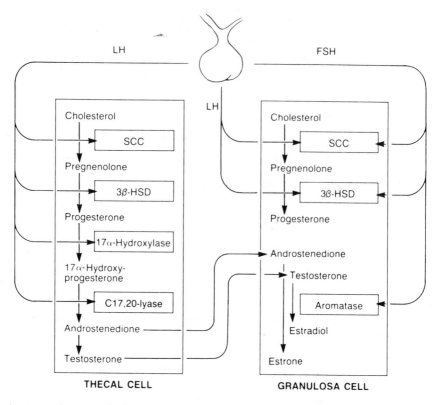

Fig. 9-3 The control of ovarian estrogen, progesterone, and androgen production by LH and FSH. LH acts on both thecal and granulosa cells. FSH acts only on granulosa cells.

sharply soon after the preovulatory surge of gonadotropins (see below, Dynamics of the Hypothalamic–Pituitary–Ovarian Relationship). This decrease appears to be brought about by LH, which, however, simultaneously increases inhibin production in smaller follicles not destined to ovulate in that cycle. It also appears that the formation of POMC peptides, vasopressin, and oxytocin, is under gonadotropin control. While gonadotropins stimulate the formation of oxytocin and β-endorphin, they decrease the production of vasopressin.

Mechanism of Action of Gonadotropins

Both LH and FSH act on the ovary by first binding to high-affinity, low capacity–specific membrane receptors. Interaction with their respective receptors activates the enzyme adenylyl cyclase that catalyzes the formation of cyclic AMP from ATP. Cyclic AMP then activates specific protein kinases that induce the phosphorylation of proteins involved in the synthesis of enzymes, such as the cholesterol side chain cleavage enzyme and 3β-hydroxysteroid dehydrogenase.

Actions of Ovarian Steroids

Ovarian steroids exert intra- and extraovarian actions. The best known intraovarian actions are those exerted on follicular development. Among the extraovarian actions are those exerted on the hypothalamic–pituitary unit (negative and positive feedback on gonadotropin secretion), on reproductive tissues such as fallopian tubes, uterus, vagina, and mammary gland, and on other tissues such as bone, kidney, and liver.

Intraovarian Actions

Estrogens. Estradiol (E_2) induces proliferation of the granulosa cells, increases E_2 receptors in these cells, and facilitates the action of FSH.

Androgens and Progestins. Testosterone and progesterone have been shown to affect ovarian steroidogenesis through binding to their respective receptors, which are present in granulosa cells. Progesterone inhibits FSH-induced estradiol production. Androgens, on the other hand, enhance both the progesterone and estradiol secretion induced by FSH.

Extraovarian Actions

The Uterus. The endometrium or uterine mucosa consists of a superficial layer of epithelial cells and a deeper stromal layer. The epithelial layer is interrupted by tubular invaginations or glands that penetrate into the stromal layer and that are lined by both epithelial cells and columnar secretory cells. The stroma is permeated by blood vessels (spiral arteries) and contains characteristically spindle-shaped cells.

During the first half of the menstrual cycle (i.e., before ovulation), estrogen stimulates proliferation of both the epithelial and stromal layers. The thickness of the endometrium increases 3- to 5-fold and the uterine glands become enlarged, but they maintain their straight shape. The spiral arteries also elongate. Because of such changes this has been called the *proliferative phase* of the endometrium.

The uterine cervix contains endocervical glands that elaborate a mucus which, under the influence of estrogen, changes from a scanty, viscous substance to a much more abundant, watery material. The elasticity of the mucus, usually referred to as *spinnbarkeit*, is greatly increased, a change that can be easily determined by stretching a small drop of mucus between two fingers. When estrogen production is maximal, that is, close to the time of ovulation, the elasticity of the cervical mucus is such that the drop can be stretched maximally without rupture. If a drop of such mucus is spread on a slide, upon drying, the mucus forms a typical palm-leaf arborization or ferning pattern.

Under the influence of progesterone, which is produced in increasing amounts after ovulation, proliferation of the endometrium decreases and the uterine glands lose their straight configuration, becoming tortuous. Glycogen accumulates in large vacuoles at the basal portion of the cells. As the corpus luteum becomes firmly established, the glycogen vacuoles move toward the apical end of the cells and the secretory activity of the gland increases pronouncedly. The stroma becomes edematous and the elongating spiral arteries become coiled. These transformations establish the *secretory phase* of the endometrium. Coinciding with the loss of corpus luteum function and the resulting decline in plasma levels of estrogen and progesterone the endometrium undergoes focal necrosis, probably because of vasospasm of the spiral arteries. The areas of necrosis increase progressively and the mucosa exfoliates with loss of all superficial cells except those lining the base of the glands.

Since necrosis also affects the endometrial blood vessels blood loss occurs as menstrual bleeding.

Vasospasm of the endometrial blood vessels is believed to be mediated by prostaglandins, in particular prostaglandin $F_{2\alpha}$. It appears that the loss of luteal progesterone and estrogen destabilizes the lysosomal membranes of endometrial cells, which enhances phospholipid hydrolysis and subsequently prostaglandin formation. Reinitiation of follicular growth during menstruation is accompanied by an increased production of estrogen stimulating the endometrium to proliferate again.

In addition to its effects on the endometrium, progesterone decreases both the quantity and the elasticity of the cervical mucus and inhibits its ferning pattern.

The Vagina. The human vagina is lined by a stratified multilayered squamous epithelium that consists of superficial, intermediate, inner parabasal, and basal layers. In the presence of low levels of estrogen, such as before puberty or at menopause, the vaginal epithelium is thin and dry, and more susceptible to infections. Estrogens induce both proliferation and keratinization of the epithelium. As a result, mucosa thickness increases and there is increased exfoliation of superficial cells that can be readily identified under the microscope because of their acidophilia and pyknotic nuclei. Interestingly, the epithelium that lines the urethra is also sensitive to ovarian steroids. Thus, examination of urethral cells collected from a fresh urinary sediment can be used to assess alterations in circulating estrogen and progesterone levels in both prepubertal and mature women.

In general, progesterone has an effect on the vaginal epithelium that is opposite to that of estrogen. Under its influence the number of cornified cells decreases and the number of polymorphonuclear leukocytes increases.

The Mammary Gland. The human breast is composed of several lactiferous duct systems that are embedded in adipose and connective tissue. Each duct system contains thousands of sac-like milk-secreting alveoli, which are surrounded by myoepithelial cells and which drain into an increasingly larger milk-transporting system comprised of ductules, ducts, lactiferous sinuses, and a single ampulla that opens into the nipple at the center of the areola. At puberty, when estrogen secretion rises, growth of the ducts accelerates and the size of the areola increases. Furthermore, estrogen causes selective accumulation of adipose tissue around the lactiferous duct systems and this contributes significantly to the overall growth of the breast.

While estrogen stimulates the development of the duct system, progesterone induces the formation of secretory alveoli, an effect more clearly observed during pregnancy. Progesterone also facilitates the proliferating effect of estrogen on the duct epithelium. However, in the absence of priming by estrogen, progesterone has little effect on the mammary gland.

Metabolic Effects. One of the best known metabolic effects of estrogen is to increase the amount of plasma proteins produced by the liver that bind estradiol, testosterone, cortisol, progesterone, and thyroxine. Estrogens also increase the plasma proteins that bind iron and copper, and the levels of high-density lipoproteins (HDL) and very low–density lipoproteins (VLDL). The latter results in increased circulating levels of triglycerides, which are the lipids preferentially associated with these lipoproteins. In contrast to these effects, estrogen lowers the circulating levels of cholesterol and low-density lipoproteins (LDL). Estrogen also exerts appreciable effects on the electrolyte balance because of its stimulatory effects on the formation of angiotensinogen and aldosterone. Awareness of these metabolic effects has been heightened by the use of oral contraceptives that, as will be discussed later, are a combination of estrogen and progesterone.

Progesterone itself has little effect on plasma triglycerides. However, some synthetic progestins used in oral contraceptives have been shown to decrease triglyceride levels. More importantly, progesterone increases the levels of cholesterol and LDL and decreases the levels of HDL. Progesterone can also transiently increase sodium excretion in the urine, due to its capacity to antagonize aldosterone action.

The best known metabolic effect of progesterone is to increase basal body temperature. This is observed after ovulation and can serve as an index that ovulation has occurred.

Effects on Nonreproductive Tissues. We have already mentioned the effect of estrogen on the liver. Two additional target tissues are the bone and the kidney. Estrogen acts on bone both to accelerate linear growth and to induce closure of the epiphyseal plates. In addition, it prevents bone resorption. The importance of this effect becomes clearly evident at menopause. At this time, estrogen secretion decreases resulting in osteoporosis and bone fragility in some women. It is now clear that at least part of this supportive effect of estrogen is exerted directly via binding of the steroid to specific receptors, previously believed not be present in bone.

In the kidney, estradiol enhances the reabsorption of sodium from the renal tubules.

THE MENSTRUAL CYCLE

Reproductive function in the female is cyclic. A series of functional interrelationships between the hypothalamus, the anterior pituitary, and the ovaries leads to the monthly rupture of an ovarian follicle and extrusion of an ovum (i.e., ''ovulation''), which is then transported to the fallopian tubes to be fertilized. Should fertilization fail to occur, menstruation ensues within 14 days and the hormonal and morphological events that led to ovulation are repeated. This coordinated series of events is known as the *menstrual cycle*. The normal menstrual cycle has an average duration of about 30 days, with a range that extends from 25 to 35 days. It can be divided into four phases: menstruation, which lasts about 4–5 days; a follicular phase of the ovary, which corresponds to the proliferative phase of the endometrium and lasts 10–16 days; an ovulatory phase, which lasts about 36 hours; and a luteal phase, which corresponds to the secretory phase of the endometrium. The latter is usually very constant, lasting about 14 days; thus, the length of the menstrual cycle is determined by the follicular phase, which can be highly variable. The greatest variability occurs during the few years that follow the first menstruation (menarche) and during the years preceding the loss of reproductive function (menopause).

An understanding of the menstrual cycle requires knowledge of the functional components responsible for its occurrence. These are (1) the gonadotropin-releasing system and its neural (hypothalamic) control, (2) the ovarian follicle, including its rupture at ovulation and its subsequent transformation into a corpus luteum, and (3) the ovarian negative and positive feedback control of gonadotropin secretion.

The Gonadotropin-Releasing System and Its Neural Control (Fig. 9-4)

As discussed in Chapter 6 the adenohypophysis has a population of cells—the gonadotrophs—that secretes both LH and FSH. Some cells secrete only LH, others FSH, and still others secrete both hormones.

Release of LH and FSH to the bloodstream is pulsatile. This is particularly noticeable in the case of LH, which has episodes of release every 70–100 minutes. Because the interpulse intervals are about an hour long, this rhythm has been called

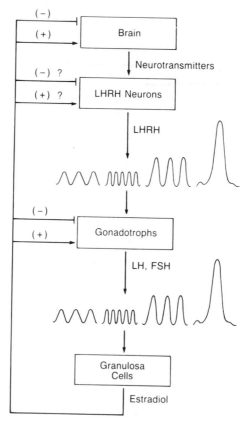

Fig. 9-4 The control of LH and LHRH secretion by ovarian estradiol. Both LH and LHRH are released in pulses. Estrogen (and combinations of estrogen plus progesterone) can alter the frequency and/or amplitude of the pulses by acting at both the hypothalamus and anterior pituitary. Within the brain, estradiol may inhibit (−) or stimulate (+) the activity of neurotransmitters that affect LHRH secretion. It is not clear whether estradiol can act directly on LHRH neurons(?). Estradiol can also induce a large, preovulatory surge of LH and LHRH secretion.

circhoral. LH pulses are less frequent during the luteal phase of the menstrual cycle than during the follicular phase, presumably because of an inhibitory effect of progesterone that is secreted in large amounts by the corpus luteum.

In addition to this circhoral rhythm, LH secretion also exhibits a sleep-related rhythm. This mode of secretion is not clearly manifested in adult women, but it constitutes a fundamental feature at the onset of puberty. At this time, an increased amount of LH is released during sleep, the first overt neuroendocrine manifestation of the initiation of puberty (see p. 185).

Both types of release (circhoral and diurnal) are a function of luteinizing hormone-releasing hormone (LHRH), the hypothalamic peptide responsible for the control of gonadotropin secretion. Experiments performed in rats, sheep, and monkeys have demonstrated that LHRH is released into the portal circulation, and hence reaches the adenohypophysis, in a pulsatile fashion.

In nonhuman primates, surgical isolation of the medial basal hypothalamus fails to disturb the circhoral release of LH, indicating that the rhythm originates within this area of the brain. Lesion of the arcuate nucleus, which is located in the foremost ventral portion of the medial basal hypothalamus, abolishes basal release of both LH and FSH, indicating that the most essential neural center controlling gonadotropin secretion resides within the arcuate nucleus of the hypothalamus. It is also clear, however, that the function of this "pulse generator" is subject to modifying influences emanating from both extra- and intrahypothalamic loci. An increase in electrical activity in the arcuate nucleus area precedes, with astounding regularity, the discharge of LH into the bloodstream.

The involvement of several neurotransmitter systems in the control of LHRH secretion has also been demonstrated. Among them, the most important appear to be those neuronal networks that utilize excitatory amino acids and γ-amino butyric acid (GABA) as neurotransmitters. While excitatory amino acids (most predominantly glutamate) stimulate LHRH secretion, GABAergic neurons exert a powerful inhibitory influence. Other stimulatory neurotransmitters include the catecholamine norepinephrine (NE), and the peptide neuropeptide Y (NPY). Neurotransmitters, such as the amines dopamine (DA) and serotonin, can either stimulate or inhibit LHRH secretion, depending on the prevalent steroid milieu present.

In addition to GABA, opioid peptides are also able to inhibit LHRH release. β-Endorphin, synthesized in neurons of the medial basal hypothalamus, is a prominent example of this class of inhibitory inputs to LHRH neurons. If naloxone, a blocker of opioid receptors, is administered during the luteal phase of the menstrual cycle the frequency of LH discharges increases to levels similar to those seen during the early follicular phase. Since progesterone increases β-endorphin production in the hypothalamus it has been suggested that the inhibitory effect exerted by progesterone during the luteal phase on LH pulse frequency is mediated by β-endorphin.

Despite the complexity of the transsynaptic control of LHRH secretion, it is becoming increasingly evident that most of the above-mentioned neuronal networks play roles secondary to that of the glutamate–GABA regulatory systems. Recent experiments in rodents and nonhuman primates have implicated glial cells as an additional component of the regulatory complex that controls LHRH secretion. Glial cells of the astrocytic lineage were shown to affect LHRH release via production of trophic molecules, which, acting in a paracrine–autocrine fashion, stimulate the glial secretion of neuroactive substances able to stimulate LHRH release. TGF-α was identified as one of these glial regulatory molecules.

Follicular Development, Ovulation, and Atresia (Fig. 9-5)

Follicular Development

The primordial germ cells originate in the endoderm of the yolk sac, allantois, and hindgut of the embryo. By 5–6 weeks of gestation they migrate to the genital ridge and begin to multiply rapidly, so that by 24 weeks of gestation the total number of oogonia is about 7 million. They concentrate mostly within the cortical portion of the primitive gonad. As multiplication of the oogonia proceeds, some of them reach the prophase stage of meiosis and are called *primary oocytes*. Many others degenerate and die so that at birth only 2 million primary oocytes remain, a number that is further reduced before puberty to about 400,000. The first meiotic division is not completed until the moment of ovulation; at this time the oocyte has grown from a size of 10–25 μm to 150 μm in diameter. The factors determining their long suspension in meiotic prophase are unknown.

A single layer of flattened epithelial cells surrounds an oogonium as it enters into meiosis and a *primordial follicle* is formed. These cells, which will later generate the granulosa cells, provide nutrients to the oocyte via protoplasmatic processes that reach the oocyte plasma membrane. The primordial follicle becomes separated from surrounding spindle-shaped mesenchymal cells by a basement membrane called the *basal lamina*. When the flattened epithelial cells become cuboidal and begin to divide, forming several layers of granulosa cells around the oocyte, the follicle becomes a *primary follicle*. The oocyte secretes a mucoid substance rich in glycopro-

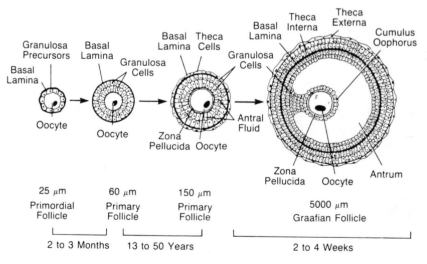

Fig. 9-5 Schematic representation of the development of an ovarian follicle (not to scale). (Redrawn from Berne RM, and Levy MN: *Physiology,* Mosby, St. Louis, p. 1069–1115, 1983, with permission.)

teins that forms a band that separates the oocyte from the granulosa cells. This band is called the *zona pellucida*.

The granulosa cells located next to the zona pellucida continue to extend protoplasmatic processes to the oocyte, presumably to maintain a nutrient flow and chemical signals to the growing oocyte. As these changes occur, the stromal cells adjacent to the basal lamina become differentiated in concentric layers around the primary follicle, constituting the *follicular theca*. The portion of the theca contiguous to the basal lamina is called the *theca interna:* the outer portion is called the *theca externa.*

Development of primary follicles is a very slow process lasting several years. At puberty, and throughout the reproductive years, cohorts of primary follicles enter into a new phase of development that is completed in approximately 3 weeks. During this phase, granulosa cells proliferate more rapidly, and some of the cells of the theca interna become cuboidal and filled with lipid droplets, indicative of enhanced steroidogenesis. In addition, the theca interna is invaded by capillaries that terminate at the basal lamina. In some follicles the continuity of the granulosa cell layers is disrupted by the appearance of spaces filled with fluid. Gradually, these spaces become confluent forming a single central cavity called the *antrum.* The follicles with these characteristics receive the name of *antral* or *graafian follicles.* At this stage, the oocyte reaches its maximum size (150 μm).

The antrum is filled with a fluid that contains steroids, pituitary hormones, local growth factors, etc., at concentrations several times greater than in plasma. As the granulosa cells continue to proliferate and the antrum enlarges, the oocyte is displaced from the center of the follicle and becomes surrounded by a hillock of granulosa cells called the *cumulus oophorus.* By the end of this stage, the antral follicle has reached a diameter of 5 mm. Since the granulosa cells and the oocyte are surrounded by the basal lamina throughout follicular development, they remain avascular; hormones, neurotransmitters, and blood-borne nutrients must reach them by diffusion.

During the second half of the follicular phase, one follicle becomes dominant and undergoes further growth to finally expel its oocyte at ovulation. The factors responsible for the selection of a single follicle (from a cohort of 6 to 10 follicles that become antral) in every cycle are poorly understood.

Ovulation

As discussed below, at the end of the follicular phase a large increase in plasma levels of gonadotropins takes place. This greatly increases the production of antral fluid in the dominant follicle, and further proliferation of the granulosa cells occurs. The follicle enlarges markedly reaching a diameter of 10–20 mm in 48 hours. Depolymerization of the mucopolysaccharides of the antral fluid increases the colloid osmotic pressure inside the antrum. The granulosa became less cohesive and the cumulus oophorus loosens. The follicular wall deteriorates at a specific site on its surface called the *stigma.* Formation of the stigma begins with a local decrease of blood flow, followed by a gradual thinning and depolymerization of the connective tissue in the area, and reorganization of the vasculature around the edge of the stigma. Sixteen to 24 hours after the gonadotropin surge the follicular wall at the stigma becomes exceedingly thin and finally appears to dissolve, allowing the extrusion of the oocyte surrounded by the cumulus oophorus. The focal dissolution of the follicular membrane is apparently due to enzymatic activity, the nature of which is not fully understood. It is now clear that follicular rupture at ovulation is preceded by a number of inflammatory-like events that include the increased production of prostaglandins and lipooxygenase products of arachidonic acid metabolism, as well as the production of a series of proteases (such as plasminogen activator) able to disrupt the continuity of the follicular wall. It appears that an important factor in the dissolution of the collagenous connective tissue that forms the follicular wall is the activation of thecal fibroblasts from a resting into a proliferating condition. Disruption of thecal fibroblast cell–cell contacts may precipitate the ovulatory rupture. The mechanism by which these changes occur remains unknown.

The first meiotic division is completed at the time of ovulation. The first polar body is eliminated and the secondary oocyte is captured by the fimbria of the oviduct, which is closely applied to the surface of the ovary. The oocyte is then transported to the ampulla of the oviduct via ciliary movements. The second meiotic division occurs at the time of fertilization, yielding the ovum with a haploid number of chromosomes and a second polar body that is discarded.

Atresia

Since the human female has approximately 30 years of reproductive life, only 300–400 follicles reach maturity within a lifetime. The rest undergo a degenerative process called atresia. Atresia is initiated very early in life (as soon as the first primordial follicles develop in the fetal ovary) and occurs at any stage of follicular maturation. It takes place throughout prepubertal development and at every menstrual cycle. Morphologically, it is characterized by necrosis of both the oocyte and the granulosa cells. Their nuclei become pyknotic and the cells degenerate. In antral follicles the last granulosa cells to die are those of the cumulus oophorus. When this occurs the zona pellucida disintegrates and the oocyte degenerates. In other follicles death of the oocyte is one of the first events to occur. Interestingly, some oocytes are stimulated to resume meiosis during the initial phases of atresia and they extrude the first polar body before dying. In marked contrast to the granulosa cells, thecal cells lose their differentiated condition, and instead of dying, return to the pool of interstitial cells not associated with follicles. The atretic follicle, on the other hand, is invaded by fibroblasts and becomes an avascular, nonfunctional scar.

The Corpus Luteum

Following ovulation the collapsed follicle becomes reorganized to form the corpus luteum consisting of "luteinized" granulosa cells, and thecal cells, fibroblasts, and capillaries that invade the new structure. Luteinization of the granulosa cells involves the appearance of lipid droplets in the cytoplasm, development by the mitochondria of a dense matrix with tubular cristae, and hypertrophy of the endoplasmic reticulum. The outer portion of the corpus luteum is made up of thecal cells that are also luteinized. The basal lamina of the follicle, however, regresses. Should fertilization fail to occur the corpus luteum remains functional for 13–14 days and then undergoes luteolysis. The luteal cells become necrotic, progesterone secretion ceases, and the corpus luteum is invaded by macrophages and then by fibroblasts. Endocrine function is rapidly lost and the corpus luteum is replaced by a scar-like tissue called the *corpus albicans.*

Ovarian Feedback Control of Gonadotropin Secretion

The ovary regulates the secretion of gonadotropins through two basic mechanisms: an inhibitory or negative feedback effect, and a stimulatory or positive feedback loop (Fig. 9-4). The former is exerted mainly by estrogen, although other steroids such as progesterone and androgens and the gonadal protein inhibin also play a role. Positive feedback is exerted by estradiol, although progesterone can amplify the E_2 effect. In addition to these hormones, it is likely that the newly discovered ovarian FSH-releasing protein (activin) has a physiological role to play in maintaining FSH secretion.

Negative Feedback

Ablation of the ovaries produces a prompt increase in plasma levels of LH and FSH. Conversely, administration of estradiol reverts the levels toward normal values. Estradiol acts on both the hypothalamus and the anterior pituitary to inhibit gonadotropin secretion. At the hypothalamus it suppresses LHRH secretion; at the pituitary it inhibits the release of LH induced by LHRH. The course of action of estrogen is fairly rapid. Within minutes of its intravenous injection LH levels begin to decline. FSH levels, however, are not brought to normal values by administering estradiol to ovariectomized women since, in addition to estradiol, the secretion of FSH is also under the inhibitory control of inhibin. Inhibin controls the secretion of FSH via a direct inhibitory action on the pituitary gonadotrophs.

The mechanism by which estradiol inhibits LHRH secretion in the hypothalamus appears to involve activation of inhibitory neurotransmitter systems coupled to the LHRH neuronal network. GABA neurons have been implicated as a major inhibitory neuronal group mediating the negative feedback effect of estradiol on LHRH secretion. At the pituitary level, estradiol decreases gonadotropin synthesis by suppressing gene expression of the LH β-subunit, and to a lesser extent, that of the FSH β-subunit (for an understanding of gonadotropin structure, see Chapter 6).

Positive Feedback

As with the negative feedback loop, positive feedback is for the most part estrogen dependent. Progesterone, however, can markedly potentiate the stimulatory effect of estradiol on gonadotropin release when given several hours after estradiol. In contrast to the negative feedback that occurs within minutes, the positive effect of estradiol has a latency of several hours. Typically, administering estradiol results in a rapid decline in plasma LH levels that last for at least 24 hours (negative feedback); this is followed by an abrupt, marked increase in the secretion of both LH and FSH (positive feedback). During the normal menstrual cycle this discharge of gonadotropins is a consequence of increases in plasma estrogen levels produced by the growing follicles (see below); the gonadotropin surge at the end of the follicular phase is essential for ovulation.

The anatomical site(s) where estradiol exerts its stimulatory effect have been a matter of controversy. In the Rhesus monkey estradiol can elicit gonadotropin release even if the pituitary gland is disconnected from the hypothalamus provided that LHRH is administered in a pulsatile manner. An invariable pulsatile LHRH regimen administered to these animals is capable of sustaining the complete sequence of endocrine events characteristic of a normal menstrual cycle, suggesting that the hypothalamus—through the release of LHRH—plays a permissive rather than a commanding role in the genesis of the preovulatory surge of gonadotropins. In conflict with this view, however, are the more recent demonstrations by other investigators that administration of both estradiol or estradiol plus progesterone to monkeys increases the release of LHRH from the hypothalamus.

It thus appears that estradiol can act at both the hypothalamus and the anterior pituitary to affect gonadotropin secretion. At both loci the steroid is able to exert first inhibitory effects, and, then, a stimulatory action. Whereas the stimulatory effect of estradiol on LHRH release appears to be mediated by the activation of stimulatory neurotransmitter systems functionally coupled to LHRH neurons (glutamatergic, noradrenergic, and NPY neurons), at the pituitary level, estradiol may act by both increasing the releasable pool of gonadotropins and by stimulating the expression of the β-subunit genes. This stimulation only occurs in the presence of LHRH.

An involvement of activin in stimulating the preovulatory discharge of FSH has not been demonstrated but is suggested by the increased expression of activin βA subunit mRNA in Graafian follicles close to the time of ovulation. Activin βA subunit mRNA has also been found in pituitary gonadotrophs where it may regulate FSH production through paracrine–autocrine actions.

Dynamics of the Hypothalamic–Pituitary–Ovarian Relationship

The hormonal changes during the menstrual cycle are depicted in Fig. 9-6. During menstruation and the early follicular phase, plasma FSH levels are 1.5- to 2-fold higher than at the end of the preceding luteal phase. LH values are low. Thereafter, plasma FSH falls and LH increases moderately, so that by the end of the follicular phase the LH/FSH ratio has increased to a value of about 2. This increase in mean plasma LH appears to be mainly due to an increased frequency of episodic LH discharges, probably due to the rising estradiol levels.

Estrogen levels (estradiol and estrone) increase very little during the first half of the follicular phase, while progesterone, 17α-hydroxyprogesterone, and the aromatizable androgens, androstenedione and testosterone, remain essentially unchanged.

During the second half of the follicular phase, and as follicles grow, plasma estrogen levels begin to rise, a change that becomes more pronounced toward the ovulatory phase, at which time estrogen levels rise up to 9-fold over basal values. Estradiol is produced directly by the ovaries, but most of the circulating estrone originates from peripheral conversion of estradiol and androstenedione. Progesterone and 17α-hydroxyprogesterone levels remain unchanged until the ovulatory phase; they are mostly derived from the ovary. Indeed, the concentration of progesterone in the follicular fluid increases markedly during the second half of the follicular phase, but not sufficiently to alter the systemic levels. Plasma androstenedione and testosterone also remain at relatively constant levels throughout the follicular phase.

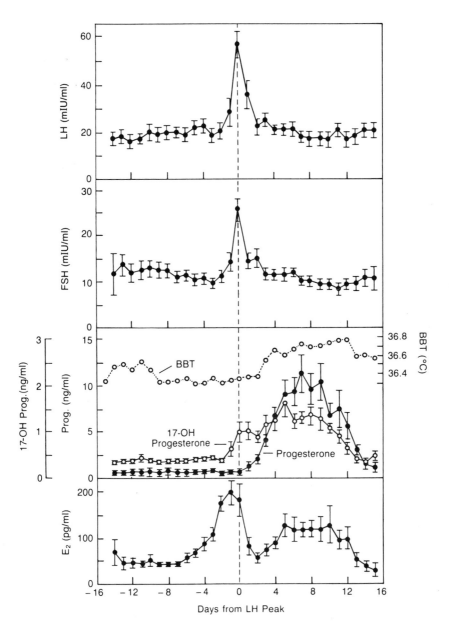

Fig. 9-6 Changes in plasma levels of LH, FSH, progesterone, estradiol, and 17-hydroxy-progesterone during the human menstrual cycle. The day of the LH peak is considered as day 0. The changes in basal body temperature (BBT) induced by progesterone are also represented. (Redrawn from Thorneycroft IH, et al: Am J Obstet Gynecol 111:947–951, 1971, and from Ross GT, et al: Recent Prog Horm Res 26:1–62, 1970, with permission.)

Normally, circulating androgens derive from both the ovary and the adrenal gland. The relative contribution of these glands, however, varies both during the day and during the cycle.

The initiation of the ovulatory phase is characterized by elevated estrogen levels that trigger an abrupt discharge of gonadotropins. This preovulatory gonadotropin surge is more pronounced in the case of LH than FSH, and lasts about 24 hours. As gonadotropins begin to rise, there is a concomitant increase in 17α-hydroxyprogesterone and a smaller rise in testosterone and androstenedione. Progesterone, on the other hand, increases transiently during the LH surge, but not to the same extent as 17α-hydroxyprogesterone. Although LH and FSH are still elevated, there is a precipitous fall in circulating estrogen levels. Plasma levels of progesterone and androgens also fall, but this occurs after gonadotropin levels have returned to basal values.

Ovulation occurs 16–20 hours after the peak of the gonadotropin surge. At this time gonadotropins, estrogen, progesterone, and androgen levels have fallen. In contrast, 17α-hydroxyprogesterone values remain elevated.

Both LH and FSH remain at low levels during the luteal phase. At the beginning of this phase the frequency of episodic LH discharges decreases, but they have a greater amplitude. Later on there is a progressive decrease in both amplitude and frequency of the LH pulses. A sustained elevation of progesterone and 17α-hydroxyprogesterone secreted by the corpus luteum becomes a predominant feature of this phase. Estrogens and androstenedione of luteal origin are also increased throughout the length of the luteal phase. After about 8–10 days, progesterone values begin to decrease, falling rapidly thereafter so that by day 14 after ovulation they have reached very low levels and menstruation ensues.

There is little doubt that the selective elevation in FSH levels observed during the early follicular phase plays a decisive role in initiating the growth of primary follicles. Although the formation of primordial and primary follicles is independent of the pituitary, development of primary follicles to the antral stage cannot proceed in the absence of FSH. A first wave of follicular development is initiated by the midcycle preovulatory surge of FSH. Upon reaching the antral stage, however, these follicles undergo atresia in the midst of the luteal phase. A second crop of follicles that is recruited by the elevation of FSH levels at the beginning of the follicular phase progresses beyond the antral stage, but only one of them, the "dominant" follicle, reaches the ovulatory condition.

As the follicular phase progresses, aromatase activity increases under FSH control. As more estradiol is produced, the steroid further facilitates the proliferation of granulosa cells, enhances synthesis of its own receptor, and synergizes with FSH to amplify the stimulatory effect of the hormone on granulosa cell differentiation. More specifically, it facilitates the ability of FSH to stimulate estrogen secretion and to induce the formation of both FSH and LH receptors. A gradual increase in plasma LH levels promotes the differentiation of thecal cells and induces the synthesis of the 17,20-lyase enzyme, thereby increasing available androgens for estrogen biosynthesis.

At the end of the follicular phase, estrogen secretion is further increased as the dominant follicle reaches the preovulatory stage. Plasma FSH levels are depressed, perhaps as a consequence of the increase in estrogen and inhibin secretion (the latter stimulated by FSH itself). Pituitary responsiveness to LHRH increases as a consequence of estrogen action, and within 48 hours the preovulatory surge of gonadotropins occurs.

It was previously mentioned that during the second half of the follicular phase a follicle is selected to reach the ovulatory stage. The factors involved in this selection are unknown; however, recent findings demonstrating the presence of noradrenergic and peptidergic nerves in the ovary of several species, including the human, raise the possibility that selection of a dominant follicle may have a neural component.

Once this selection has been made, the rest of the growing follicles undergo atresia. Exactly how atresia is induced is not known, but again, a neural component may be involved. An important characteristic of follicles undergoing atresia is their increased androgen content and reduced estrogen and progesterone levels in comparison with healthy graafian follicles. This is associated with a decreased concentration of FSH and increased levels of prolactin in the follicular fluid. It seems evident, therefore, that the fate of follicles is individually determined by the action of local factors. The nature of these factors remains obscure.

After ovulation, loss of the oocyte from the ruptured follicle appears to initiate luteinization of the granulosa cells. However, the LH discharge itself contributes significantly to the process. The mechanism involved in maintaining the function of the corpus luteum for 14 days and in precipitating its regression at the end of this period is incompletely understood. It is clear, however, that in humans LH is luteotropic, that is, it maintains the functional and morphological integrity of the corpus luteum. Prolactin has been found to be luteotropic in rodents, but its role in humans is unclear. The capacity of LH to maintain corpus luteum function declines as the luteal phase progresses. This phenomenon may be related to the influence of luteolytic factors that acquire predominance as the corpus luteum becomes older. Among these factors estrogen, oxytocin, and prostaglandin $F_{2\alpha}$ deserve mention. The luteolytic effect of both estrogen and oxytocin appears to be mediated at least in part by local formation of prostaglandin $F_{2\alpha}$.

DEVELOPMENT AND MAINTENANCE OF REPRODUCTIVE FUNCTION

Acquisition of reproductive competence is a long and extraordinarily complex process that takes an average of 12–14 years in the human female. Understanding of this process necessitates knowledge of both the initiation and progression of the neuroendocrine changes within the hypothalamic–pituitary–ovarian axis that lead to adult reproductive function.

Fetal, Neonatal, and Infantile Phases

The Hypothalamic–Pituitary Unit
The fetal pituitary gland synthesizes both FSH and LH as early as the fifth week of gestation and circulating levels of these two hormones are detected by the end of the third month (the earliest fetus studied). Pituitary content of both LH and FSH rises from this gestational age, reaching maximum levels at 25–29 weeks, and then

declines as pregnancy progresses. Newborn infants have two to five times less pituitary gonadotropin content than fetuses at 25–29 weeks.

Serum gonadotropins, and in particular FSH, begin to increase around the third month of gestation reaching very high values (in the castrate range) by 150 days, that is, before the peak in pituitary gonadotropins. From then to birth serum levels fall.

This pattern of gonadotropin release is thought to be related to the maturation of the negative feedback mechanism exerted by gonadal steroids. In the absence of this inhibitory feedback, FSH and LH secretion is relatively unrestrained. Later in fetal development the inhibitory feedback mechanism matures and becomes operative, leading to suppression of the synthesis and release of FSH and LH.

After birth serum FSH (and to a lesser extent LH) levels increase again and remain elevated for at least the first 2 postnatal years. This increase seems to result from the loss of placental steroids, which greatly contributes to the inhibition of fetal gonadotropin release during the second half of pregnancy. It is unclear, however, why FSH levels remain elevated for such a long period after birth when the infantile ovary is capable of producing estrogens, which can keep gonadotropin secretion restrained. In fact, in ovariectomized neonatal nonhuman primates or agonadal patients plasma gonadotropin levels are markedly increased (Fig. 9-7).

Two to 4 years after birth, gonadotropin secretion begins to decline and remains low throughout childhood (Fig. 9-7). This reduced output of gonadotropins is most likely due to events that occur within the central nervous system. Recent findings suggest that a loss of excitatory inputs coupled with an increase of inhibitory influences on LHRH neurons may be an important factor determining the decline in gonadotropin secretion that characterizes the juvenile period.

By the eighth to tenth year of life, gonadotropin secretion begins to increase. FSH levels rise before LH levels do and attain adult values while LH levels are still

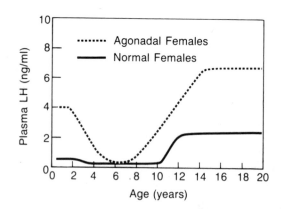

Fig. 9-7 Changes in circulating gonadotropin levels during postnatal development in normal and agonadal human females. Notice that gonadotropin levels in the agonadal subjects decrease during childhood and increase at puberty similar to normal subjects, indicating that these changes occur independently from the ovary. (Redrawn from Grumbach MM: The neuroendocrinology of puberty. In: *Neuroendocrinology*, DT Krieger and JC Hughes, eds., Sinauer, Sunderland, MA, p. 249–258, 1980, with permission.)

increasing. In fact, mean plasma LH may not increase noticeably until secondary sexual development is quite advanced. As before, the mechanism responsible for the pubertal reactivation of gonadotropin release is of central nervous system origin and independent of the ovary (see pp. 188–189).

The Ovaries

The prepubertal ovary shows morphological signs of activity such as follicular maturation to the stage of graafian follicle and subsequent atresia. The cellular residue of atretic follicles contributes substantially to the mass of the human ovary, which increases rectilinearly from birth to puberty.

Ovarian steroidogenic capability is present from birth. Estradiol is the major ovarian secretory product, although estrone, androstenedione, testosterone, and other steroids are also produced. Umbilical artery levels of estradiol, reflecting fetal production of the steroid, are more than 100 times higher than in adult females despite considerable metabolism of estrogen by the fetus. Estrogen levels decline rapidly in the first postnatal week and remain at less than 10 pg/ml until the onset of puberty.

Puberty

Hormonal Changes

The first endocrinological manifestation of the onset of puberty is amplification of a sleep-related pattern of LH release. During childhood, LH is secreted in a pulsatile manner, but the magnitude of the pulses is small. In addition, greater LH levels are seen during sleep than during waking hours. At the end of childhood, before mean gonadotropin levels begin to increase, the difference in LH secretion becomes more pronounced and the magnitude of the LH pulses is enhanced (Fig. 9-8). These changes not only reflect the initial activation of the central mechanism governing LH secretion, but are also believed to be the initial event leading to the attainment of puberty.

As puberty progresses and plasma gonadotropin levels increase, there is a wide range of serum LH and FSH values, probably reflecting superimposed menstrual cyclicity. Episodic increases in LH levels (ovulatory spikes) have occasionally been recorded in girls within several months of the first menstruation. A short rise in progesterone levels following these LH peaks has also been found in these girls, suggesting corpus luteum formation and ovulation. The luteal phase appears to be shortened. Accordingly, the infertility of the young adolescent girl may be related either to anovulatory cycles or to inadequate corpus luteum function.

The rising levels of plasma gonadotropins stimulate the ovary to produce increasing amounts of estradiol. Estradiol is responsible for the development of secondary sexual characteristics, that is, growth and development of the breasts and reproductive organs, fat redistribution (hips, breasts), and bone maturation. Using an ultrasensitive assay for estradiol, evidence was recently provided that in childhood, plasma estradiol levels, though very low, are several times higher in girls than in boys. This difference may contribute to the earlier onset of puberty in girls. As puberty approaches, plasma estradiol levels fluctuate widely, probably reflecting successive waves of follicular development that fail to reach the ovulatory stage. The uterine endometrium is affected by these changes and undergoes cycles of pro-

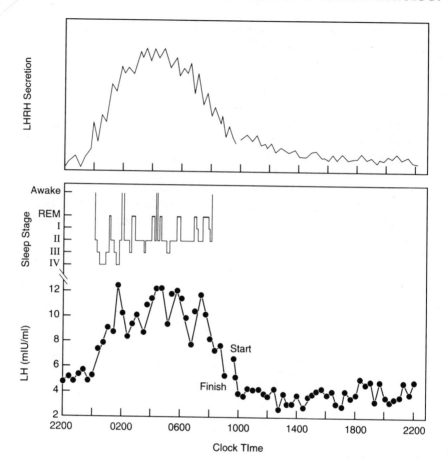

Fig. 9-8 Diurnal changes in pulsatile release of LH at puberty. The upper panel depicts the hypothetical changes in LHRH secretion assumed to occur during puberty at night. This increase in LHRH secretion does not require the presence of the gonads and is thought to be responsible for the nocturnal increase in LH secretion depicted in the lower panel. (Redrawn from Boyar RM, et al: N Engl J Med 287:582–586, 1972.)

liferation and regression, until a point is reached when substantial growth occurs so that withdrawal of estrogen results in the first menstruation. This is called *menarche*. Plasma testosterone levels also increase at puberty although not as markedly as in males. In contrast, plasma progesterone remains at low levels, even if secondary sexual characteristics have appeared. A rise in progesterone after menarche is, in general, indicative that ovulation has occurred.

Menarche occurs at an average of 12 years of age. However, the first ovulation does not take place until 6–9 months after menarche because the positive feedback mechanism of estrogen is not developed. A single estrogen injection to premenarchial subjects fails to induce LH release. However, recent experiments with Rhesus monkeys have shown that if "priming" doses of estradiol are given, the surge mechanism can be activated shortly after menarche.

Somatic Changes

Somatic changes during puberty include a rapid general increase in the growth rate of skeleton, muscles, and viscera, known as the *adolescent growth spurt*, a sex-specific increase in hip width, and changes in body composition caused by an increase in muscle and in fat tissue, the latter being more pronounced in girls than in boys. The large increase in fat (120%) between growth spurt initiation and menarche in both early and late maturing girls may have a secondary significance. The mean fat for both early and late maturers at menarche is about 11 kg, which is equivalent to 99,000 calories. The number of calories estimated to sustain a pregnancy is 80,000. Thus, although speculative, it is possible that one of the main functions of the adolescent growth spurt in females may be the storage of energy to sustain pregnancy and lactation. Figure 9-9 shows schematically the sequence of events at puberty in females. The development of the breasts is known as *thelarche* and the growth of axillary and pubic hair is known as *pubarche*.

The age at which these morphological changes occur varies from individual to individual. A classification that is widely used by clinicians to assess the normalcy of the pubertal progression bases its five stages on the degree of breast development, age at peak velocity of linear growth, and growth and distribution of the pubic hair.

In addition to estrogens and androgens, GH may play a role in inducing pubertal growth. Plasma growth hormone levels are low but pulsatile in peripubertal subjects. Immediately upon falling asleep, there is an increase in GH release; usually several episodes of GH secretion can be detected during the sleep period. The magnitude of these episodes reaches a maximum in advanced puberty.

The Critical Weight at Menarche

The mean weight of girls at the beginning of the adolescent growth spurt (around 30 kg), at the time of peak velocity of weight gain (around 39 kg), and at menarche

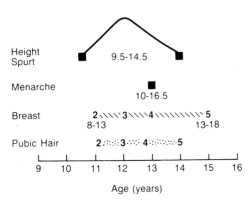

Fig. 9-9 Schematic sequence of somatic events during female puberty. The ranges of ages at which each event occurs is indicated at the beginning and end of each event. The numbers 2–5 correspond to the different stages of puberty as defined by Tanner. (Redrawn from Tanner JM: Sequence and tempo in the somatic changes in puberty. In: *The Control of the Onset of Puberty*, MM Grumbach, GD Grave, and FE Mayer, eds., Wiley, New York, p. 448–470, 1974, with permission.)

(around 47 kg) does not differ for early and late maturing girls. This has led to the notion that the attainment of a critical weight causes a change in metabolic rate per unit mass (or per unit of surface area), which, in turn, may affect the hypothalamic ovarian feedback by decreasing the sensitivity of the hypothalamus to gonadal steroids so that gonadotropin release increases, triggering the pubertal process.

The hypothesis of a critical weight has recently been modified to state that a particular ratio of fat to lean mass is normally necessary for puberty and the maintenance of female reproductive ability. The metabolic signals that influence the timing of puberty have not yet been identified, but recent evidence suggest that IGF-I may represent one of these signals. In addition, experiments with nonhuman primates have demonstrated that LH pulsatile release is dramatically decreased by underfeeding and activated by administration of calorie-rich food, suggesting that caloric intake may be a mechanism by which the diet could affect the timing and/or progression of puberty. It seems clear that undernutrition delays the onset of puberty and disrupts menstrual cyclicity.

Of interest in this regard are the observations that in the past century, girls in the United States reached a weight of 46 kg at about 14 years, which was the age of menarche. In 1970, this body weight was reached at the age of 12.9 years, at which age menarche also occurred. Belgian girls who in the past century reached a weight of 46 kg at about age 16.5 years also attained menarche at that age.

Mechanisms Involved in the Regulation of the Onset of Puberty (Fig. 9-10)

Change in the Setpoint of Negative Feedback Mechanism. As previously discussed, the negative feedback of gonadal steroids is poorly developed during fetal life but becomes fully operative during childhood. Several years ago, from experiments in rats, the concept was advanced that a decrease in hypothalamic sensitivity to circulating sex steroids ("resetting of the gonadostat") results in the increased gonadotropin secretion seen at puberty.

Nevertheless, more recent experiments in rats, Rhesus monkeys, and humans have demonstrated that the change in hypothalamic setpoint to steroid negative feedback is a late phenomenon in puberty. Thus the "resetting" mechanism can no longer be considered as the cause of puberty, but rather a consequence of it.

Activation of the Hypothalamic "Pulse Generator." As discussed above, pulsatile release of LH becomes more prominent at the end of childhood. If juvenile monkeys are treated with LHRH, delivered in a pulsatile manner for several weeks, they show repetitive, normal ovulatory cycles. These cycles cease upon removal of the LHRH treatment. Direct measurement of LHRH released from the mediam eminence of the hypothalamus in nonhuman primates demonstrated that LHRH secretion does indeed increase as puberty approaches. These and other observations have led to the concept that the onset of puberty is determined by the enhancement of endogenous pulsatile LHRH release (Fig. 9-10). Such an activation would then drive the pituitary to secrete larger amounts of LH, particularly during sleep, and puberty would be initiated.

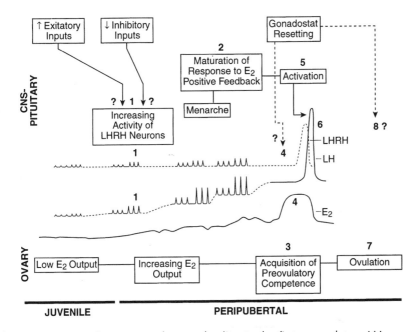

Fig. 9-10   Proposed sequence of events leading to the first preovulatory LH surge in humans. The numbers indicate the sequence in which the events may occur. The sleep-related increase in plasma LH levels (1), which can be regarded as the first neuroendocrine manifestation of puberty, is the result of an increase in LHRH release caused by an enhancement of excitatory inputs, withdrawal of an inhibitory tone, or, more likely, the simultaneous occurrence of both changes. Estradiol positive feedback becomes operative several months after menarche but it can be activated shortly after menarche (2). Present evidence does not permit a firm conclusion as to the timing of the gonadostat resetting. It may occur shortly before the first ovulation (4) or after it (8). A preovulatory LHRH surge precedes the first preovulatory gonadotropin surge. (Modified from Ojeda SR, et al: The onset of female puberty: Underlying neuroendocrine mechanisms. In: *Neuroendocrine Perspectives,* Vol. 3, EE Muller and RM MacLeod, eds., Elsevier, Amsterdam, p. 225–278, 1984, with permission.)

The pubertal activation of LHRH release may be brought about by the coordinated interplay of transsynaptic inputs and glial influences on LHRH neurons. According to this concept, LHRH release would increase as a consequence of an activation of stimulatory transsynaptic inputs (glutamate, but also NE and NPY), a decrease in GABA inhibition, and a growth factor–dependent increase in glial production of neuroactive substances, such as prostaglandins.

Another factor that operates at the time of puberty is an enhanced pituitary responsiveness to LHRH. The release of LH following the administration of LHRH is minimal in prepubertal children but increases markedly at puberty and is even greater in adult males. There is a sex difference in the rise of serum FSH: prepubertal and pubertal females release more FSH than males at all stages of sexual maturation.

This increased gonadotropin response to LHRH is probably due to the elevated circulating levels of gonadal and adrenal steroids present at that time. Since the response can be elicited before the development of secondary sex characteristics in both girls and boys, this change in pituitary responsiveness probably plays an important role in the progress of the pubertal process.

The Adrenal Gland and Puberty

Circulating levels of aromatizable androgens such as dehydroepiandrosterone and its sulfate, and androstenedione, are mostly of adrenal origin. Their concentrations in the bloodstream begin to increase at about 7–8 years of age, reaching maximum values by ages 13–15. This phenomenon is called *adrenarche*. It begins before the rise in gonadotropin secretion. Adrenal androgens are responsible to a significant extent for the growth of pubic and axillary hair; however, they do not appear to play a decisive role in determining the initiation of puberty. Premature adrenarche is not associated with advancement of gonadarche (i.e., the activation of the hypothalamic–pituitary–ovarian axis). Conversely, gonadarche occurs in the absence of adrenarche in children with chronic primary adrenal insufficiency, and in children with true idiopathic precocious puberty. It is evident, therefore, that adrenarche and gonadarche are separately controlled, albeit temporally associated phenomena.

Reproductive Cyclicity (Fig. 9-11)

This reproductive phase has already been discussed. Basically it entails the orderly repetition of the menstrual cycle, a sequence that may be interrupted by pregnancy. It begins with the first ovulation at puberty and extends on the average to about 50 years of age.

Menopause

Menopause can be defined as the time at which the final menstrual bleeding occurs. Like ''puberty,'' however, the term is used in a much broader sense, and describes a period of the female climacteric during which reproductive cyclicity gradually disappears. The menstrual cycles become irregular, and the intervals between menses become shorter because of a reduction in the duration of the follicular phase.

Menopause is regarded as the result of the loss of ovarian function. There is an exhaustion of ovarian follicles, mainly due to the repeated cycles of atresia that occur twice in every menstrual cycle. Histologically, the stroma hypertrophies and primordial follicles disappear. Ovarian weight declines as menopause progresses, and plasma gonadotropin levels increase when estradiol production falls as follicles fail to mature (Fig. 9-11). The loss of estrogen is accompanied by atrophic changes of the breasts, uterus, and vagina, vasomotor instability (hot flashes), and in many instances, osteoporosis.

The occurrence of hot flashes is temporally related to pulses of LH secretion. Since such pulses appear to be the consequence of an increase in noradrenergic impulses arriving at hypothalamic LHRH neurons, the inference has been made that hot flashes are caused by a change in hypothalamic noradrenergic activity.

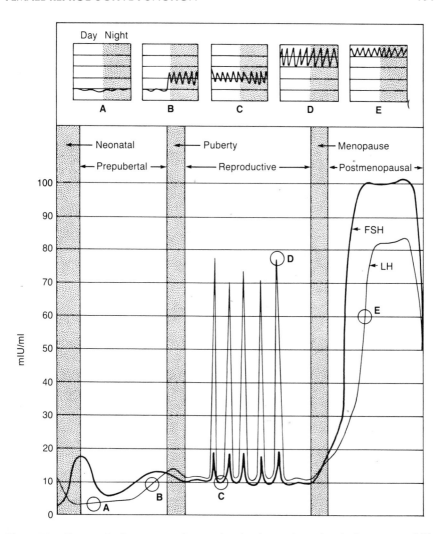

Fig. 9-11 Changes in the pattern of serum levels of gonadotropins during postnatal life in the human female. During neonatal and prepubertal phases, FSH levels are greater than LH, and pulsatile LH secretion is minimal (A). As puberty approaches, LH secretion increases during sleep (B). When puberty is completed, LH secretion is greater than FSH secretion (C) and cyclic gonadotropin release occurs (D). At menopause, cyclic gonadotropin release ceases (E) and plasma levels of both gonadotropins increase. (Redrawn from Yen SSC: In: *Neuroendocrinology*, DT Krieger and JC Hughes, eds., Sinauer, Sunderland, MA, p. 259–272, 1980, with permission.)

THE MAMMARY GLAND

Lactation is an evolutionarily conserved mammalian physiological function. The offspring of placental mammals are entirely dependent in early postnatal life on nutrients provided by the mother through the secretion of milk. The mammary gland replaces the nourishing function of the placenta after birth.

Development and Hormonal Control

During early fetal development, epithelial cells cluster in areas that will later give rise to the areola of the breast and proliferate to form cord-like structures that penetrate the underlying mesenchyme. These epithelial cords give origin to the duct system of the mammary gland that converges into a single ampulla that opens to the nipple. During the last part of gestation, the distal portion of the ducts develops into alveolar structures that at birth are able to secrete. This secretion, known as *witch's milk*, can be observed in most infants by the first week of postnatal life and lasts for 4–6 weeks. Once the effect of placental steroids subsides, the secretory alveoli regress so that in children only scattered ducts without alveoli are observed embedded in stromal tissue.

As previously indicated, when estrogen secretion increases at puberty, breast development is markedly stimulated. When the ovulatory cycles begin and the corpus luteum secretes progesterone, the estrogen-dependent branching and lengthening of the ducts are enhanced, and alveoli development is accelerated.

During pregnancy the mammary gland undergoes further growth under the concerted influence of estrogen, progesterone, glucocorticoids, prolactin, and human placental lactogen (hPL). The duct system develops markedly, the alveoli form clusters called *lobules*, and their epithelial cells acquire vacuoles indicative of intense secretory activity. Actual secretion of milk (lactation) does not occur, however, until after birth, due to an inhibitory effect of placental estrogen and progesterone. Following delivery of the placenta, estrogen and progesterone levels drop precipitously and secretion of milk begins.

A pituitary hormone that plays a critical role in the process of breast development during pregnancy is prolactin. In combination with estrogen, prolactin causes primarily ductal but also lobuloalveolar growth. In the presence of progesterone its effect on alveolar growth is greatly enhanced. Prolactin not only induces cell proliferation but also controls the synthesis of key components of milk such as casein and α-lactoalbumin. Its mechanism of action is not completely understood, but it is clear that it involves binding of prolactin to specific membrane receptors. Recent experiments have shown that the interaction of prolactin with its receptor leads to tyrosine phosphorylation of a protein that may act as a transcription factor as it binds to regulatory DNA sequences present in genes encoding milk proteins. Interestingly, this mechanism of action appears to be similar to that described for the cytokine interferon (INF)-α.

Growth hormone produced by the anterior pituitary shares many of the actions of prolactin and, in fact, is as effective as prolactin in inducing duct development. Nevertheless, its presence is not essential for breast development or lactation to occur. The human placenta produces hPL, which has actions similar to growth hormone.

Insulin and glucocorticoids are also required for most phases of breast development. Their function, however, is of a permissive rather than a regulatory nature.

Lactation

Lactation can be divided into four stages: *milk synthesis,* which is initiated during the last part of pregnancy by prolactin and hPL action; *lactogenesis,* which comprises he synthesis of milk by the alveolar cells and its secretion into the alveolar lumen—t is initiated by the loss of placental steroids after birth; *galactopoiesis*, or maintenance of established lactation, which is mainly controlled by prolactin, whose release increases because of suckling by the infant; and *milk ejection,* which is controlled by the neurohypophyseal hormone oxytocin and comprises the passage of milk from the alveolar lumen to the duct system, its collection in the ampulla and larger ducts, and its delivery to the infant (see below).

Oxytocin and prolactin play fundamental roles in the maintenance and dynamics of lactation. The act of nursing itself is a powerful stimulus for oxytocin release. Sensory impulses arising from the nipple travel via the spinal cord through a multisynaptic pathway to the hypothalamic neurons that produce oxytocin and evoke its release. Oxytocin then reaches the mammary gland via the bloodstream and induces the contraction of the myoepithelial cells that surround the alveoli and the ducts. Contraction of these cells mobilizes milk from the alveoli and duct system to the nipple, producing the sensation of "milk let-down" in the mother. Oxytocin release is also induced by audiovisual stimuli such as the sight of the baby or hearing its cry, or even just by the anticipation of nursing.

Prolactin secretion is stimulated by the act of suckling (Fig. 9-12), but not by audiovisual or psychological stimuli. Throughout the lactational period plasma prolactin levels vary as follows: basal values are quite high during the first few weeks postpartum and each nursing episode further elevates the levels. After about 2 months of lactation, basal prolactin levels return to normal values (10–20 ng/ml), but nursing

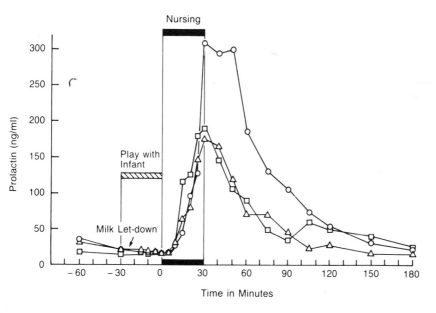

Fig. 9-12 Changes in plasma prolactin levels during anticipation of nursing and nursing itself in women. The women played with their infants for 30 minutes before suckling began. Milk let-down, which results from an action of oxytocin on the breast, occurred before suckling. (Redrawn from Noel GL, et al: Prolactin release during nursing and breast stimulation in postpartum and nonpostpartum subjects. J Clin Endocrinol Metab 38:413–423, 1974, with permission.)

is still able to elicit a rise in prolactin levels. Such a response persists for many months postpartum. It is clear that episodes of prolactin secretion are important for the maintenance of lactation since prolactin-inhibiting drugs such as ergot alkaloids (see below) result in cessation of lactation.

The mechanism by which nursing increases prolactin release appears to involve both secretion of prolactin-releasing substances from the hypothalamus into the portal vasculature (e.g., vasoactive intestinal polypeptide [VIP], oxytocin) and suppression of dopamine release from the tuberoinfundibular dopaminergic neuronal system. Since dopamine itself is a prolactin-inhibiting factor, suppression of its release enhances the secretion of prolactin. On the other hand, dopamine may stimulate the release of a prolactin-inhibiting peptide to further depress prolactin secretion.

A common abnormality of breast function is galactorrhea, that is, the persistent discharge of milk in the absence of parturition or beyond 6 months postpartum in a nonnursing mother. In most of these cases, plasma prolactin levels are elevated and imaging of the pituitary occasionally reveals the presence of microadenomas. Most of these patients suffer from reproductive malfunction ranging from ovulatory failure to amenorrhea, that is, failure to menstruate.

The discovery that ergot alkaloids such as bromocriptine are potent suppressors of prolactin release provided a useful tool in the treatment of hyperprolactinemia. Bromocriptine treatment not only suppresses prolactin secretion, but often results in significant regression of the pituitary prolactin-secreting tumor. Ergot alkaloids suppress prolactin release because they are dopamine agonists.

Major Milk Components

The major milk proteins are casein, α-lactalbumin, and β-lactoglobulin. The protein content in the human colostrum (the first milk secreted at parturition) is 2.7 g/dl as compared to 1.2 g/dl in mature milk. The major carbohydrate is lactose (5.3 g/dl in the colostrum, 6.8 g/dl in mature milk). Milk is also rich in fat (2.9 vs 3.8 g/dl) and contains many other substances including minerals (calcium, magnesium, phosphorus, iron), electrolytes (chloride, potassium, sodium), all vitamins with the exception of vitamin B_{12}, iron-binding proteins (transferrin), immunoglobulins (IgMs, IgEs, and particularly IgAs), and growth factors such as epidermal growth factor and IGF-I.

CLINICAL ASSESSMENT OF REPRODUCTIVE FUNCTION

A thorough history and physical examination are essential in the assessment of female reproductive function. Valuable information includes the age of first menstruation, the presence of secondary sexual characteristics such as breast development, and the presence of regular, cyclic menses. The latter implies that ovulation is occurring, that the secretion of gonadotropins and ovarian steroids is adequate, and that the outflow reproductive tract is intact. Should the symptoms be ambiguous, both plasma gonadotropin and steroid concentrations can be determined by radioimmunoassay.

As stated in Chapter 10 an accurate assessment of plasma gonadotropin values generally can be obtained by drawing multiple blood samples every 20 minutes for 2 hours and pooling them to obtain a mean value. This is usually not necessary in women. There are cases (e.g., precocious puberty and delayed puberty), however, in which each blood sample must be analyzed individually and the sequential sampling needs to be extended for several hours.

Plasma estradiol levels can also be measured by sensitive immunoassays, but because of the normal variations in circulating levels of the steroid during the menstrual cycle it is preferable to use other less direct, but more informative tests. For example, adequate estrogen production can be inferred by the presence of a moist vagina and a copious, clear cervical mucus that shows a high degree of elasticity upon stretching and an arborization (ferning) pattern upon spreading on a glass slide. A test commonly used to assess the estrogen status is the progesterone withdrawal test. Administration of progesterone followed by menstrual bleeding upon termination of the treatment is indicative of adequate estrogen priming to allow the withdrawal bleeding.

The occurrence of ovulation can be assessed by measuring plasma progesterone by radioimmunoassay. A functional test that can be employed to assess progesterone status is the increase in basal body temperature associated with normal luteal function. If such an increase occurs and is maintained for 2 weeks, progesterone secretion and corpus luteum function can be considered to be adequate.

One very important diagnosis is that of pregnancy. Again, the history and physical examination are critical for a correct diagnosis. In this case, however, the laboratory measurement of the placental product human chorionic gonadotropin (hCG) is recommended because of its reliability and sensitivity.

DERANGED REPRODUCTIVE FUNCTION

There are many forms of reproductive dysfunction, but only the most common will be considered.

Sexual Precocity and Delayed Puberty

Early advent of sexual maturity can be isosexual, that is, feminization in girls and virilization in boys, or heterosexual, that is, feminization in boys and virilization in girls.

Isosexual Precocity

Isosexual precocity can be manifested in three forms: true precocious puberty, precocious pseudopuberty, and incomplete isosexual precocity.

True precocious puberty is characterized by an early but normal sequence of pubertal development, including the initiation of menstrual cycles in girls. In 90% of cases the factors determining the early advent of sexual maturation are unknown; the syndrome is called *constitutional* or *idiopathic precocious puberty*. Prompt treatment is essential and has usually relied on the use of the progesterone derivative medroxyprogesterone acetate to suppress gonadotropin release. More recently, LHRH agonists have proved to be extremely effective in reverting the progression of precocious puberty. These analogs act by down-regulating pituitary LHRH receptors and desensitizing the gland to LHRH.

Precocious pseudopuberty is defined as a dysfunction in which girls feminize but fail to ovulate. In general, precocious pseudopuberty is caused by an enhanced secretion of estradiol, which in most cases originates from granulosa-theca cell tumors of the ovary. These tumors can be detected by rectoabdominal examination, sonography, or laparoscopy.

Incomplete isosexual precocity is defined as the premature development of a single clinical pubertal event. This event may be breast enlargement before the age of 8, known as *premature thelarche*, or appearance of axillary and pubic hair, known as *premature pubarche*. The former is thought to be caused by a transient increase in estrogen levels, or to a heightened breast sensitivity to normal plasma levels of estrogen. Premature pubarche is caused by a premature increase in androgens of adrenal origin. Neither syndrome requires treatment since in both cases these changes are transient and often regress, the patient entering puberty at the normal age.

Heterosexual Precocity

Heterosexual precocity of girls, that is, virilization, is usually caused by congenital adrenal hyperplasia (see Chapter 8) or to androgen-secreting tumors of the ovary or adrenal.

Delayed Puberty

Delayed puberty is much less common in girls than in boys. The onset of puberty may be regarded as delayed if breast development has not begun by age 13. The progression of puberty is regarded as delayed if menses have not occurred 5 years after the onset of breast development. Delayed puberty may be due to disorders of the ovary or the hypothalamic–pituitary unit. In the former case, gonadotropin levels are elevated because of the absence or insufficiency of steroid secretion. The syndrome is called *hypergonadotropic hypogonadism*; the most common form in girls is gonadal dysgenesis known as the *Turner syndrome* (see Chapter 8). The term *hypogonadotropic hypogonadism* is applied to disorders of the hypothalamic–pituitary unit that result in decreased secretion of gonadotropins. It must be differentiated from a constitutional delay in the onset of puberty that is characterized by a pattern of slow maturation but an otherwise normal attainment of puberty at a late age. Patients with hypogonadotropic hypogonadism do not reach puberty because of their inability to produce gonadotropins. Because failure of gonadotropin secretion is usually the only pituitary hormone deficiency observed in these patients, the syndrome is termed *isolated gonadotropin deficiency*; it is the result of defective synthesis or release of hypothalamic LHRH (see also Chapter 10).

Abnormal Uterine Bleeding

Abnormal uterine bleeding is a disorder of the menstrual cycle that may be associated with either ovulatory or anovulatory cycles. In the first case, abnormal bleeding may occur in the face of regular cycles, but it deviates from a normal menstruation because it is too abundant and prolonged (hypermenorrhea) or too light and interrupted (hypomenorrhea). Hypermenorrhea may be caused by polyps, submucus myomas, or adenomyosis of the uterus. Hypomenorrhea may result from obstruction of the uterinecervical lumen. Bleeding may also occur between normal menstruations and is usually due to cervical and/or endometrial lesions.

Uterine bleeding associated with anovulatory cycles is painless bleeding, unpredictable with respect to onset, amount, and duration. The syndrome is called *dysfunctional uterine bleeding* and is caused by alterations of ovarian function rather than to abnormalities of the uterus. As previously discussed, the sequential changes in estradiol and progesterone secretion during the menstrual cycle determine both the cyclic pattern of endometrial activity and the onset of menses upon their withdrawal.

Dysfunctional uterine bleeding may result from three different mechanisms: (1) estrogen withdrawal bleeding that results from interruption of ovarian estrogen production in the absence of progesterone. For example, if estrogen is given to a postmenopausal woman and the treatment is then interrupted, bleeding will occur. (2) Estrogen breakthrough bleeding occurs when the corpus luteum fails to form (or function) so that progesterone levels are low and the uterus is chronically exposed to continuous estrogen. It is the most common type of dysfunctional uterine bleeding. (3) Progesterone breakthrough bleeding occurs when the progesterone/estrogen ratio to which the endometriun is exposed is abnormally high, as in women on continuous progesterone-only contraceptives (see below).

Amenorrhea

Amenorrhea is the failure to menstruate. It may be classified as primary amenorrhea (i.e., a woman who has never menstruated) and secondary amenorrhea (i.e., a woman who had menses, but they ceased). Operationally, amenorrhea has been defined as the failure of menarche by age 16, irrespective of the presence of secondary sexual characteristics, or the absence of menstruation for 6 months in a woman with previous periodic menses.

Amenorrhea may result from *anatomical defects* (e.g., pseudohermaphroditism, congenital defects of the vagina, imperforate hymen), *ovarian failure* (the most frequent being gonadal dysgenesis in which the ovary does not develop normally), or *chronic anovulation* with or without estrogen present. More than 80% of all amenorrhea is the result of chronic anovulation. This disorder is characterized by the failure to ovulate spontaneously with the possibility of ovulating upon therapy. In the most common type of chronic anovulation estrogen is present. Women with this condition have withdrawal bleeding after progesterone administration, indicating the prior presence of estrogen. However, most of the estrogen produced is estrone formed by peripheral aromatization of androstenedione. In most cases the ovaries are sclerotic and show a thick capsule, multiple follicular cysts, and hyperplastic thecal and stromal tissue. The disorder is known as *polycystic ovarian disease* and is usually characterized by hirsutism, obesity, and amenorrhea. Its etiology is complex but it does not appear to be caused by a primary defect in gonadotropin secretion as plasma LH and FSH are variable in patients suffering from the disease. Recent findings suggest that alterations in the ovarian microenvironment (e.g., growth factor dysfunction) may play a more predominant role.

Hirsutism

Hirsutism, or excessive growth of hair, is the result of an abnormally high production of androgens. The androgens may be produced by neoplastic tissue; in this case, the onset of hair growth is rapid. They may also have an adrenal origin in the absence of a tumor; in most such cases the excessive androgen production results from heritable defects in adrenal steroidogenesis (congenital adrenal hyperplasia), which can occur as "late onset" forms. An excess of circulating androgens may also result from intake of some drugs or steroids with androgenic capacity.

A more common cause of hirsutism, however, is ovarian hyperandrogenism, and, in particular, polycystic ovarian disease. Its pubertal onset is associated with chronic anovulation.

The treatment of hirsutism depends on the source of excessive androgen production. If it is a tumor, the tumor should be removed. If the source is the ovary, suppression of ovarian function is attempted by administering oral contraceptives; adrenal hyperfunction is reduced by suppressing ACTH secretion with dexamethasone.

FERTILITY CONTROL

Few things are more important in today's world than birth control. There are several methods of contraception, including (1) natural rhythm, (2) withdrawal, (3) barriers such as condoms, foams, and diaphragms, (4) intrauterine devices known as IUDs, (5) sterilization, (6) abortion, and (7) oral and subdermal steroid contraceptives.

The most effective preventive method is that of oral and subdermal contraceptives. The least effective are the rhythm and withdrawal methods. Barrier methods are much more effective than the "natural" methods, but not as efficient as IUDs, which have a success rate of more than 95%.

Use of an IUD, however, carries significantly greater health risks than barrier methods. It is believed that IUDs prevent pregnancy by inducing a chronic inflamation of the endometrium, which then becomes unable to allow implantation of the blastocyst. The most serious side effect of the IUD is pelvic infection.

Oral contraceptives are the most widely used method of fertility control. When properly used their success rate is more than 99%, they are easy to administer, and in general the incidence of side effects is low. Oral contraceptives prevent pregnancy by inhibiting ovulation through suppression of gonadotropin secretion.

There are three types of oral contraceptive pills: a combination of estrogen–progestin, phasic estrogen–progestin, and progestin-only pills. The combination pill consists of a synthetic estrogen (mestranol or ethinyl estradiol) and a synthetic progestin (such as norethindrone or norgestrel). They are taken continuously for 21 days followed by 7 days of rest. During these 7 days the person may discontinue taking the pill altogether, or—as some manufacturers recommend—take a placebo pill for 7 days. Phasic pills consist of different concentrations of estrogen plus progesterone, which varies throughout a 21-day treatment schedule followed by 7 days of placebo. Progesterone-only pills are taken continuously on a daily basis.

The ideal combination pill has the least amount of steroids necessary to prevent pregnancy, but a sufficient amount to avoid breakthrough bleeding. Oral contraceptives are not free of side effects. Users of progesterone-only pills have a high incidence of breakthrough bleeding. Combination pills may in some patients increase the risk of deep-vein thrombosis, pulmonary embolism, thromboembolic stroke, and hemorrhagic stroke. Because of the stimulatory effect of estrogen on liver protein synthesis, including the renin substrate angiotensinogen, use of contraceptive pills results in a small rise in blood pressure. In some women, however, hypertension may develop after several years of use. As previously indicated, estrogen increases both HDL and VLDL levels and decreases cholesterol and LDL concentrations. Progestins increase cholesterol and LDL levels and decrease HDL. However, their combination in pills usually has minimal effects in the low doses used today. In some women glucose tolerance changes, manifested by elevated plasma insulin and glucose levels after a glucose load. Because of all these possible side effects oral contraceptives are contraindicated in women with diabetes mellitus, abnormal liver function tests, hypercholesterolemia, hypertension, or heavy smoking habits. There is no convincing evidence that oral contraceptives increase the incidence of uterine, cervix, breast, or ovarian cancer.

Although oral contraceptives are contraindicated in a small fraction of the population, they are the most reliable and effective method of birth control available to women throughout the world.

An alternative system of steroid administration is the subdermal implantation of the progestin levonorgestrel (Norplant). Implantation of six silicone capsules in the inside part of the lower arm provides continuous contraception for 5 years. The mechanism of action of the progestin is by inhibition of ovulation. It may be used by women who cannot receive estrogens or by women who do not want surgery and/or cannot use any other form of contraception. The main problem with this system is the surgery it requires; it also causes irregular menstrual bleeding and headaches in some patients.

An alternative form of contraception that provides a safe and efficient means of abortion is the administration of the antiprogestin mifepristone (known as RU486). When RU486 administration is followed 48 hours later by prostaglandin F2α given orally, intravaginally, or by injection, abortion occurs in almost 100% of the cases. Mifepristone in combination with prostaglandin F2α can be used to induce abortion in case of fetal death or for the voluntary termination of pregnancy during the first trimester of pregnancy.

SUGGESTED READING

Boyar RM, Finkelstein J, Roffwarg H, Kapen S, Weitzman E, and Hellman L: Synchronization of augmented luteinizing hormone secretion with sleep during puberty. N Engl J Med 287:582–586, 1972.

Carr BR, and Griffin JE: Fertility control and its complications. In: *Williams Textbook of Endocrinology,* 8th ed., JD Wilson and DW Foster, eds., Saunders, Philadelphia, pp. 1007–1031, 1992.

Courie AT, Forsyth IA, and Hart IC: *Hormonal Control of Lactation,* Springer-Verlag, Berlin, 1980.

DiZegera GS, and Hodgen GD: Folliculogenesis in the primate ovarian cycle. Endocr Rev 2:27–49, 1981.

Filicori M, Santoro N, Merriam GR, and Crowley WF Jr: Characterization of the physiological pattern of episodic gonadotropin secretion throughout the human menstrual cycle. J Clin Endocrinol Metab 62:1136–1144, 1986.

Frantz AG, and Wilson JD: Endocrine disorders of the breast. In: *Williams Textbook of Endocrinology,* 8th ed., JD Wilson and DW Foster, eds., Saunders, Philadelphia, pp. 953–975, 1992.

Genuth SM: The endocrine system. In: *Physiology,* RM Berne and MN Levy, eds., Mosby, St. Louis, pp. 1069–1115, 1983.

Grumbach MM, and Styne DM: Puberty: Ontogeny, neuroendocrinology, physiology, and disorders. In: *Williams Textbook of Endocrinology 8th ed., JD Wilson and DW Foster, eds., Saunders, Philadelphia, pp 1139–1221, 1992.*

Hirshfield AN: Development of follicles in the mammalian ovary. Int Rev Cytol 124:43–101, 1991.

Hsueh AJW, Adashi EY, Jones PBC, and Welsh TH Jr: Hormonal regulation of the differentiation of cultured ovarian granulosa cells. Endocr Rev 5:76–127, 1984.

Kellie AE. Structure and nomenclature. In: *Biochemistry of Steroid Hormones,* Makin HLJ, ed., Blackwell Scientific, Oxford, UK, pp. 1–9, 1984.

Knight CH, and Pecker M: Development of the mammary gland. J Reprod Fertil 65:521–536, 1982.

Knobil E, and Neill JD: *The Physiology of Reproduction,* Vol 1 and 2, Raven Press, New York, 1994.

Komm BS, Terpening CM, Benz DJ, Graeme KA, Gallegos A, Kore M, Greene GL, O'Malley BW, and Hausler MR: Estrogen binding, receptor mRNA, and biologic response in osteoblast-like osteosarcoma cells. Science 241:81–84, 1988.

Marshall WA, and Tanner JM: Variations in the pattern of pubertal changes in girls. Arch Dis Childhood 44:291–303, 1969.

Oerter Klein K, Baron J, Colli MJ, McDonnell DP, and Cutler B, Jr: Estrogen levels in childhood determined by an ultrasensitive recombinant cell bioassay. J Clin Invest 94:2475–2480, 1994.

Peters H: The human ovary in childhood and early maturity. Eur J Obstet Gynecol Reprod Biol 9:137–144, 1979.

Ross GT, Cargille CM, Lipsett MB, Rayford PL, Marshall JR, Strott CA, and Rodbard D: Pituitary and gonadal hormones in women during spontaneous and induced ovulatory cycles. Recent Prog Horm Res 26:1–62, 1970.

Santoro N, Filicori M, and Crowley WF Jr: Hypogonadotropic disorders in men and women: Diagnosis and therapy with pulsatile gonadotropin-releasing hormone. Endocrine Rev 7:11–23, 1986.

Shoupe D, Mishell DR: Norplant: Subdermal implant system for long-term contraception. Am J Obstet Gynecol 160:1286–1292, 1989.

Sidis Y, and Horseman ND: Prolactin induces rapid p95/70 tryosine phosphorylation, and protein binding to GAS-like sites in the *anx lcp35* and c-*fos* genes. Endocrinology 134:1979, 1985.

Spitz IM, and Bardin CW: Mifepristone (RU 486)—A modulator of progestin and glucocorticoid action. N Engl J Med 404–412, 1993.

Terasawa E, Claypool LE, Gore AC, and Watanabe G: The timing of the onset of puberty in the female rhesus monkey. In: *Control of the Onset of Puberty III,* HA Delemarre-van de Waal, TM Plant, GP van Rees, and J Schoemaker, eds., Excerpta Medica, New York, pp. 123–136, 1989.

Yen SSC: Neuroendocrine regulation of the menstrual cycle. In: *Neuroendocrinology,* DT Krieger and JC Hughes, eds., Sinauer, Sunderland, MA, pp. 259–274, 1980.

10

Male Reproductive Function

JAMES E. GRIFFIN

Like the ovaries, the testes are the source of both germ cells and hormones important for reproductive function. The production of both sperm and steroid hormones is under complex feedback control by the hypothalamic–pituitary system.

STRUCTURAL ORGANIZATION OF THE TESTIS AND MALE REPRODUCTIVE TRACT

The testis consists of a network of tubules for the production and transport of sperm to the excretory ducts and a system of interstitial or Leydig cells that contain the enzymes necessary for the synthesis of androgens. The spermatogenic or seminiferous tubules are lined by a columnar epithelium composed of germ cells and Sertoli cells and surrounded by peritubular tissue made up of collagen, elastic fibers, and myofibrillar cells (Fig. 10-1). Tight junctions between Sertoli cells at a site between the spermatogonia and the primary spermatocyte form a diffusion barrier that divides the testis into two functional compartments, the basal and adluminal. The basal compartment consists of the Leydig cells surrounding the tubule, the peritubular tissue, and the outer layer of the tubule containing the spermatogonia. The adluminal compartment consists of the inner two-thirds of the tubules including primary spermatocytes and more advanced stages of spermatogenesis. The base of the Sertoli cell is adjacent to the basement membrane of the spermatogenic tubule (Fig. 10-1) with the inner portion of the cell engulfing the developing germ cells so that spermatogenesis actually takes place within a network of Sertoli cell cytoplasm. The mechanism by which spermatogonia pass through the tight junctions between Sertoli cells to begin spermatogenesis is not known. The close proximity of the Leydig cell to the Sertoli cell with its imbedded germ cells is thought to be critical for normal male reproductive function.

The seminiferous tubules empty into a network of ducts termed the rete testis. Sperm are then transported into a single duct, the epididymis. Anatomically, the epididymis can be divided into the caput, the corpus, and the cauda regions. The caput epididymidis consists of 8 to 12 ductuli efferentes, which have a larger lumen tapering to a narrower diameter at the junction of the ductus epididymidis. The diameter then remains constant through the corpus, or body, of the epididymis. In

201

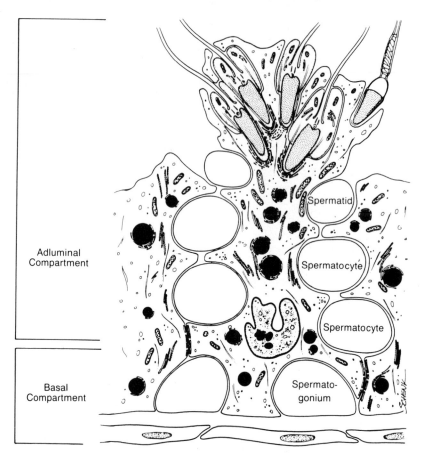

Fig. 10-1 Diagram of a Sertoli cell showing the relation between Sertoli cell cytoplasm and developing spermatocytes. (From Griffin JE, and Wilson JD: In: *Williams Textbook of Endocrinology,* 8th ed., JD Wilson and DW Foster, eds., Saunders, Philadelphia, p. 802, 1992.)

the cauda epididymis the diameter of the duct enlarges substantially and the lumen acquires an irregular shape. The entire epididymal length is 5 to 6 m. The epithelium exhibits regional differences of ciliated and nonciliated cells with evidence for a blood–epididymis barrier.

The vas deferens is a tubular structure about 30 to 35 cm in length beginning at the cauda epididymis and terminating in the ejaculatory duct near the prostate. In cross section there is a middle circular muscle layer surrounded by inner and outer longitudinal muscle layers. The pseudostratified epithelium lining the lumen of the vas deferens is composed of basal cells and three types of tall thin columnar cells, all three of which have cilia.

The normal adult prostate weights about 20 g. The gland is composed of alveoli that are lined with tall columnar secretory epithelial cells. The acini of these alveoli drain into the floor and lateral surfaces of the posterior urethra. The alveoli and the ducts draining them are embedded with a stroma of fibromuscular tissue. The seminal vesicles are paired pouches 4–5 cm long that join the ampulla of the distal vas deferens to form the ejaculatory ducts. The seminal vesicles are composed of tubular alveoli containing viscous secretions. The ejaculatory ducts pass through the prostate and terminate within the prostatic urethra.

ANDROGEN PHYSIOLOGY

Testosterone Formation

The pathway of testosterone synthesis from cholesterol and the conversion of testosterone to active androgen and estrogen metabolites is depicted in Fig. 10-2. Cholesterol, the precursor steroid, can be either synthesized *de novo* or derived from the plasma pool by receptor-mediated endocytosis of low-density lipoprotein particles. In contrast to the ovary which uses lipoprotein-derived cholesterol as its major source of substrate for steroid hormone formation (see Chapter 9), the testis appears to derive about half of the cholesterol required for testosterone formation from *de novo* synthesis. Five enzymatic steps are involved in the conversion of cholesterol to testosterone: 20,22-desmolase, 3β-hydroxysteroid dehydrogenase, 17α-hydroxylase, 17,20-lyase, and 17β-hydroxysteroid oxidoreductase (17β-HSOR). The initial reaction in the process is the side-chain cleavage of cholesterol by the 20,22-desmolase complex of enzymes in mitochondria to form pregnenolone. This is the rate-limiting reaction in testosterone synthesis and a specific cytochrome $P\text{-}450_{scc}$ under regulatory control of luteinizing hormone (LH). The remaining four enzymes in the pathway of testosterone synthesis are located in the microsomes. The 3β-hydroxysteroid dehydrogenase/isomerase (3β-HSD) enzyme complex oxidizes the A ring of the steroid to the Δ^4-3-keto configuration. Evidence indicates that the 17α-hydroxylase and 17,20-lyase activities are accomplished by a single cytochrome $P\text{-}450_{17\alpha}$. Whereas the first four enzymatic activities in the pathway of testosterone synthesis are also common to the adrenals, the 17β-hydroxysteroid oxidoreductase activity is present primarily in the testes.

Although testosterone is the major secretory product of the testis, some of the precursors in the pathway as well as small amounts of dihydrotestosterone and estradiol are also directly secreted by the testis. The major sites of formation of dihydrotestosterone and estradiol are androgen target tissues and adipose tissue, respectively (see below). Concentrations of testosterone in testicular lymph and testicular venous blood are similar; however, because of its greater flow, the major route of testosterone secretion is via the spermatic vein. About 25 μg of testosterone is present in the normal testes so that the total hormone content must turn over several hundred times each day to provide the 5–10 mg of testosterone secreted daily.

Testosterone Transport and Tissue Delivery

Testosterone circulates in the plasma largely bound to plasma proteins, primarily albumin and testosterone-binding globulin (TeBG, also called *sex hormone-binding globulin* [SHBG]). TeBG is a β-globulin composed of nonidentical subunits with

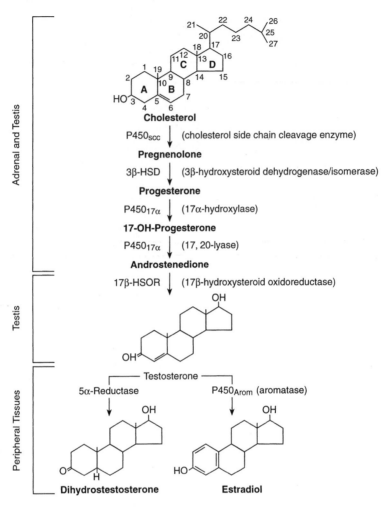

Fig. 10-2   Pathway of testosterone formation in the testis and the conversion of testosterone to active metabolites in peripheral tissues. (Redrawn from Griffin JE, and Wilson JD: Disorders of the testes. In: *Harrison's Principles of Internal Medicine*, 13th ed., KJ Isselbacher, E Braunwald, JD Wilson, JB Martin, AS Fauci, and DL Kasper eds., McGraw-Hill, New York, p. 2006–2017, 1994.)

a size of 95,000 Da containing 30% carbohydrate. In normal men only about 2% of testosterone is unbound in *in vitro* tests of peripheral blood whereas 44% is bound to TeBG and 54% is bound to albumin and other proteins. Albumin has about 1000-fold lower affinity for testosterone than does TeBG, but albumin has an approximately 1000-fold greater binding capacity than does TeBG so that the capacity × affinity product is similar. The proportion of testosterone (or estradiol) bound to the TeBG fraction in serum is directly proportional to the TeBG concentration. As discussed in Chapter 5, we now know that debinding of steroid hormones occurs in the microcirculation *in vivo* and that tissue availability of steroid hormones is greater than the "free" fraction measured *in vitro*. The amount of hormone available for entry into cells depends on the given organ as a function of capillary transit time, dissociation rate from the given binding protein, and endothelial membrane permeability. In studies of *in vivo* tissue delivery, nearly all of the albumin-bound testosterone is available for brain uptake while the TeBG-bound testosterone is not significantly transported into the brain. (Brain is used as a representative nonhepatic organ.) Although capillary transit time in liver is longer than in brain, it is still short in relation to the half-time of dissociation from TeBG so that testosterone available for transport into hepatocytes is similar to that in brain or about 40–50% of the total plasma testosterone in normal men (i.e., the free plus the albumin-bound fraction). In contrast, since estradiol dissociates more rapidly from TeBG, both albumin-bound and TeBG-bound estradiol are available for transport into liver cells.

This difference between testosterone and estradiol delivery to certain tissues from TeBG increases the significance of changes in TeBG levels. TeBG levels in men are one-third to one-half those in women, and they decrease from the prepubertal level to the lower level in normal adult men as a result of sexual maturation and the associated testosterone secretion. Decreased TeBG levels occur in hypothyroidism, and thyroid hormone excess increases TeBG levels, possibly because of increased estrogen formation. (The increased estrogen levels of normal pregnancy increase the TeBG level by 5- to 10-fold.) In men with an intact hypothalamic–pituitary–testicular axis, increases or decreases in TeBG levels do not affect tissue delivery of testosterone in the steady state because of temporary compensatory changes in testosterone formation. However, since the level of plasma estradiol is not directly regulated in men (see below), changes in TeBG levels may have a more profound effect on estradiol tissue delivery. Since TeBG binds to estradiol less avidly than testosterone or dihydrotestosterone, increases in TeBG amplify the amount of estradiol cleared by the liver relative to the amount of testosterone. For example, increases in TeBG cause decreased hepatic clearance of testosterone but have no effect on the hepatic clearance of estradiol. Thus, even in normal men, changes in TeBG levels can produce alterations in the ratios of androgens to estrogens that persist even when androgen levels themselves are not permanently altered.

Metabolism of Androgens

Testosterone and its metabolites are excreted primarily in the urine. About half of the daily turnover is recovered in the form of urinary 17-ketosteroids (androsterone and etiocholanolone) and the other half in polar metabolites (diols, triols, and conjugates). These excretory metabolites are thought to be largely inactive. A small fraction of testosterone metabolism is initially to two types of active metabolites, which in turn mediate many androgen actions. These metabolites are formed in peripheral tissues, and in this regard testosterone serves as a prohormone (Fig. 10-2). The first category of active metabolites is 5α-reduced steroids, primarily dihydrotestosterone, that mediate many androgen actions in target tissues. The second category of active testosterone metabolites is estrogens formed by the aromatization of androgens in a number of peripheral tissues. Estrogens in some instances act in concert with androgen to affect physiological processes, but they may also exert independent effects or oppose the actions of androgens. Each of these reactions to form active metabolites (5α-reduction and aromatization) is physiologically irreversible; furthermore, 5α-reduced androgens cannot be converted to estrogens. Thus, physiological actions of testosterone are the result of the combined effects of testosterone itself plus those of estrogen and the active androgen metabolites. In some instances these metabolites are active only in the tissues in which they are formed, whereas in other cases the 5α-reduced and estrogen metabolites reenter the circulation to act in other tissues.

5α-Reductase accepts a number of Δ^4-3-ketosteroids as substrate and requires reduced nicotinamide-adenine dinucleotide phosphate (NADPH) as cofactor. There are two steroid 5α-reductase isozymes. Enzyme 1 is encoded by a gene on chromosome 5, and enzyme 2 is encoded by a gene on chromosome 2. 5α-Reductase has a distinct tissue localization with the most activity found in the accessory organs of reproduction, liver, and skin. In skin the highest activity is in the genital skin, but hair follicles from all anatomical sites contain the enzyme. 5α-Reductase 2 has a lower Km for testosterone and is primarily localized in androgen target tissues. In normal men about 6–8% of the total testosterone produced is metabolized to dihydrotestosterone by 5α-reductase. Testosterone can also be converted to 5β-reduced metabolites in the liver; however, these 5β-reduced metabolites do not have androgenic activity. A key observation in the history of androgen physiology was the discovery in 1968 that 5α-dihydrotestosterone was the main intracellular androgen and the predominant androgen concentrated in the nucleus of the rat prostate. These findings, together with the observation that dihydrotestosterone is about twice as potent as testosterone in most bioassay systems, indicated that dihydrotestosterone is a major cellular mediator of androgen action. The importance of dihydrotestosterone in normal androgen physiology has been confirmed by studies of human mutations in which the steroid 5α-reductase 2 enzyme responsible for dihydrotestosterone formation is defective (see Chapter 8).

The aromatization of androgens to estrogens in testes and in extraglandular tissues of men is accomplished by the same enzyme complex present in placenta and ovary. Estrogen formation involves sequential hydroxylation, oxidation, and removal of the C-19 carbon and aromatization of the A ring of the steroid. Three moles of NADPH and 3 moles of oxygen are required to convert each mole of androstenedione or testosterone to estrone or estradiol, respectively. The oxidations are of the mixed function type (see Chapter 9) involving a specific cytochrome P-450. The enzymes involved appear to be bound in a microsomal complex that includes NADPH–cytochrome P-450 reductase as well as cytochrome. Of the 45 μg of estradiol produced per day on average in normal young men, only about 10–15% comes from direct secretion by the testes while the remaining 85–90% is derived from peripheral aromatization of secreted androstenedione (via estrone) and testosterone. Estrogen formation in the testes is enhanced by increased levels of LH or human chorionic gonadotropin (hCG). The rate of overall aromatase in nongonadal tissue does not appear to be influenced by gonadotropins but rises with advancing age.

Androgen Action

The current concepts of androgen action in target cells are diagrammed in Fig. 10-3. Among the major functions of androgen are regulation of gonadotropin secretion by the hypothalamic–pituitary system, stimulation of spermatogenesis, formation of the

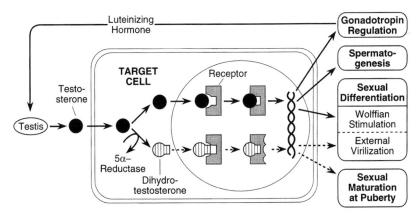

Fig. 10-3   Normal androgen physiology. Testosterone, secreted by the testis, binds to the androgen receptor in a target cell, either directly or after conversion to dihydrotestosterone. Dihydrotestosterone binds more tightly than testosterone, and the complex of dihydrotestosterone and the androgen receptor can bind more efficiently to the chromatin. The major actions of androgens, shown on the right, are mediated by testosterone (solid lines) or by dihydrotestosterone (broken lines). (Redrawn from Griffin JE: Androgen resistance—the clinical and molecular spectrum. N Engl J Med 326:612, 1992).

male phenotype during sexual differentiation, and promotion of sexual maturation at puberty. Inside the cell, testosterone can be converted to dihydrotestosterone by 5α-reductase. The two hormones then bind to the same high-affinity androgen–receptor protein. The androgen receptor is predominantly in the nucleus in the unbound state. The binding of androgen to the hormone-binding domain of the receptor to form the hormone–receptor complexes is a prerequisite to the transformation reaction in which the hormone–receptor complexes acquire increased affinity for nuclear components. The DNA-binding domains of the receptors then bind to androgen response elements (AREs) in the promotor region of target genes activating transcription (see Chapter 3). Estradiol, as discussed above, may be either secreted directly by the testis or formed in peripheral tissues. The mechanisms by which estrogens augment or block androgen action are not fully understood.

Based on studies in humans and animals, it appears that the testosterone–receptor complex is responsible for gonadotropin regulation, stimulation of spermatogenesis, and the virilization of the wolffian ducts during sexual differentiation, whereas the dihydrotestosterone–receptor complex is responsible for external virilization during embryogenesis and for most male secondary sexual characteristics of the adult (see below). Androgen receptor concentrations are highest in the accessory organs of male reproduction that depend on androgens for their growth. Other tissues such as skeletal muscle, liver, and heart have smaller amounts of the receptor. In the testis, androgen receptors are present both in isolated Sertoli cells and in interstitial cells. Whether the presence of androgen receptors identifies a tissue as androgen responsive is not clear, although androgen receptors are present in greater numbers in tissues known to respond to androgen. Single gene mutations in humans and animals (see Chapter 8) indicate that a single androgen receptor binds both testosterone and dihydrotestosterone and that the receptor protein is coded by a locus on the X chromosome. Dihydrotestosterone formation appears to be required for normal androgen action even though there is only a single receptor. In the case of the human androgen receptor, a partial explanation for this requirement of dihydrotestosterone may be the severalfold greater affinity of the receptor for dihydrotestosterone than for testosterone. Dihydrotestosterone may also facilitate transformation of the receptor for transactivation more efficiently (Fig. 10-3). Thus, a difference in the interaction of the two hormones with the androgen receptor may serve as an amplifying mechanism for androgen action in certain target tissues.

SPERMATOGENESIS, SEMINAL FLUID FORMATION, AND CAPACITATION

Stages of Spermatogenesis

After migration of the germ cells to the gonadal ridges during embryogenesis (see Chapter 8), the total number of germ cells is approximately 3×10^5 per gonad. This number increases to about 6×10^6 spermatogonia per testis by the time of puberty. With sexual maturation, sperm production of about 2×10^8 sperm per day develops. As depicted in Fig. 10-4, each spermatogonium undergoing differentiation gives rise to 16 primary spermatocytes, each of which enters meiosis and gives rise to four spermatids and ultimately four spermatozoa. Thus 64 spermatocytes can develop from each spermatogonium. It appears that contiguous groups of spermatogonia

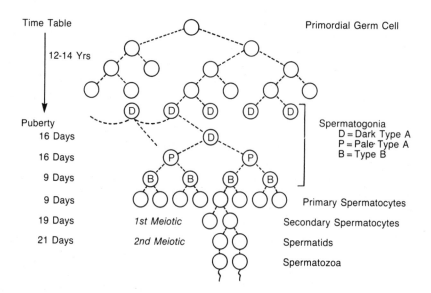

Fig. 10-4 Cell divisions during spermatogenesis. (From Griffin JE, and Wilson JD: In: *Williams Textbook of Endocrinology*, 8th ed., JD Wilson and DW Foster, eds., Saunders, Philadelphia, p. 810, 1992.)

begin the differentiation process simultaneously, resulting in certain typical cellular associations in seminiferous tubules.

The conversion of the round spermatid to the mature spermatogonia requires the reorganization of nucleus and cytoplasm and development of a flagellum. The nucleus comes to occupy an eccentric position near what will become the head of the spermatozoon and is covered by an acrosomal cap. The cilial structure that serves as the core of the sperm tail consists of nine outer fibers and two inner fibers. The mitochondria form a helix around this cilia in the middle piece of the spermatozoon, and most of the cytoplasm is lost. The differentiation of the spermatocyte to a motile sperm takes about 70 days, and the transport of the sperm through the epididymis to the ejaculatory ducts takes about 14 days. Sperm maturation occurs during passage through the epididymis, as evidenced by the development of capacity for sustained motility. The mechanism of sperm motility is believed to involve sliding action of microtubules (fibers) contained in the axial structure of the tail. The microtubules are attached to each other by dynein arms that contain the protein dynein, an ATPase. The energy for the process is derived from hydrolysis of ATP generated in the mitochondrial sheath.

Regulation of Spermatogenesis

Normal spermatogenesis requires follicle-stimulating hormone (FSH), testosterone, and normal Sertoli cell function. FSH binds to the surface of Sertoli cells, stimulating adenylyl cyclase and thus activating protein kinase. Testosterone interacts with androgen receptors in Sertoli cells to activate specific genes necessary for the differentiation process. Testosterone may also act indirectly by stimulating peritubular cells to produce peritubular factors that modulate Sertoli cell function (P-Mod-S). These factors are yet to be purified proteins. Following removal of the pituitary gland in the adult, no spermatocytes are formed. Spermatogenesis can be restored by treatment with the combination of hCG plus human menopausal gonadotropin (hMG), which contains FSH, and once restored, spermatogenesis can be maintained by hCG alone. This might seem to imply that FSH is required only for initiation of spermatogenesis but not its maintenance. But in studies of normal men in whom FSH is selectively suppressed by the sequential administration of exogenous testosterone followed by hCG, FSH administration is required for quantitatively normal spermatogenesis. Thus, it is likely that both FSH and luteinizing hormone (LH) play a continuing role in human spermatogenesis.

Based on animal studies, it appears that at least three other factors in addition to FSH and testosterone play an essential role in modulating spermatogenesis: vitamin-A, c-fos, and stem cell factor. Rats fed on a vitamin A–deficient diet develop defective spermatogenesis with germ cell arrest. The nuclear protooncogene c-fos was found to be necessary for spermatogenesis by studies of mice homozygous for germ-line mutations at the locus. Stem cell factor is the ligand for c-kit. Stem cell factor is made in Sertoli cells, and the c-kit receptor is present in germ cells. Mutations in either stem cell factor or c-kit genes result in abnormal spermatogenesis.

A number of other factors have been studied in vitro and postulated to have paracrine/autocrine roles in modulating spermatogenesis. These include transferrin, insulin-like growth factor (IGF)-I, β-fibroblast growth factor (FGF), nerve growth factor (NGF), interleukin (IL)-1, opioids, transforming growth factors (TGF), growth hormone-releasing hormone (GHRH), inhibin/activin (see below), and others.

Seminal Fluid Formation

During the approximately 12 days of transit through the epididymis, sperm undergo maturation with development of the capacity for sustained motility, modification of the structural state of the nuclear chromatin and tail organelles, and loss of the remnant of spermatid cytoplasm. The epididymis also serves as reservoir for sperm. After transport of the sperm and secretory products of the testis and epididymis through the vas deferens, the fluid reaching the ejaculatory ducts is enriched by the secretions from the seminal vesicles. The seminal vesicles are the source of seminal fluid fructose and prostaglandins and contribute about 60% of the total volume of the seminal fluid. Approximately 20% of the total seminal fluid volume is derived from prostatic secretions added to the semen when the ejaculatory ducts terminate in the prostatic urethra. The prostatic fluid is the source of seminal fluid spermine, citric acid, zinc, and acid phosphatase.

Capacitation

Fertilization normally takes place within the fallopian tube. Spermatozoa usually require a period in the female reproductive tract before they can fertilize. This functional change, termed *capacitation*, is thought to have at least two components: (1) increased rate of flagellar beat and acceleration of sperm movement, and (2) an acrosome reaction in sperm that allows the underlying plasma membrane of the sperm to fuse with the ovum. The time required for optimal capacitation of normal sperm may vary from 2 to more than 6 hours. Whether capacitation is an absolute requirement in the human or serves only to enhance fertilizing capabilities has not been established.

The elements of the capacitation reaction that promote motility may include a change in the intracellular concentration or metabolism of calcium or cAMP. The acrosome reaction appears to be more complex but also involves calcium. Neither the fallopian tube nor the egg itself appears to be essential for this process. The acrosome reaction involves fragmentation and loss of the acrosome with release of a variety of hydrolytic enzymes and proteases allowing the sperm to penetrate and fuse with the ovum. Understanding of the mechanism of sperm penetration is largely based on studies of fertilization of human eggs *in vitro*. Ovulated eggs are surrounded by layers of cumulus cells embedded in a matrix of hyaluronic acid. The mechanism by which spermatozoa tunnel through the cumulus is not known. Possibly, hyaluronidase is released by the degenerating acrosome, and the mechanical agitation of the flagellum may disperse cumulus cells (see Chapter 11).

THE HYPOTHALAMIC–PITUITARY–TESTICULAR AXIS

LHRH and Gonadotropins and Their Actions

As discussed in detail in Chapter 6, peptidergic neurons in the hypothalamus secrete luteinizing hormone-releasing hormone (LHRH, also called gonadotropin-releasing hormone [GnRH]). LHRH is transported to the pituitary by a portal vascular system and interacts with cell surface receptors on pituitary gonadotrophs to stimulate the release of LH and FSH (Fig. 10-5). LHRH is a decapeptide that is widely distributed in the central nervous system and may also be present in other tissues. However, a physiological role for LHRH in sites other than the pituitary has not been established. The amount of LH and FSH released in response to LHRH depends on age and hormonal status. In a primate model the sensitivity of the gonadotrophs to LHRH is high during the first few months of life and then declines and remains low until the onset of puberty when it increases and attains an adult level of response. The secretion of FSH in response to LHRH is relatively greater than that of LH before puberty.

LH and FSH are secreted by the same basophilic cells in the pituitary. Like thyrotropin (TSH) and hCG, LH and FSH are glycoproteins composed of two polypeptide chains designated α and β. The α-subunit of each of the four hormones is identical, and distinct immunological and functional characteristics are determined by unique β-subunits. The turnover of FSH is slower than that of LH. LH interacts with specific high-affinity cell surface receptors on Leydig cells and utilizes a cAMP

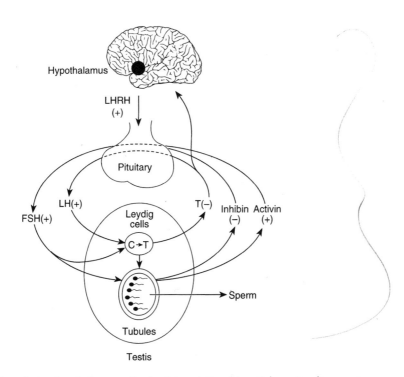

Fig. 10-5 Hypothalamic–pituitary–testicular interrelationships. Schematic diagram to indicate feedback relationship of testosterone, inhibin, and activin produced by testes on gonadotropin secretion by the hypothalamic–pituitary complex, and site of action of FSH and LH on testis. C, cholesterol; T, testosterone; FSH, follicle-stimulating hormone; LH, luteinizing hormone; LHRH, LH-releasing hormone. (From Griffin JE, and Wilson JD: Disorders of the Testis. In: *Harrison's Principle of Internal Medicine*, 13th ed., KJ Isselbacher, E Braunwald, JD Wilson, JB Martin, AS Fauci, and DL Kasper, eds., Mc-Graw-Hill, New York, pp. 2006–2017, 1994.)

second messenger (see Chapter 3). Activation of Leydig cell protein kinase via un specified intermediates eventually results in stimulation of testosterone formation by enhancing side-chain cleavage of cholesterol (see above). The rate of testosterone synthesis correlates more closely with the degree of occupancy of the regulatory subunits of the protein kinase by cAMP than with the total amount of cAMP in the cells. Whether the LH–receptor complex undergoes internalization for degradation or receptor recycling is unclear.

In the intact testis and in cultured Leydig cells, the receptors for LH decrease in number following administration of LH or hCG. This down-regulation of LH receptors is maximal at 24 hours and is followed by return to control levels several days later. It is associated with a decreased responsiveness to subsequent LH ad ministration, but this desensitization is not solely due to decreased receptor number Rather, the diminished steroidogenic response appears to be due to inhibition of some postreceptor event since cAMP is ineffective in overcoming the desen sitization. In fact, the desensitization can be produced by cAMP stimulation of ste roidogenesis under conditions that do not alter LH receptor number. The dimin ished response of the Leydig cell to LH that follows LH administration appears to be a component of an intratesticular control system for regulating testosterone pro duction.

As mentioned above the second messenger for FSH action following its binding to receptors on the basal aspects of Sertoli cells is also cAMP. FSH is known to stimulate the rate of synthesis of androgen-binding protein (ABP) and the aromatase enzyme complex in Sertoli cells. FSH may also play a role in steroidogenesis by influencing Leydig cell maturation and modifying effects of LH on steroidogenesis by stimulating paracrine–autocrine factors. Based on *in vitro* studies, IGF-I, TGF β, epidermal growth factor (EGF), TGF-α, FGF, inhibin/activin, cytokines, GHRH corticotropin-releasing hormone (CRH), arginine vasopressin (AVP), endothelin and other agents have been postulated to have effects on Leydig cells. Their phys iological significance is unclear.

Regulation of Secretion of LHRH and Gonadotropins

The secretion of LHRH is episodic resulting in the intermittent secretion of both immunoreactive and bioactive LH in 8–14 pulses/24 hours in adult men. Pulsatile secretion of FSH also occurs but is of smaller amplitude, in part because of the longer half-life of FSH in the circulation. The rate of LH secretion is controlled by the action of sex steroids on the hypothalamus and pituitary. The control of LH in men operates primarily by negative feedback since normal levels of gonadal steroids inhibit secretion (Fig. 10-5). Both testosterone and estradiol can inhibit LH secretion Testosterone and its metabolites act on the central nervous system to slow the hy pothalamic pulse generator and consequently decrease the frequency of LH pulsatile release. Testosterone can be converted to estradiol in the brain and pituitary, but the two hormones are thought to act independently. The fact that the testosterone me tabolite dihydrotestosterone, which cannot be converted to estrogen, exerts negative feedback on LH secretion suggests that testosterone does not require conversion to estradiol to inhibit LH secretion. Testosterone also appears to have a negative feed back on LH secretion at the pituitary level, since moderate elevations of plasma testosterone levels result in diminished LH response to acute LHRH stimulation in the presence of normal levels of plasma estradiol. In normal men endogenous estrogens also act to restrain tonically the hypothalamic release of LHRH.

The negative feedback inhibition of testicular hormones on FSH secretion is less well understood. Serum FSH levels increase selectively in proportion to the loss of germinal elements in the testis. A nonsteroidal inhibitor of FSH is present in the testis, semen, and cultured Sertoli cells, and a similar material is present in ovarian follicular fluid. The inhibitor is termed *inhibin*. Inhibin is a heterodimeric protein consisting of α- and β-polypeptide subunits connected by disulfide bonds (see Chapter 9). Interestingly, significant structural homology of the β-subunit of inhibin to a subunit of human TGF-β has been found. As discussed in Chapter 12, transforming growth factors may not only affect tumor growth in an autocrine manner but also participate in the regulation of nonneoplastic processes. There are two different forms of the β-subunit of inhibin $β_A$ and $β_B$. Thus the α-subunit can combine with either β-subunit to form inhibin A (α-$β_A$) or inhibin B (α-$β_B$). An additional hormone identified during the purification of inhibin, activin, is formed by the dimerization of the inhibin β-subunits. Activin A ($β_A$-$β_A$), activin B ($β_B$-$β_B$), and activin AB ($β_A$-$β_B$) stimulate FSH secretion from anterior pituitary cells in culture. Sertoli cells are the site of inhibin production, and inhibin production is enhanced by FSH and androgen. Studies of pituitary cells cultured with purified bovine inhibin indicate that inhibin acts to block LHRH-stimulated FSH release. Whether inhibin also has effects at the hypothalamic level are not known. Testosterone and estradiol also have direct effects on FSH secretion, and a decrease in the pulse frequency of LHRH release may produce a selective increase in FSH. A primate model has been useful in defining the relative importance of inhibin and gonadal steroids in physiological feedback control of FSH secretion. When episodic gonadotropin secretion was maintained by chronic, intermittent LHRH infusion, neither circulating testosterone nor estradiol could account for the testicular inhibition of FSH, implying by exclusion that inhibin must be the gonadal hormone of greatest importance for feedback control at the level of the pituitary.

DEVELOPMENT AND MAINTENANCE OF REPRODUCTIVE FUNCTION

The development and maintenance of reproductive function in the male can be correlated with alterations in the plasma testosterone. During embryogenesis testosterone levels rise in the late first trimester in association with male sexual differentiation (see Chapter 8). In the first 6 months of life, the testosterone level rises to about half of adult levels, falling back to low levels before age 1 year. This neonatal surge in testosterone secretion results from a rise in plasma LH levels, but the function of the temporary increase in testosterone secretion is not well understood. Temporary inhibition of the pituitary–testicular axis in the neonatal primate is associated with impaired testicular function at puberty. The concentration of testosterone remains low in the childhood years until the onset of puberty when the level rises to adult levels by about age 17. Plasma testosterone concentration remains more or less constant in the adult until it declines somewhat during the later decades of life.

Puberty

Plasma gonadotropins and testosterone are low during the prepubertal years. In boys around age 6 or 7 there is an increased secretion of adrenal androgens termed *adrenarche*. This increased secretion of dehydroepiandrosterone and androstenedione is probably under the control of corticotropin (ACTH) and is in part responsible for the prepubertal growth spurt and the initial development of axillary and pubic hair. These adrenal androgens are thought to act via the androgen receptor after conversion to testosterone. Total plasma testosterone increases only slightly during this time, and plasma LH levels are low but show some low-amplitude pulsations. However, the low levels of gonadotropins in prepubertal boys are under feedback control by the prepubertal testis and low levels of plasma testosterone (5–20 ng/dl), since castration elevates both gonadotropins. The degree of elevation of gonadotropins in the absence of gonadal feedback is less during the late childhood years than at other times, suggesting that more than exquisite sensitivity of the hypothalamic–pituitary system to feedback inhibition is involved in the low gonadotropin levels of normal prepubertal boys (see also Chapter 9).

The pituitary hormonal changes associated with the onset of puberty are sleep-associated surges in LH secretion. Later, as puberty advances, the increased plasma gonadotropin levels are present throughout the day and are accompanied by increased plasma levels of testosterone. The increase in gonadotropin secretion is thought to result from both increased LHRH secretion and enhanced responsiveness of the pituitary to LHRH. Plasma levels of bioactive LH increase even more than those of the immunoreactive hormone. Thus, it appears that with sexual maturation the hypothalamic-pituitary system becomes less sensitive to the feedback inhibition of testosterone on gonadotropin secretion.

The changes in the testes at puberty include the maturation of Leydig cells and the initiation of spermatogenesis. The anatomical changes characteristic of puberty are a result of testosterone secretion. Usually the first sign of male puberty is testis enlargement with some reddening and wrinkling of the scrotal skin. Pubic hair growth begins first at the base of the penis. On average, this initiation of pubertal development occurs between ages 11 and 12, but occasionally begins as early as age 9 or as late as age 13 or 14. About a year after the onset of testis growth, the penis begins to increase in size. The prostate, seminal vesicles, and epididymis increase in size over a period of several years. Characteristic hair growth includes development of mustache and beard, regression of the scalp line, appearance of body and extremity hair, and extension of pubic hair upward into the upper pubic triangle. The larynx enlarges and the vocal cords become thickened resulting in a lower pitch of the voice. There is an enhanced rate of linear growth resulting in a height spurt somewhat later in puberty to a rate of about 3 inches per year. At this time the androgen-sensitive muscles of the pectoral region and the shoulder also increase in their characteristic male pattern, the hematocrit increases, and libido and sexual potency develop. These various maturation processes take place over about 4 years and reach some normal limit for the individual based on genetic and nutritional factors. Linear growth is halted by fusion of the epiphyses which requires normal estrogen formation or action. Administration of excess exogenous androgens has little effect on the parameters of sexual development once puberty is completed.

Adulthood

At the completion of puberty, usually between ages 16 and 18, plasma testosterone levels are in the adult range of 300–1000 ng/dl, sperm production is normal, and plasma gonadotropins are in the 5–20 mIU/ml range. Most anatomic changes are completed at this time except for androgen-mediated hair growth, which is usually not maximal until the mid- to late 20s. Some effects of androgens, such as the development of the larynx, are permanent and do not reverse if androgen production decreases. Other effects of androgen, such as stimulation of erythropoietin, reverse with androgen deprivation. Beard growth slows following postpubertal androgen deficiency but usually does not completely cease. Two results of postpubertal androgen deficiency are clear-cut. There is a negative nitrogen balance probably resulting from regression of accessory organs of reproduction and to a certain extent skeletal muscle. And complete androgen deficiency is followed by a progressive decline in male sexual drive so that most such men are unable to have intercourse after a few years.

Old Age

Men have a gradual decrease in free plasma testosterone levels and an increase in TeBG levels, beginning around age 40. In elderly men there is a decrease in total plasma testosterone levels, a decrease in the ratio of androgen to estrogen, and some elevation in plasma LH and FSH. Sperm production decreases about 30% between the ages of 50 and 80. It is likely that any decline in sexual activity with age is the result of nonendocrine factors. Healthy elderly men commonly maintain a healthy sex life and reproductive capacity.

CLINICAL ASSESSMENT OF REPRODUCTIVE FUNCTION

History and Physical Exam

The assessment of androgen status should include inquiry about defects of the urogenital tract at birth, sexual maturation at puberty, rate of beard growth, and current libido, sexual function, strength, and energy. If Leydig cell failure occurs before puberty, sexual maturation will not occur. In contrast, the detection of Leydig cell failure beginning after puberty requires a high index of suspicion of androgen deficiency and, usually, laboratory assessment. Many men without hormonal abnormalities complain of decreased sexual function. In addition, even when Leydig cell function is impaired, the rate of beard growth may not decrease for many months or even years.

The prepubertal testis measures about 2 cm in length and increases in size to a range of 3.5–5.5 cm in length in the normal adult. When damage to the seminiferous tubules occurs before puberty, the testes are small and firm. Following postpubertal damage the testes are characteristically small and soft. However, considerable damage must occur before the overall size is decreased below the lower limits of normal.

Breast enlargement is the most consistent feature of feminizing states in men and may be an early sign of androgen deficiency.

Gonadotropins and Testosterone

Because of the pulsatile secretion of LH, the consequent pulsatile secretion of testosterone and the need to interpret the LH in light of the testosterone, it is usually appropriate to measure gonadotropins and testosterone on a pool formed by combining equal quantities of blood obtained from three or four samples at 15- to 20-minute intervals. The normal plasma immunoreactive LH and FSH values must be established for each laboratory based on the antibody and standard used. The normal range for plasma testosterone in adult men is 300–1000 ng/dl. Free testosterone concentrations can be estimated by equilibrium dialysis, but as discussed above, a more accurate assessment of available testosterone *in vivo* can be obtained by measuring the non-TeBG-bound fraction (or free plus albumin-bound fraction). TeBG concentrations can also be measured directly.

Leydig cell function may be assessed before puberty by measuring the response to hCG administration. Single or multiple injections of hCG may be given. The testosterone response increases at the start of puberty, peaking in early puberty.

In certain circumstances the change in plasma LH after the administration of LHRH is measured to assess the functional integrity of the hypothalamic–pituitary–Leydig cell axis. When 100 μg of LHRH is given subcutaneously or intravenously to normal men, there is, on average, a 4- to 5-fold increase in LH, with the peak level at 30 minutes. However, the range of response is broad, some normal men having less than a doubling of LH levels. In general, the peak LH following a single LHRH injection correlates with basal levels. In patients with primary testicular failure, measurement of basal LH is usually sufficient, and measurement of LHRH response adds little to the diagnosis. Men who have either pituitary disease or hypothalamic disease may have either a normal or abnormal LH response to an acute dose of LHRH. Therefore, a normal response is of no diagnostic value, whereas a subnormal response is of value in determining that an abnormality exists, even though the site is not determined. Men with secondary hypogonadism and a subnormal response to an acute infusion of LHRH who develop a normal response to an acute dose after daily infusion of LHRH for a week usually have a hypothalamic cause of the hypogonadism (see Chapter 5 and Fig. 5-4).

Seminal Fluid Examination

Routine evaluation of seminal fluid is largely dependent on tests that do not assess the functional capacity of sperm. Although methods to measure sperm penetration of bovine cervical mucus and zona-free hamster ova have been developed, they are not sufficiently standardized to permit general use. Seminal fluid should be obtained after masturbation into a clean glass or plastic container. The volume of the normal ejaculate is 2–6 ml. The specimen should be analyzed within an hour. Estimation of motility is made by examining a drop of undiluted seminal fluid and recording the percentage of motile forms. Normally 60% or more of the sperm should be motile with forward progression. Sperm density may be determined by diluting seminal fluid 20-fold with an appropriate solution and estimating density in a hemocytometer or with the aid of an electronic particle counter. The normal value is usually considered to be greater than 20 million/ml with total sperm per ejaculate of greater than 60 million.

After the first 2 days, the total sperm in the ejaculate per the number of days since the last ejaculation is relatively constant in normal men who ejaculate daily. Random sampling of sperm density in men is complicated by variable extragonadal sperm reserves and by effects of toxic factors such as hot baths, acute febrile illnesses, and unknown medications. The net result is that it is difficult to define the minimally adequate ejaculate. Ordinarily, three ejaculates are required to establish inadequacy of sperm number or cytology.

Seminal cytology is a useful index of fertility, and the seminal fluid smear is prepared in the same way as a blood smear but with special stains. Some abnormal spermatozoa are present in all semen. The best correlations between histological abnormalities and infertility are seen when a single anomaly is found in a large percentage of the sample. It is generally believed that 60% or more of the spermatozoa should have a normal morphology. When it is available the details of sperm structure can be studied by electron microscopy. Such studies are useful in identifying the abnormalities in immotile sperm.

DERANGED REPRODUCTIVE FUNCTION

Abnormalities of male reproductive function may occur at any stage of life from embryogenesis through old age. Disorders of testicular function during embryogenesis result in abnormalities of sexual differentiation (see Chapter 8). This section will consider examples of deranged reproductive function occurring at puberty and in adulthood.

Deficient Puberty: Isolated Gonadotropin Deficiency

Deranged reproductive function related to pubertal events in boys more commonly involves delayed or deficient puberty rather than sexual precocity. This is not so in girls (see Chapter 9). It is often difficult to decide whether puberty is merely being delayed or is actually incomplete or deficient. A common cause of deficient puberty is isolated gonadotropin deficiency, which occurs in both sporadic and familial forms. It is second only to Klinefelter's syndrome (see Chapter 8) as a cause of hypogonadism in men. The disorder was originally described as a familial syndrome associated with an impaired sense of smell (Kallmann's syndrome). The defect in most affected individuals is detected because of a failure to undergo puberty. A subset of patients, particularly familial cases, have associated congenital defects commonly involving midline facial and head structure. There is no sense of smell because of defective development of the olfactory bulbs. Less severely affected individuals may have a negative family history (half of all subjects with the disorder) and only partial defects in FSH and/or LH secretion. Such individuals may undergo partial sexual maturation, develop partial testicular enlargement, and not come to medical attention until adulthood.

The underlying defect is at the hypothalamic level: absence of or inadequate release of LHRH. A defect in a neural adhesion molecule may be responsible for abnormal migration of neurons leading to both the impaired sense of smell and defective LHRH release. Randomly obtained gonadotropin levels may be undetectable, low, or apparently normal in the face of a low testosterone level. The fact that the underlying defect is at the hypothalamic level was established when LHRH became available; following short-term administration of LHRH, plasma LH and FSH increase in about half of subjects. After repetitive infusion of LHRH for 5 days or longer, plasma gonadotropins rise to the normal range in virtually all affected patients but not in individuals with panhypopituitarism.

One group of investigators has attempted to define some of the clinical heterogeneity in patients with isolated gonadotropin deficiency by studying the patterns of LHRH secretion as inferred by frequent measurements of plasma LH. Men with isolated gonadotropin deficiency displayed several abnormal patterns of LHRH secretion. Whereas the majority of men had an apulsatile pattern of LH release, other individuals demonstrated episodic LH secretion only during sleep (as in normal pubertal subjects). In these latter subjects a history of early but arrested pubertal development was often obtained, and testicular size was larger than in those subjects with an apulsatile pattern of LH release. In another family with isolated gonadotropin deficiency, affected men consistently demonstrated a diminished LH pulse amplitude compared to normal men or their unaffected siblings. In yet another family, a diminished frequency of LHRH pulsations was associated with a variably low to low normal testosterone in two affected subjects but not in their normal brother. These findings imply that maintenance of a physiological amplitude and frequency of endogenous LHRH secretion appear to be essential for normal reproductive function.

Adult Reproductive Failure: Infertility

Approximately 15% of couples attempting their first pregnancy are unsuccessful. Couples who have been unable to achieve a pregnancy after 1 year of unprotected intercourse are usually considered to have primary infertility. In approximately one-third of such cases a significant abnormality impairing fertility is thought to be in the man alone, and in an additional fifth of couples both the man and the woman have some abnormality. Thus, in about half of infertile couples some abnormality in the male is in part responsible for the failure to conceive. Any man who has been unsuccessful in achieving fertility with a regularly menstruating woman after 1 year should be considered potentially subfertile.

Infertility in men may be an isolated problem with normal androgen production or one manifestation of an abnormality of testicular function that involves both Leydig cells and seminiferous tubules. In contrast, since sperm production depends on normal Leydig cell function, disorders that primarily affect androgen formation or action are usually associated with infertility. Therefore, it is essential to exclude the presence of subtle Leydig cell dysfunction in every man with infertility.

Adult abnormalities of testicular function can be due to hypothalamic–pituitary defects, testicular disorders, or abnormalities in sperm transport. The hypothalamic–pituitary defects include not only isolated gonadotropin deficiency mentioned above, but also a variety of destructive and infiltrative lesions in the pituitary, and functional impairments of gonadotropin secretion resulting from excess prolactin, glucocorticoids, or endogenous or exogenous androgens.

Testicular disorders known to be associated with infertility can be grouped into several categories. Developmental and structural defects include Klinefelter's syndrome and its variants (see Chapter 8), abnormal testicular descent, and a structural defect of sperm tails leading to immotile sperm. Acquired testicular defects include viral and other infections, trauma, radiation, drugs including alcohol and marijuana, and self-directed immune responses affecting the entire testis or primarily sperm. Systemic diseases may be associated with testicular abnormalities as in chronic liver disease, renal failure, and certain neurological disease. Finally, androgen resistance due to a receptor defect may be manifested by underandrogenization with infertility or infertility alone in men with normal development of the external genitalia.

Disorders of sperm transport lead to infertility associated with normal Leydig cell function. The abnormality may be unilateral or bilateral, congenital or acquired. In men with unilateral obstruction infertility may be due to antisperm antibodies. Obstruction at the level of the epididymis may occur in association with chronic infections of the paranasal sinuses and lungs. Or the vas deferens may be congenitally absent in association with an absence of the seminal vesicles.

Unfortunately, the above elaboration of the known causes and of conditions associated with male infertility does not account for the problem in the majority of men. When a large series of consecutive patients seen by referral groups is evaluated, more than half of such men are classified as having ''idiopathic'' infertility. Since as many as 1 in 20 men are infertile, this represents a major area for improvement in medical knowledge.

FERTILITY CONTROL

Vasectomy

Other than the condom, the only proven effective means of fertility control in men is surgical interruption of the vas deferens (vasectomy). In the United States about 50% of couples adopting surgical sterilization for fertility control choose vasectomy, and about 1 million vasectomy procedures are performed each year. Bilateral partial vasectomy is a relatively simple operative procedure usually performed with local anesthesia. Vasectomy is considered successful when sperm cannot be demonstrated in the semen on two consecutive specimens. On average, this takes 24 ejaculations following vasectomy. In less than 1% of men, the procedures fails due to recanalization, which occurs at a median time of 6 months. The changes in the testis following occlusion of the vas have been studied in animals. In the primate, spermatogenesis continues following vasectomy with the sperm being resorbed or stored in distended ducts and cysts. Testicular volume does not change after vasectomy in men. There appear to be no significant changes in peripheral hormone levels following vasectomy as assessed by measuring plasma testosterone and gonadotropins. Moreover, Leydig cell reserve, as assessed by response to hCG, is normal several years later. Although immune complex-mediated atherosclerosis has been described in vasectomized monkeys, no increased prevalence of atherosclerotic cardiovascular disease has been found in vasectomized men. Vasectomy has no deleterious effects on potency or sexual performance. Antisperm antibodies do develop following vasectomy, and the presence of these antibodies may limit the success of attempted vasectomy reversal. Thus, vasectomy reversal (or vasovasostomy) has a success rate for appearance of sperm in the ejaculate of 80–90%, but the associated pregnancy rate is only 30–40%.

Search for a Male Contraceptive

Whereas the oral contraceptive is effective and relatively safe in controlling ovulation and fertility in women (see Chapter 9), there is no readily reversible and effective pharmacological contraceptive for men. One explanation for this delay in pharmacological contraception for men is that it intuitively seems easier to prevent the production of only one ovum per month in the female than to prevent the production of millions of sperm each day in the male. In addition, sperm migration, capacitation, fertilization and implantation all take place in women. Thus, even some measures designed to affect the sperm can be used in women but not in men.

In searching for a male contraceptive, pharmacological and other methods have been directed at inhibiting hypothalamic–pituitary function, directly inhibiting spermatogenesis, or inhibiting epididymal function. It was hoped that the injection of long-acting testosterone esters might suppress gonadotropin secretion, resulting in defective spermatogenesis and at the same time replace endogenous testosterone. However, this proved successful in completely eliminating sperm from the ejaculate in a little over half of men. The inhibition of spermatogenesis in men who responded was reversible. Gonadotropin levels are decreased in men given injections of long-acting analogues of LHRH with a resultant decrease in testosterone levels and associated impaired sexual function. Combining testosterone injections with LHRH agonist treatment results in a return of testosterone levels to normal but inconsistent success in totally inhibiting spermatogenesis. Trials of LHRH antagonists suggest that inhibition of the hypothalamic–pituitary system does not appear to be an optimal approach to reversible contraception. However, men with low but measurable numbers of sperm may have significant subfertility.

A number of anticancer drugs and other relatively toxic compounds impair sperm production by direct effects on the testes. Most of these compounds produce additional unacceptable side effects. An antifertility agent in cottonseed oil discovered in China is a naphthalphenol termed *gossypol*. It has been given to more than 10,000 men in clinical trials over a decade. Administration of this compound orally for 60 days causes the sperm in the ejaculate to become immotile and to decrease in number. Since the drug affects motility it may not be necessary to have absence of sperm for an antifertility effect. Unfortunately, significant toxicity in regard to lowering the serum potassium has been observed, and the reversibility of the effect of the drug is uncertain. Other drugs with antispermatogenic activity have been studied less extensively. Physical methods such as heat and ultrasound have been tried to a limited extent, and immunological methods with antibodies directed at various testicular components have an uncertain duration of action.

Selective inhibition of epididymal function to cause impairment of sperm maturation would theoretically control fertility without the risk of impaired testicular function. In addition, the time required to achieve an effect on fertility with such agents should be less than the 2 to 3 months necessary for agents affecting the pituitary or testis. Since the epididymis is a target organ for androgens, antiandrogens were used in an attempt to inhibit sperm maturation in the epididymis. Both cyproterone and the nonsteroidal antiandrogen flutamide appear to be ineffective in inhibiting epididymal function. A number of chlorinated sugars have been shown to be effective in producing reversible infertility in animal models by inhibiting glycolysis in sperm and impairing sperm motility. Unfortunately, all such compounds studied thus far have significant toxicity to preclude their trials in men.

In summary, the search for a male contraceptive has not been rewarding.

SUGGESTED READING

Baker HWG, Burger HG, de Kretser DM, and Hudson B: Relative incidence of etiological disorders in male infertility. In: *Male Reproductive Dysfunction,* RJ Santen and RS Swerdloff, eds., Marcel Dekker, New York, pp. 341–372, 1986.

Carr BR, and Griffin JE: Fertility control and its complications. In: *Williams Textbook of Endocrinology,* 8th ed., JD Wilson and DW Foster, eds., Saunders, Philadelphia, pp. 1007–1031, 1992.

DeKretser DM, Risbridger GP, and Kerr JB. Basic endocrinology of the testis. In *Endocrinology,* LJ DeGroot, M Besser, HG Burger, JL Jameson, DL Loriaux, JC Marshall, WD Odell, JT Potts Jr, and AH Rubenstein, eds., Saunders, Philadelphia, pp. 2307–2335, 1995.

Ewing LL: Leydig cell. In: *Infertility in the Male,* LI Lipshultz and SS Howards, eds., Churchill Livingstone, New York, pp. 43–69, 1983.

Griffin JE, and Wilson JD: Disorders of the testes and male reproductive tract. In: *Williams Textbook of Endocrinology,* 8th ed., JD Wilson and DW Foster, eds., Saunders, Philadelphia, pp. 799–852, 1992.

Johnson L: A reevaluation of daily sperm output of men. Fertil Steril 37:811–816, 1982.

Lieblich JM, Rogol AD, White BJ, and Rosen SW: Syndrome of anosmia with hypogonadotropic hypogonadism (Kallmann syndrome). Clinical and laboratory studies in 23 cases. Am J Med 73:506–519, 1982.

Matsumoto AM, Karpas AE, and Bremner WJ: Chronic human chorionic gonadotropin administration in normal men: Evidence that follicle-stimulating hormone is necessary for the maintenance of quantitatively normal spermatogenesis in man. J Clin Endocrinol Metab 62:1184–1192, 1986.

Matsumoto AM, Paulsen CA, and Bremner WJ: Stimulation of sperm production by human luteinizing hormone in gonadotropin-suppressed normal mem. J Clin Endocrinol Metab 59:882–887, 1984.

Mayo KE: Inhibin and activin. Trends Endocrinol Metab 5:407–415, 1994.

Pardridge WM: Serum bioavailability of sex steroid hormones. Clin Endocrinol Metab 15:259–278, 1986.

Pescovitz H, Srivastava CH, Breyer PR, and Monts BA: Paracrine control of spermatogenesis. Trends Endocrinol Metab 5:126–131, 1994.

Saez JM: Leydig cells: Endocrine, paracrine, and autocrine regulation. Endocr Rev 15:574–626, 1994.

Santoro N, Filicori M, and Crowley WF Jr: Hypogonadotropic disorders in men and women: Diagnosis and therapy with pulsatile gonadotropin-releasing hormone. Endocr Rev 7:11–23, 1986.

Smith EP, Boy D, Frank GR, Takahashi H, Cohen RM, Specker B, Williams TC, Lubahn DB, and Korach KS: Estrogen resistance caused by a mutation in the estrogen-receptor gene in man. N Engl J Med 331:1056–1061, 1994.

Swerdloff RS, Overstreet JW, Sokol RZ, and Rajfer J: Infertility in the male. Ann Intern Med 103:906–919, 1985.

Verhoeven G: Local control systems within the testis. Bailliere's Clin Endocrinol Metab 6:313–333, 1992.

Wilson JD: Metabolism of testicular androgens. In: *Handbook of Physiology.* Section 7: *Endocrinology,* Vol. 5, *Male Reproductive System,* RO Greep and EB Astwood, eds., American Physiological Society, Washington, DC, pp. 491–508, 1975.

11

Fertilization, Implantation, and Endocrinology of Pregnancy

BRUCE R. CARR

The complex and coordinated set of events leading to sperm and egg maturation and transport in the female genital tract that culminates in fertilization is one of the most remarkable phenomena in nature. This set of events is followed by the equally important and unique process of implantation, fetal maturation, and parturition. The hormonal changes that regulate these events are dependent upon the close interaction of the fetal–placental–maternal unit.

FERTILIZATION AND IMPLANTATION

Ovum Maturation and Transport

Just before ovulation, the egg, which has been arrested in the diplotene stage, completes the first meiotic division and forms the first polar body. The second meiotic division starts at the time of ovulation but ends only after fertilization by a sperm.

The process of egg maturation is regulated through a closely interrelated set of hormonal events, most notably involving follicle-stimulating hormone (FSH), luteinizing hormone (LH), and estrogen. At the time of ovulation the fimbria of the oviduct are closely applied to the surface of the ovary. The extruded oocyte and adherent granulosa cells, known as the *cumulus oophorous*, are collected by the ciliated fimbrial end of the fallopian tube. The transport of the egg into the end of the fallopian tube occurs within minutes and is regulated primarily by ciliary action. The cumulus cells are able to communicate with one another via a network of intercellular bridges through the zona to the perivitelline space. The cumulus cells have also been reported to play a role in nutrition and maintenance of the ovum.

There are three different stages of passage of the ovum through the fallopian tube. The first stage includes the transfer of the ovum from the fimbriated end of the fallopian tube until the egg reaches and is retained at the ampullary–isthmic junction. The ampullary–isthmic junction is a functional block but is not a clearly defined anatomical structure. The ovum remains at this junction for 1–2 days during which time fertilization occurs. The second stage begins soon after fertilization when the

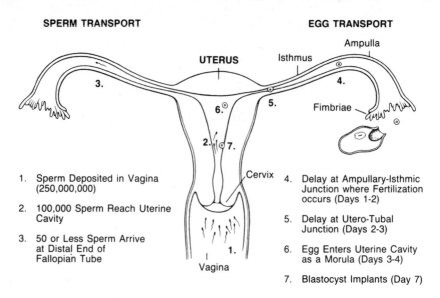

Fig. 11-1   The transport of sperm and the egg in the female reproductive tract.

egg traverses the isthmic portion of the tube where it again is retained at another functional block, the isthmic–utero or utero–tubal junction. The length of time from the process of ovulation until the release of the egg from the isthmus is species dependent and averages 3 days in women. The detection of the egg at the utero–tubal junction appears to be influenced by ovarian steroid hormones, namely estrogen and progesterone. During this stage, estrogen and progesterone also act on the endometrium to prepare it for implantation of the fertilized egg. The final stage of egg transportation occurs during the 3–4 days after ovulation when the fertilized egg leaves the isthmus and arrives in the uterine cavity (Fig. 11-1).

Sperm Transport and Capacitation

The sperm are required to traverse an even greater distance in the female genital tract than is the ovum. Spermatozoa leave the vagina after ejaculation and pass through the cervix, the entire length of the uterine cavity, the utero–tubal junction, the isthmus, and finally the ampullary–isthmic junction where fertilization occurs. The process of sperm transport is very rapid in comparison to egg transport. Spermatozoa have been found in the distal end of the fallopian tube 5 minutes after ejaculation. However, the rate of attrition of sperm is high. Of an estimated 250 million sperm deposited in the vagina, only 50 or less ever reach the oviduct (Fig. 11-1). The principal mechanism that controls sperm transport is flagellar movement; thus, semen characterized by low sperm motility is usually associated with infertility.

In most species, the process of sperm capacitation requires sperm to reside in or be exposed to the female tract before they are capable of fertilizing an ovum. This process is not yet clearly defined, but it includes the acrosome reaction that involves a breakdown and merging of the plasma membrane and acrosomal membrane of the sperm head. This is followed by a release of enzymes thought to play a role in sperm penetration. During the process of capacitation, the sperm also become hypermotile. However, exposure of sperm to the female genital tract does not appear to play a significant role in humans, since sperm incubated in defined medium are capable of fertilizing an ovum. Furthermore, it has been shown that direct intraovum injection of a single sperm can result in fertilization and successful human pregnancy.

Process of Fertilization

The first stage of fertilization requires the equitorial region of the sperm head to adhere to the zona pellucida surrounding the egg (Fig. 11-2). The sperm must first penetrate the zona, which requires 15–25 minutes. After passage through the zona pellucida, the sperm moves rapidly across the perivitelline space (in less than 1 second) where it attaches to the perivitelline membrane and penetrates it (in less than 1 minute).

After the sperm head penetrates the vitelline membrane, the male pronucleus is formed. The penetration of the vitelline membrane initiates two important events: the release of cortical granules into the perivitelline space, which prevents further penetration of the egg by other sperm, and the development of polyspermia. The second event is the triggering of the final stage of meiosis of the oocyte. The second polar body is extruded from the egg and a haploid number of chromosomes (23) are present in the egg pronuclei just before fertilization.

The male and female pronuclei are visible about 2–3 hours after the sperm has penetrated the vitelline membrane. Within 4 hours, the sperm tail is incorporated with the egg. Within 24 hours, the two pronuclei have moved toward the center of the egg. Next, their respective haploid chromosomes replicate and a mitotic spindle forms. The fertilized egg then divides and forms two blastomeres.

The initial cleavages of the egg occur in the fallopian tube and the rate of cleavage remains remarkably constant in mammals. The rate of cleavage of fertilized human eggs has been determined primarily from *in vitro* studies. The timing of

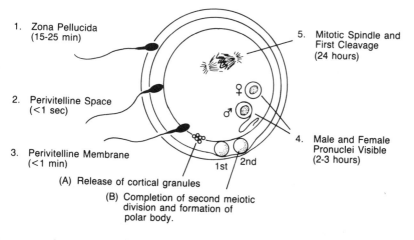

1. Zona Pellucida (15-25 min)
2. Perivitelline Space (<1 sec)
3. Perivitelline Membrane (<1 min)
5. Mitotic Spindle and First Cleavage (24 hours)
4. Male and Female Pronuclei Visible (2-3 hours)
1st   2nd
(A) Release of cortical granules
(B) Completion of second meiotic division and formation of polar body.

Fig. 11-2   The process of fertilization of a human egg.

cleavage of the human egg is as follows: 2 cells (38 hours), 4 cells (46–48 hours), 8 cells (51–62 hours), morula formation (111–135 hours), and the formation of a blastocyst (123–147 hours).

The life span of an unfertilized egg is less than 20 hours following ovulation, after which it undergoes cell death and cytolysis. The life span of a sperm is about 24 hours after ejaculation.

In Vitro Fertilization

The initial report of a delivery of a live born infant by the method of *in vitro* fertilization in 1978 stimulated further clinical investigations. The knowledge and methods that were developed from these studies are now available for the treatment of infertile couples with a variety of disorders including damaged or blocked fallopian tubes, male infertility, cervical factors, and even unexplained infertility.

In vitro fertilization and embryo transfer starts with the administration of exogenous human menopausal gonadotropins (hMG) or pure human FSH to stimulate growth and development of multiple ovarian follicles. Follicles are aspirated 32 hours after the injection of human chorionic gonadotropin (hCG), which has LH-like activity. The ova are aspirated from ripe follicles by transvaginal ultrasonographic techniques. The eggs are then transferred to a culture dish containing medium and serum.

Freshly ejaculated semen is washed in buffer and centrifuged to remove the seminal fluid. The sperm are preincubated in defined medium for 1–2 hours to initiate capacitation. Next, approximately 1 million sperm are added to each culture dish containing an ovum. Fertilization occurs within 6–8 hours. During the next 2–3 days, the egg divides and reaches the 4- to 8-cell stage. Then, up to a maximum of 4–6 fertilized, 2- to 8-cell stage eggs are transferred by a catheter to the uterine cavity. The rest of the embryos that are not initially transferred are frozen and placed in the uterus during a subsequent menstrual cycle. The pregnancy rate is about 25% of those embryos transferred with a rate of live births of about 10% worldwide.

Implantation and Placentation

In the human the egg enters the uterine cavity as a morula on the third to fourth day after ovulation. The morula is transformed into a blastocyst during the fifth to sixth day over ovulation. At this stage of development, the embryo appears as a hollow sphere with two clearly distinguishable cell types: the outer cells are the trophectoderm cells that will form the placenta and the inner cell mass from which the fetus will develop.

On the seventh day after ovulation, the blastocyst implants on the endometrial lining of the uterine cavity. The embryo attaches to the endometrium with the embryonic pole facing the uterine cavity. The endometrium, under the influence of progesterone secreted by the corpus luteum, is transformed into a decidua. The decidua consists of large polyhedral endometrial cells that are laden with glycogen and lipid and are often multinucleated. In the absence of progesterone, the development of decidua and implantation fail to occur. Continued progesterone secretion ensures further development of the decidua and maintenance of pregnancy. The corpus luteum is the main source of progesterone during the first 6–8 weeks of pregnancy. During the remainder of gestation, the principal source of progesterone is the placenta. Decidual development also appears to respond to signals from the invading embryo.

Next, endometrial cells begin to interdigitate with the microvilli of the trophectodermal cells. The layer of trophoblast cells develops into an inner cytotrophoblast layer and an outer syncytiotrophoblast layer. The syncytiotrophoblasts secrete proteolytic enzymes that errode the endometrium and allow the syncytiotrophoblast cells to invade further. The human embryo undergoes interstitial implantation in which the embryo is deeply embedded and enclosed by the endometrium by the eleventh day after ovulation (Fig. 11-3).

The placenta develops from the trophoblast. Lacunar spaces form among the syncytiotrophoblasts and these spaces are contiguous with the maternal capillary circulation. Within these spaces the functional and structural compartment of the placenta, the chorionic villi, develops. In the absence of vascularization of the villi, further development into secondary and tertiary villi do not occur. Instead, the villi become cystic and fill with fluid (hydatidiform mole). The hydatidiform mole is often spontaneously aborted and may result in extensive maternal hemorrhage and shock. This process often precedes the development of invasive trophoblastic disease known as *choriocarcinoma*.

Placental development varies widely between different animal species. The human placenta is hemochorial: fetal endothelium and fetal connective tissues are directly bathed by maternal blood. The maternal side of the placenta is composed of the chorionic villi fused into separate lobes known as *cotyledons*. The cotyledons and their chorionic villi are exposed to maternal blood from the decidual spiral arterioles. The fetal side of the placenta is composed of the amnion and chorion. The amnion is a thin avascular membrane that surrounds the fetus and contains amniotic fluid. Adjacent to the amnion is the chorion tissue layer and these cells are

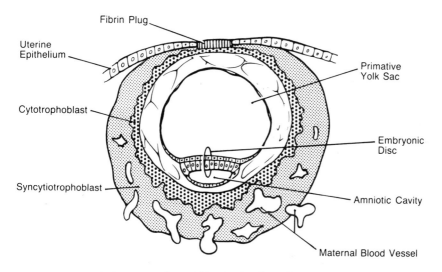

Fig. 11-3   The early human embryo following implantation.

next to the decidua lining the uterine cavity. It has been proposed that hormones produced in the fetus, amnion, chorion, and decidua communicate information between each compartment. This information may play a role in the onset of parturition.

ENDOCRINOLOGY OF PREGNANCY

The Fetal–Placental–Maternal Unit

The hormonal change and maternal adaptation that occur during human pregnancy are extensive. During pregnancy the placenta, supplied with precursor hormones from the maternal–fetal unit, synthesizes large quantities of steroid hormones as well as various protein and peptide hormones and in turn secretes these products into the fetal and maternal circulation. Near the end of pregnancy a woman is exposed daily to large quantities of estrogen, progesterone, mineralocorticoids, and glucocorticosteroids. The mother and, to a lesser extent, the fetus are also exposed to large quantities of human placental lactogen (hPL), hCG, prolactin (PRL), relaxin, prostaglandins, and smaller amounts of proopiomelanocortin (POMC)-derived peptides such as corticotropin (ACTH) and endorphin, gonadotropin-releasing hormone (GnRH), also called luteinizing hormone-releasing hormone (LHRH), thyroid-stimulating hormone (TSH), corticotropin-releasing hormone (CRH), somatostatin, and other hormones.

Implantation, the maintenance of pregnancy, parturition, and finally lactation depend on a complex interaction of hormones in the maternal–fetal–placental unit. In addition, protein and peptide hormones produced by the placenta act via a paracrine mechanism to regulate the secretion of placental steroid hormones.

Placental Compartment

In mammals, particularly humans, the placenta has evolved into a complex structure that delivers nutrients to the fetus, produces numerous steroid and protein hormones, removes metabolites from the fetus, and delivers them to the maternal compartment.

Progesterone

The main source of progesterone during pregnancy is the placenta. The corpus luteum, however, is the major source of progesterone secretion during the first 6–8 weeks of gestation. It is believed that the developing trophoblast takes over as the major source of progesterone secretion by 8 weeks gestation, since removal of the corpus luteum before, but not after, this time leads to abortion. After 8 weeks gestation the corpus luteum of pregnancy continues to secrete progesterone, but the amount of progesterone secreted is only a fraction of that secreted by the placenta. The placenta of a term pregnancy produces approximately 250 mg of progesterone each day. Maternal progesterone plasma levels rise from 25 ng/ml during the late luteal phase to 150 ng/ml at term (Fig. 11-4A). Most of the progesterone secreted by the placenta enters the maternal compartment.

The biosynthetic origin of progesterone has recently been clarified. Although the placenta produces large amounts of progesterone, under normal circumstances

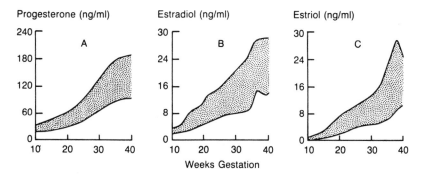

Fig. 11-4 The range (mean ± 1 SD) of (A) progesterone, (B) estradiol-17β, and (C) estriol in plasma of normal pregnant women as a function of weeks of gestation. (Adapted from Carr BR: In: *Principles and Practice of Endocrinology and Metabolism*, KL Becker, ed., Lippincott, Philadelphia, pp. 887–898, 1990.)

it has very limited capacity to synthesize cholesterol from acetate. It is now clear that maternal cholesterol in the form of low-density lipoprotein (LDL) cholesterol is the principal source of precursor substrate for biosynthesis of progesterone in human pregnancy. LDL cholesterol attaches to its receptor on the trophoblast and is taken up and degraded by the trophoblast to free cholesterol, which then is converted to progesterone and secreted.

A functioning fetal circulation is not important for the regulation of progesterone levels in the maternal unit. In fact, fetal death, ligation of the umbilical cord, and anencephaly, which are all associated with a decrease in estrogen production, have no significant effect on progesterone levels in the maternal compartment.

The physiological role of the large quantity of progesterone produced by the placenta has been an area of great interest. Progesterone binds to receptors in uterine smooth muscle, thereby inhibiting smooth muscle contractility leading to myometrial quiescence and prevention of uterine contractions. It also inhibits prostaglandin formation, which is known to be critically involved in human parturition. Progesterone is essential for the maintenance of pregnancy in all mammals, possibly due to its ability to inhibit T lymphocyte cell–mediated responses involved in graft rejection. Since the fetus is a foreign body within the uterus, the high local levels of progesterone can block cellular immune responses to foreign antigens and may be important in giving immunological privilege to the pregnant uterus.

Estrogen

During human pregnancy the rate of estrogen production and the level of estrogens in plasma increase markedly (Fig. 11-4B and C). The levels of urinary estriol have been reported to increase 1000-fold during pregnancy. The corpus luteum is the principal source of estrogen during the first few weeks of pregnancy, but afterward nearly all of the estrogen formed is from the trophoblast of the placenta. However, the mechanism by which estrogen is produced by the placenta is unique. The placenta is unable to convert progesterone to estrogens because of a deficiency of 17α-hydroxylase. Thus, the placenta has to rely on preformed androgens produced in the maternal and fetal adrenal glands. Estradiol-17β and estrone are synthesized by the placenta via conversion of dehydroepiandrosterone sulfate (DHEA-S) that reaches it from both maternal and fetal bloodstreams.

The placenta metabolizes DHEA-S to estrogens in the presence of the following enzymes: placental sulfatase, 3β-hydroxysteroid dehydrogenase, and aromatase enzyme complex. Estriol is synthesized by the placenta from 16α-hydroxydehydroepiandrosterone sulfate formed in fetal liver from circulating DHEA-S secreted by the fetal adrenal gland. At least 90% of urinary estriol is ultimately derived from the fetal adrenal gland. The sources of estrogen biosynthesis in the maternal–fetal–placental unit is presented in -Fig. 11-5. The major source of fetal adrenal DHEA-S is cholesterol circulating in fetal blood. A minor source of fetal adrenal DHEA-S is formed from pregnenolone secreted by the placenta. It has been reported that only 20% of fetal cholesterol is derived from the maternal compartment and since amnionic fluid cholesterol levels are negligible, the main source of cholesterol appears to be the fetus itself. Since the fetal liver synthesizes cholesterol at a high rate, and if one considers the size of the fetal liver, it can be computed that the fetal liver may supply sufficient cholesterol to the fetal adrenal to maintain steroidogenesis.

The measurement of estrogens, in particular, urinary estriol, has been utilized to monitor fetal well being in high-risk pregnancies. There are several disorders in addition to fetal distress that lead to low urinary excretion of estriol by the mother. The most notable of these conditions is placental sulfatase deficiency, also known as steroid sulfatase deficiency syndrome, which is an X-linked inherited metabolic disorder characterized during fetal life by decreased maternal estriol production due

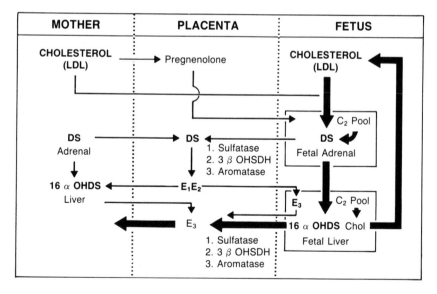

Fig. 11-5 Sources of estrogen biosynthesis in the maternal–fetal–placental unit. LDL, low-density lipoprotein; Chol, cholesterol; C_2 pool, carbon–carbon unit; DS, dehydroepiandrosterone sulfate; E_1, estrone; E_2, estradiol-17β; E_3, estriol. (Adapted from Carr BR, and Gant NF: The endocrinology of pregnancy-induced hypertension. Clin Pernatol 10:737, 1983.)

to deficient placental sulfatase (see Fig. 11-5). In this disorder, the placenta is unable to cleave the sulfate moiety from DHEA-S and, consequently, the levels of maternal estrogens, particularly estriol, are quite low. Placental sulfatase deficiency is also associated with prolonged gestation and difficulty in cervical dilatation at term, often requiring cesarean section. Steroid sulfatase deficiency is thought to occur in 1 of every 2000–6000 newborns. The male offspring are, of course, sulfatase deficient and they have a characteristic skin condition termed *ichthyosis*, which is first apparent after the first few months of life.

The ultimate destination of estrogen and progesterone secretion by the placenta is primarily the maternal compartment. The physiological role of the large quantity of estrogen produced by the placenta of women is not completely understood. It has been proposed that estrogen may regulate or fine tune the events leading to parturition since pregnancies are often prolonged when estrogen levels in maternal blood and urine are low, as in placental sulfatase deficiency or when associated with anencephaly. Estrogen stimulates phospholipid synthesis and turnover, increases incorporation of arachidonic acid into phospholipids, stimulates prostaglandin synthesis, and increases the number of lysosomes in uterine endometrium. Estrogens are known to increase uterine blood flow and may also play a role in fetal organ maturation and development.

Human Chorionic Gonadotropin (hCG)

Human chorionic gonadotropin is secreted by the syncytiotrophoblast of the placenta into both the fetal and maternal circulation. It is a glycoprotein with a molecular weight of about 38,000 that consists of two noncovalently linked subunits, α and β, and is similar to LH in structure and action. Human chorionic gonadotropin has been used extensively as a pregnancy test and can be detected in serum as early as 6–8 days following ovulation. Plasma levels rise rapidly in normal pregnancy, doubling in concentration every 2–3 days until they reach a peak between 60 and 90 days of gestation. Thereafter the concentration of hCG in maternal plasma declines, plateauing at about 120 days before delivery. The levels of hCG are higher in multiple pregnancies, in pregnancies associated with Rh isoimmunization, and in pregnant diabetic women; they are highest in pregnancies associated with hydatidiform moles and in women with a tumor known as choriocarcinoma.

There is some evidence that the regulation of the rate of secretion of hCG by the syncytiotrophoblastic cells may be a paracrine mechanism involving the release of LHRH by the cytotrophoblast. Activin and inhibin as well as the transforming growth factors α and β secreted by the trophoblastic cells also appear to regulate hCG secretion. Fetal concentrations of hCG reach a peak at 11–14 weeks gestation and thereafter fall progressively until delivery.

Various theories have been proposed as to the functional role of hCG in pregnancy. The most widely accepted theory is that hCG maintains the early corpus luteum of pregnancy to ensure continued progesterone secretion by the ovary until this function is replaced by the growing trophoblast. Some investigators have demonstrated that hCG promotes steroidogenesis (namely progesterone) by the trophoblast. It is most likely that a primary role for hCG in the fetus is to regulate the development and the secretion of testosterone by the fetal testes. Male sexual differentiation occurs at an early but critical time in development when fetal serum hCG levels are high and fetal plasma LH levels are low. It has been suggested that hCG may affect fetal ovarian development as well. An additional role may be to give immunological privilege to the developing trophoblast. Finally, several investigators have demonstrated that the excess thyrotropic activity during the clinical development of hyperthyroidism observed in some women with neoplastic trophoblast disease results from excessive hCG secretion. TSH and hCG have similar structures and purified hCG inhibits binding to thyroid membranes and stimulates adenylate cyclase in thyroid tissues.

Human Placental Lactogen (hPL)

Human placental lactogen is a single-chain polypeptide consisting of 191 amino acid residues that has a molecular weight of about 22,000. As its name suggests, hPL has both lactogenic and growth hormone-like activity and is also referred to as chorionic growth hormone or chorionic somatomammotropin. Human placental lactogen mainly exhibits lactogenic activity; it has only 3% or less of the growth-stimulating activity of human growth hormone (GH). The structure and amino acid-base residues of both hPL, prolactin, and GH are quite similar.

Human placental lactogen is secreted by syncytiotrophoblast and can be detected in serum by radioimmunoassay as early as the third week after ovulation. The plasma level of hPL continues to rise with advancing gestational age, unlike hCG, and appears to plateau at term. The concentration of hPL in serum is closely correlated with increasing placental weight. Its half-life in serum is short. Although the level of hPL in serum before delivery is the highest of all the protein hormones secreted by the placenta, it cannot be detected in serum after the first postpartum day. The time sequence and peak of hPL secretion are significantly different from that of hCG, which suggests a different regulation for each hormone. This is interesting since both are secreted by the syncytiotrophoblast rather than the cytotrophoblast.

Interestingly, hPL appears to be secreted primarily into the maternal circulation since very low levels are observed in cord blood of newborns. Thus, most of the proposed physiological roles of hPL have centered around its sites of action in maternal tissues. Human placental lactogen may have a major sparing influence on maternal glucose, providing adequate, continued nutrition for the developing fetus. It has been suggested that hPL exerts metabolic effects in pregnancy similar to those of GH. These effects include stimulation of lipolysis, resulting in increased circulating free fatty acids (which are available for maternal and fetal nutrition), inhibition of glucose uptake in the mother, resulting in increased maternal insulin levels, development of maternal insulin resistance, and an inhibition of gluconeogenesis that favors transportation of glucose and protein to the fetus.

Other Placental Peptide and Protein Hormones

In addition to hCG and hPL, several other placental hormones that are similar or closely related with respect to biological and immunological activity to hypothalamic or pituitary hormones have been reported. These include human chorionic POMC peptides, CRH, growth hormone-releasing hormone (GHRH), human chorionic thyrotropin, and human chorionic gonadotropin-releasing hormone. The regulation of the secretion of these hormones is not well understood, but classical negative feedback inhibition does not appear to exist. Furthermore, the function and significance of these hormones are speculative at best. Most of these hormones are believed to enter primarily the maternal compartment.

Fetal Compartment

During the past two decades an intense effort has been made to elucidate the human fetal endocrine system. To a large extent, this was based on the development of methods of measuring minute quantities of hormone. The regulation of the fetal endocrine system, as with that of the placenta, is not completely independent but relies to some extent on precursor hormones secreted by the placenta or maternal tissues. As the fetus develops, its endocrine system gradually matures and becomes more independent, preparing the fetus for extrauterine existence. The following section will summarize our present understanding of the fetal endocrine unit and its relationship with the maternal–placental unit.

Fetal Hypothalamic–Pituitary Axis

The fetal hypothalamus begins to differentiate from the forebrain during the first few weeks of fetal life and by 12 weeks hypothalmic development is well advanced. Most of the hypothalamic-releasing hormones, including LHRH, thyrotropin-releasing hormone (TRH), dopamine, norepinephrine, and somatostatin, as well as their respective hypothalamic nuclei have been identified as early as 6–8 weeks of fetal life. The neurohypophysis is first detected at 5 weeks and by 14 weeks the supraoptic and paraventricular nuclei are fully developed.

Rathke's pouch first appears in the human fetus at 4 weeks of fetal life. The premature anterior pituitary cells that develop from the cells lining Rathke's pouch are capable of secreting GH, PRL, FSH, LH, and ACTH *in vitro* as early as 7 weeks of fetal life. Recent evidence suggests that the intermediate lobe of the pituitary may be a significant source of POMC hormones. The intermediate pituitary lobe in the human fetus decreases in size after birth with only remnants remaining in the adult.

The hypothalamo-hypophyseal portal system is the functional link between the hypothalamus and the anterior pituitary. Vascularization of the anterior pituitary starts by 13 weeks of fetal life, but a functioning intact portal system is not present until 18–20 weeks of fetal life. There is indirect evidence, however, that hypothalamic secretion of releasing hormones may influence anterior pituitary function before 18–20 weeks by simple diffusion as a result of their close proximity in early fetal development.

Fetal growth hormone is detected in the human pituitary as early as 12 weeks; fetal pituitary GH content increases until 25–30 weeks gestation and thereafter remains constant until term. In contrast, fetal plasma GH levels rise to a peak at 20 weeks of fetal life and thereafter fall rapidly until birth. Fetal plasma concentrations of GH, however, are higher than maternal concentrations at all ages. Maternal levels may be suppressed by the high circulating levels of hPL. The regulation of GH release in the fetus appears to be more complex than in the adult. In order to explain the high levels of fetal GH at mid-gestation and the fall thereafter, unrestrained release of GHRH or inhibition of somatostatin release leading to excessive release of GH at mid-gestation has been hypothesized. As the hypothalamus matures, somatostatin may increase and GHRH levels decline so that GH release also declines. The role of GH in the fetus is unclear as well. There is considerable evidence that GH is not essential to intrauterine somatic growth in primates. In newborns with pituitary agenesis, congenital hypothalamic hypopituitarism or familial GH deficiency, birth weight, and length are usually normal. However, a class of peptides known as somatomedins, in particular insulin-like growth factors (IGF-I and IGF-II), increase in fetal plasma and IGF-I and IGF-II levels correlate better with fetal growth than GH levels. Although GH is an important trophic hormone for somatomedin production in the fetus, somatomedin regulation may be independent of GH.

Prolactin is present in pituitary lactotropes by 19 weeks of life. Fetal prolactin content in the pituitary increases throughout gestation. Fetal prolactin plasma levels increase slowly until 30 weeks gestation, after which the levels rise sharply until term and remain elevated until the third month of postnatal life. TRH and dopamine as well as estrogens appear to affect human fetal prolactin secretion. Regulation of prolactin secretion by dopamine in the fetus is supported by the observation that the dopamine agonist, bromocriptine, when ingested by the mother, crosses the placenta and inhibits prolactin release from the fetal pituitary gland lowering prolactin levels in fetal blood. It has been suggested that prolactin influences fetal adrenal growth, fetal lung maturation, and amnionic fluid volume.

Arginine vasopressin (AVP) and oxytocin are found in hypothalamic nuclei and in the neurohypophysis during early fetal development. But there have been relatively few studies of the regulation and secretion of these hormones in the fetus. The levels of AVP have been reported to be high in fetal plasma and newborn cord blood at delivery. The main stimulus to AVP release seems to be fetal hypoxia, although acidosis, hypercarbia, and hypotension also play a role. The elevated AVP level in fetal blood may lead to increased blood pressure, vasoconstriction, and the passage of meconium by the fetus. In contrast, oxytocin levels in the fetus are not affected by hypoxia but appear to increase during labor and delivery.

Fetal Thyroid Gland

Evidence now suggests that the placenta is relatively impermeable to TSH and thyroid hormone so that the fetal hypothalamic–pituitary–thyroid axis appears to develop and function independently of the maternal system.

By the end of the first trimester, the fetal thyroid has developed sufficiently to be able to concentrate iodine and synthesize iodothyronines. The levels of TSH and thyroid hormone are relatively low in fetal blood until mid-gestation. At 24–28 weeks gestation, serum TSH concentrations rise abruptly to a peak and decrease slightly thereafter until delivery. In response to the surge of TSH, thyroxine (T_4) levels rise progressively after mid-gestation until term. During this time, the pituitary becomes more responsive to TRH and an increase in hypothalamic TRH content occurs. At birth there is an abrupt release of TSH, T_4, and triiodothyronine (T_3), followed by a fall in the levels of these hormones shortly thereafter. The relative hyperthyroid state of the fetus is believed to facilitate thermoregulatory adjustments for extrauterine life. The abrupt increases in TSH and T_4 that occur at birth are thought to be stimulated by the cooling process associated with delivery. Finally, $3,3',5'$-triiodothyronine (reverse T_3) levels are high during early fetal life and begin to fall at mid-gestation and continue to fall after birth (Fig. 11-6). The difference

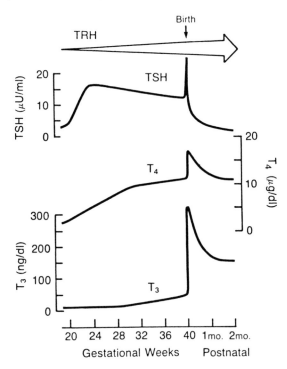

Fig. 11-6 Maturation of serum TSH, thyroxine (T_4), and triiodothyronine (T_3) during the last half of gestation and during early neonatal life. The increase in TRH effect or content is also illustrated. (From Fisher DA: Maternal–fetal thyroid function in pregnancy. Clin Perinatol 10:615, 1983. Reproduced with permission of W. B. Saunders Company.)

between the formation of T_3 and reverse T_3 is thought to be related to maturation of peripheral iodothyronine metabolism.

Fetal Gonads

Biologically active and immunoreactive LHRH has been detected in the fetal hypothalamus by 9–12 weeks of age. The hypothalamic content of LHRH has been reported to increase with fetal age, maximal content being noted between 22 and 25 weeks in females and between 34 and 38 weeks in males. The dominant gonadotropin fraction in the fetal pituitary is the α-subunit. However, the fetal pituitary is capable of secreting intact LH *in vitro* by 5–7 weeks. The plasma concentration of FSH in the human fetal plasma slowly rises and reaches a peak near the twenty-fifth week of fetal life and falls to low levels by term. Plasma FSH levels parallel the pituitary content of FSH with respect to sexual dimorphism, circulating FSH levels being higher in females than in males. The pattern of LH levels in fetal plasma parallel those of FSH. The fall in gonadotropin pituitary content and plasma concentration after mid-gestation is believed to result from the maturation of the hypothalamus. The hypothalamus also becomes more sensitive to sex steroids circulating in fetal blood, originating from the placenta.

In the male, fetal testosterone secretion begins soon after differentiation of the gonad into a testis and formation of Leydig cells at 7 weeks of fetal life. Maximal levels of fetal testosterone are observed at about 15 weeks and decrease thereafter. The early secretion of testosterone is important in initiating sexual differentiation in the male (see Chapter 8). It is believed that the primary stimulus to the early development and growth of Leydig cells and subsequent peak of testosterone is hCG. Thus, it appears that sexual differentiation of the male does not rely solely on fetal pituitary gonadotropins. However, fetal LH and FSH are still required for complete differentiation of the fetal ovary and testes. For example, male fetuses with abnormal brain formation (anencephaly), with low levels of circulating LH and FSH, secrete testosterone normally at 15–20 weeks because of adequate levels of hCG, but they have a decreased number of Leydig cells, exhibit hypoplastic external genitalia, and often have undescended testes. Likewise, male fetuses with congenital hypopituitarism often have small phalluses. These observations suggest that around and after mid-gestation, fetal pituitary gonadotropins affect testosterone secretion from the fetal testes.

The fetal ovary is involved primarily in the formation of follicles and germ cells. Although follicular development in the ovary appears to be relatively independent of gonadotropins, the anencephalic female fetus with small ovaries and a decreased number of ovarian follicles suggests some pituitary involvement (however, the fetal ovaries do not contain hCG [LH] receptors, at least by 20 weeks of gestation). The ovaries appear to be relatively inactive with respect to steroidogenesis during fetal life but are capable of aromatizing androgens to estrogens *in vitro* as early as 8 weeks of gestation.

Fetal Adrenal Gland

Of all the endocrine glands in the human fetus, the adrenal has aroused the greatest interest. The human fetal adrenal glands secrete large quantities of steroid hormones, up to 200 mg of steroid daily near term. This rate of steroidogenesis may be five times that observed in the adrenal glands of adults at rest. The principal steroids found in the fetus are C-19 steroids (mainly DHEA-S), which serve as precursor substrate for estrogen biosynthesis in the placenta.

At the beginning of this century, investigators observed that the human fetal adrenal gland contained a unique fetal zone that accounts for the rapid growth of the fetal adrenal and that this zone disappears during the first few weeks after birth. This fetal zone differs histologically and biochemically from the neocortex (also known as the *definitive* or *adult zone*). The uniqueness of a transient fetal zone has been reported in certain higher primates and some other rare species, but only humans possess the extremely large fetal zone that involutes after birth.

The cells of the adrenal cortex arise from coelomic epithelium. The cells comprising the fetal zone can be identified in the 8 to 10 mm embryo and before the appearance of the cells of the neocortex (14 mm embryo). Growth is most rapid during the last 6 weeks of fetal life. By 28 weeks gestation, the adrenal gland may be as large as the fetal kidney, and by term may be equal to the size of the adult adrenal. The fetal zone accounts for the largest percentage of growth, and after birth the adrenal gland decreases in size due to involution and necrosis of fetal zone cells (Fig. 11-7).

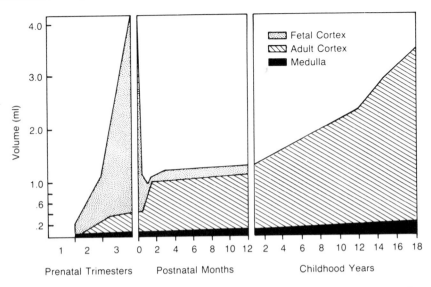

Fig. 11-7 Size of the adrenal gland and its component parts during fetal life, infancy, and childhood. (Adapted from Carr BR, and Simpson ER: Endocr Rev 2:306, 1981.)

Studies of the fetal adrenal gland have attempted to determine what factors stimulate and regulate its growth and steroidogenesis and why the fetal zone undergoes atrophy after delivery. All investigations have shown that ACTH stimulates steroidogenesis *in vitro*. Furthermore, there is clinical evidence that ACTH is the major trophic hormone of the fetal adrenal gland *in vivo*. In anencephalic fetuses the plasma levels of ACTH are very low and the fetal zone is markedly atrophic. Maternal glucocorticosteroid therapy effectively suppresses fetal adrenal steroidogenesis by suppressingfetal ACTH secretion. Further evidence that ACTH regulates steroidogenesis early in fetal life is provided by the observation that elevated levels of 17α-hydroxyprogesterone in amnionic fluid are found in fetuses with congenital adrenal hyperplasia due to the absence of 21-hydroxylase activity. Despite these observations, other ACTH-related peptides, POMC derivatives that are formed in the fetal pituitary or placenta, have been proposed as possible trophic hormones for the fetal zone, but the evidence for this proposal is weak. Other hormones or growth factors, including prolactin, hCG, GH, hPL, and epidermal and fibroblast growth factor, have no consistent significant effect on steroidogenesis or adenylyl cyclase activity in preparations of fetal zone organ cultures, monolayer cells, or membrane preparations *in vitro*. However, the role of these or other hormones in promoting growth of adrenal cells is unclear.

After birth the adrenal gland shrinks by over 50% due to regression of fetal zone cells. This suggests that a trophic substance other than ACTH is withdrawn from the maternal or placental compartment or that the secretion rates of some other trophic hormone is altered to initiate regression of the fetal zone.

The medulla of the fetal adrenal gland is formed by 10 weeks of gestation. In contrast to the fetal adrenal cortex, the adrenal medulla is relatively immature at term. The secretion of cortisol by the adrenal cortex that surrounds the medulla stimulates the formation of epinephrine from norepinephrine. However, most of the chromaffin tissue in the human fetus is formed in paraaortic paraganglia rather than in fetal adrenal medullary tissue. Both epinephrine and norepinephrine can be detected in the human fetal medulla by 10–15 weeks gestation. Except for the effect of cortisol, the regulation of catecholamine secretion in the human fetus has not been fully elucidated. In other species, hypoxia and trauma, as well as advancing gestation and maturation, are related to the release of catecholamines by the adrenal medulla.

Fetal Parathyroid Gland and Calcium Homeostasis

The level of calcium in the fetus is regulated largely by the transfer of calcium from the maternal compartment across the placenta. The maternal compartment undergoes a number of adjustments that ultimately allow for a net transfer of sufficient calcium to the fetus to sustain fetal bone growth.

The changes in the maternal compartment that permit fetal accumulation of calcium include an increase in maternal dietary intake, an increase in circulating maternal 1,25-dihydroxyvitamin D [+1,25-$(OH)_2$D]+, and an increase in circulating parathyroid hormone (PTH). No significant changes are observed in maternal calcitonin levels. The levels of total calcium and phosphorus decline in maternal serum, but ionized calcium levels remain unchanged. The "placental calcium pump" allows for a positive gradient of calcium and phosphorus to the fetus. Circulating fetal calcium and phosphorus levels increase steadily throughout gestation. Fetal levels of total and ionized calcium as well as phosphorus exceed maternal levels at term.

The fetal parathyroid gland contains PTH, and the gland is capable of hormone secretion by 10–12 weeks gestation. Fetal plasma levels of PTH reportedly are low but increase after delivery. The fetal thyroid contains calcitonin, and in contrast to maternal plasma levels, calcitonin levels in the fetus are elevated. Since there is no transfer of PTH or calcitonin across the placenta, the consequences of the observed change in these hormones on fetal calcium are consistent with an adaptation to conserve and stimulate bone growth within the fetus.

Plasma levels of the various forms of vitamin D are lower in the fetus than in the mother. The placenta and decidua are capable of 1α-hydroxylation and of formation of the active metabolite 1,25-$(OH)_2$D. However, the role, if any, of this hormone in the fetus is unknown since its major effect is on intestinal absorption of calcium.

After birth, serum calcium and phosphorus levels fall in the neonate. PTH levels begin to rise 48 hours after birth, and calcium and phosphorus levels gradually increase over the following several days, depending on dietary intake of milk.

Fetal Endocrine Pancreas

The human fetal pancreas appears during the fourth week of fetal life. The α-cells containing glucagon and the δ-cells containing somatostatin develop early, before β-cell differentiation, although insulin can be recognized in the developing pancreas before apparent β-cell differentiation. Total human pancreatic insulin and glucagon content increase with fetal age and are higher than the concentrations of the adult human pancreas.

In contrast to the pancreatic content of insulin, fetal insulin secretion is low and relatively unresponsive to acute changes in glucose in *in vitro* studies of pancreatic cells, in cord blood at delivery, and in blood samples obtained from the scalp of the fetus at term. In contrast, fetal insulin secretion *in vitro* is responsive to amino acids and glucagon as early as 14 weeks of gestation. Although the acute response to glucose is impaired in the fetal pancreas, β-cells that are chronically exposed to elevated glucose levels, as may occur in maternal diabetes mellitus, undergo hypertrophy so that the rate of insulin secretion increases.

Glucagon has been detected in human fetal plasma as early as 15 weeks gestation. Although secretion of glucagon is stimulated in late pregnancy by amino acids and catecholamines, acute changes in glucose appear to have little effect on fetal pancreatic glucagon secretion.

Maternal Compartment

A variety of maternal adaptations involving the endocrine system occur during pregnancy. Many diseases of the maternal endocrine system, if untreated, are associated with infertility and reduced conception rates. If conception occurs, the more serious the disorder, the more likely it will affect the fetus adversely, as in diabetes mellitus. Hormones or drugs used to treat the endocrine disorders may be transported across the placenta and alter the environment and development of the fetus.

Hypothalamus and Pituitary

Little is known about the endocrine alterations of the maternal hypothalamus during pregnancy. Tumors of the hypothalamus or functional disorders of the hypothalamus commonly result in infertility due to amenorrhea and resulting chronic anovulation. The anterior pituitary undergoes a 2- to 3-fold enlargement during pregnancy due primarily to hyperplasia and hypertrophy of the lactotrophs (prolactin-secreting cells) thought to result from estrogen stimulation. Thus, prolactin plasma levels parallel the increase in pituitary size throughout gestation.

In contrast to lactotrophs, the number of somatotropic cells in the pituitary decreases during pregnancy. Maternal levels of GH are low and do not change during pregnancy. Maternal levels of LH and FSH levels are also low during pregnancy. The response of gonadotropins to an infusion of LHRH is severely blunted during pregnancy. The loss of responsiveness to LHRH is thought to be caused by a negative feedback inhibition from the elevated levels of estrogen and progesterone during pregnancy. The levels of TSH are within the normal nonpregnant adult range throughout pregnancy. Furthermore, the response of TSH to a dose of TRH is similar to that in nonpregnant women.

Numerous studies have examined the levels of ACTH- and POMC-related peptides in maternal blood throughout pregnancy. The maternal plasma levels of β-endorphin remain relatively low throughout pregnancy. β-Endorphin levels do increase in maternal blood with advancing labor, but β-endorphin levels in cord blood are similar in infants delivered vaginally and delivered by elective cesarean section from women not in labor. However, fetal hypoxia is associated with significant increases in β-endorphin levels in cord blood. The plasma levels of maternal ACTH have been reported to increase from early to late gestation, but values were either in the normal range or lower than in nonpregnant women, suggesting that ACTH may be suppressed by estrogen and progesterone. The slight increases in ACTH levels that appear to occur during the course of gestation might be explained by the increased secretion of placental ACTH that is not subject to feedback control. Maternal ACTH rises to very high levels during labor and delivery. The levels of ACTH in umbilical cord plasma parallel those of β-endorphin and, since neither crosses the placenta, offer further evidence that in the fetus as in adults these two peptides are processed from a common precursor. CRH increases in maternal serum throughout pregnancy. This is due to an increase of CRH synthesis by the placenta. However, since there is also an increase in carrier protein which binds CRH, CRH probably exhibits little biologic activity.

Maternal plasma AVP levels remain low throughout gestation and are not believed to play a role in human parturition. Maternal oxytocin levels are reported to be low and do not vary throughout pregnancy but increase during the later stages of labor.

Thyroid

The thyroid gland increases slightly in size during pregnancy as a result of increased vascularity and mild glandular hyperplasia, but a true goiter is not present. There is a modest increase in oxygen consumption (basal metabolic rate) during pregnancy secondary to fetal requirements.

During pregnancy the mother is in a euthyroid state. Serum total T_4 and T_3 increase markedly but do not indicate a hyperthyroid state since there is a parallel increase in thyroxine-binding globulin (TBG). The elevation of total T_4 and T_3 is due to increased TBG from estrogen exposure. A similar finding is observed in women on oral contraceptives. A reduced T_3-resin uptake is also observed in pregnancy and in women on oral contraceptives. The levels of free T_3 and T_4, however, are not increased in pregnancy. There is little if any transfer of T_4, T_3, or TSH across the placenta. TRH is capable of crossing the placenta, but under normal physical conditions there is little TRH circulating in peripheral maternal plasma.

Adrenal

Compared to changes in the fetal adrenal, the maternal adrenal gland does not change morphologically during pregnancy. During pregnancy plasma adrenal steroid hormones increase with advancing gestation.

The increase in total plasma cortisol is due principally to a concomitant increase in cortisol-binding globulin (CBG), also known as *transcortin*. There is a slight increase in free plasma cortisol and urinary free cortisol, but pregnant women do not exhibit any overt signs of hypercortisolism.

Aldosterone levels increase with advancing gestation, reflecting an increased secretion rate of aldosterone by the zona glomerulosa. The levels of renin and angiotensinogen rise in pregnancy, which leads to elevated angiotensin II levels and markedly elevated levels of aldosterone.

Parathyroid

Calcium and parathyroid hormone changes during normal pregnancy were discussed in the fetal parathyroid section of this chapter.

Endocrine Pancreas

The metabolic adaptation of pregnancy in which glucose is spared for the fetus is related to an appropriate bihormonal secretion by the maternal endocrine pancreas. In response to a glucose load, there is a greater release of insulin from the β-cells and a greater suppression of glucagon release from the α-cells compared to the nonpregnant state. In association with the increased release of insulin, the maternal pancreas undergoes β-cell hyperplasia and islet cell hypertrophy, accompanied by an increase in blood flow to the endocrine pancreas. During pregnancy, fasting blood glucose levels fall, but rise more in response to a glucose load than in nonpregnant women. The increased release of insulin is related to a relative insulin resistance. The insulin resistance is due to hormones secreted by the placenta that spare transfer of glucose to the fetus. Glucagon levels are suppressed in response to a glucose load, with the greatest suppression occurring near term.

Endocrinology of Parturition

The mechanism by which labor is initiated in women is not completely understood. It is of vital importance that the etiology of the onset of parturition is clarified in order to eventually prevent preterm birth. The sequelae of prematurity includes not only increased fetal and neonatal mortality, but significant morbidity that can result in severe physical and mental impairment of the newborn. Several theories involving endocrine changes in the fetal–placental unit have been proposed to explain the onset of labor. However, our present knowledge in human pregnancy is fragmented.

Oxytocin

Oxytocin has been used in clinical obstetrics for years to induce or augment labor. When sensitive radioimmunoassays became available to detect low levels of oxytocin in maternal and fetal blood, it was determined that oxytocin levels do not rise before the onset of labor. The role of oxytocin is believed to be the regulation of the expulsive phase of labor and contraction of the uterus to reduce blood loss following delivery.

Relaxin

Relaxin is a large polypeptide that is synthesized by the corpus luteum and the decidua. Relaxin ripens the cervix in various animal species. In many mammals, a significant increase in circulating level of relaxin occurs prior to parturition. In humans, the level of plasma relaxin is highest in the first trimester of pregnancy but declines and thereafter remains constant until delivery. Relaxin has been reported to decrease myometrial contractility *in vitro.*

Progesterone Withdrawal

In most, if not all mammalian species, a decline in maternal levels of progesterone precedes the onset of parturition. In both primates and humans, however, there is no significant decrease in plasma progesterone levels before labor. Some investigators have reported a decrease in the number of myometrial progesterone receptors just prior to delivery.

Fetal Adrenal

The cause of labor in certain mammalian species, such as sheep, appears to be regulated by signals first occurring in the fetal pituitary (ACTH), which lead to increased cortisol secretion by the fetal sheep adrenal gland. In turn, increased fetal cortisol levels stimulate 17α-hydroxylase activity in sheep placenta leading to decreased progesterone and increased estrogen production, which triggers the onset of labor. Administration of corticoids or ACTH to the sheep fetus stimulates labor and the removal of the fetal pituitary or adrenal prolongs labor.

In the human, however, the role of fetal adrenal steroid secretion on the onset of labor is less clear. For example, cortisol or ACTH administration to mother or fetus does not stimulate parturition. Fetal plasma levels of cortisol do not increase markedly before labor. On the other hand, conditions that result in low levels of estrogen in the maternal compartment are often accompanied by prolonged or post-date pregnancy, failure of cervical ripening, or poor cervical dilation during labor. Such conditions include anencephaly or failure of brain development, congenital absence or hypoplasia of the fetal adrenal, and steroid sulfatase deficiency. In each of these cases, inadequate fetal adrenal hormone precursor is available for placental aromatization to estrogen. In addition, local application of estrogen to the cervix leads to ripening.

It is believed that estrogen derived from fetal adrenal precursors does not directly cause the onset of labor, but instead may modify or fine-tune other events leading to the onset of parturition. One of the effects of estrogen is the stimulation of prostaglandin synthesis in the tissues of the uterus and fetal membranes.

Prostaglandin

Because of the complexities and discrepancies of parturition in different mammalian species, a paracrine mechanism has been proposed to explain the onset of parturition in humans. This mechanism is related to the close proximity of the fetus, amnionic fluid, fetal membranes (amnion and chorion), and contiguous layer of uterine decidua and myometrium. Signals not yet identified, originating in the fetus, may by a paracrine mechanism be transmitted to fetal membranes and uterine tissues leading to increased production of prostaglandin. Accumulating evidence suggests that prostaglandins and prostanoids play a crucial role in human parturition. The trigger for release of prostaglandins is unknown, but is thought to be of fetal origin and regulated by hormone changes in the fetal–placental–maternal unit.

Prostaglandins, when administered to pregnant women, cause cervical ripening, softening, dilation, and the onset of uterine contractions resulting in labor. The administration of prostaglandin inhibitors to pregnant women prevents or averts preterm labor. Moreover, there are marked increases in the levels of prostaglandins in amnionic fluid and in fetal membranes before the onset of labor. Injury to the fetal membranes resulting in increased prostaglandin release results from uterine infections, stripping of the membranes from the cervix, and the installation of hypertonic saline into the amnionic sac.

SUGGESTED READING

Albrecht E, and Pepe GJ (eds): *Research in Perinatal Medicine (IV). Perinatal Endocrinology.* Perinatology Press, Ithaca, NY, 1985.

Brody SA, and Ueland K: *Endocrine Disorders in Pregnancy.* Appleton and Lange, Norwalk, CT, 1989.

Buster JE, and Simon JA: Placental hormones, hormonal preparation and control of parturition, and hormonal diagnosis of pregnancy. In: *Endocrinology,* Vol. 3, LJ Degroot, ed., Saunders, Philadelphia, pp. 2043–2073, 1989.

Carr BR: The endocrinology of pregnancy: The maternal–fetal–placental unit: In: *Principles and Practice of Endocrinology and Metabolism,* KL Becker, ed., Lippincott, Philadelphia, pp. 887–898, 1990.

Carr BR, and Rainey WE: The adrenal. In: *Infertility and Reproductive Medicine Clinics of North America* JP Bruner, ed., Saunders, Philadelphia pp 749–764 Vol. 5, 1994.

Carr BR, and Simpson ER: Lipoprotein utilization and cholesterol synthesis by the human fetal adrenal gland. Endocr Rev 2:306–326, 1981.

Cunningham FG, MacDonald PC, Gant NF, Levano KJ, and Gilstrap LC: *Williams Obstetrics,* Appleton and Lange, Norwalk, CT, pp. 81–246, 1993.

Davis OK, and Rosenwaks Z: Assisted Reproductive Technology. In: *Textbook of Reproductive Medicine,* B Carr and R Blackwell, eds., Appleton andLange, Norwalk, CT, pp. 571–586, 1993.

Fuchs F, and Klopper A (eds): *Endocrinology of Pregnancy,* Harper & Row, New York, 1983.

Harper MJK: Gamete and zygote transport. In: *The Physiology of Reproduction,* Vol. 1, E Knobil and JD Neill, eds., Raven Press, New York, pp. 103–134, 1988.

Laufer N, Grunfeld L, and Garrise C: In-vitro fertilization. In: *Infertility,* MM Seibil, ed., Appleton and Lange, Norwalk, CT, pp. 481–511, 1990.

Meldrum DR, Chetkouski R, Steingold KA, de Zeigler D, Cedars MI, and Hamilton M: Evolution of a highly successful in vitro fertilization embryo transfer program. Fertil Steril 48:86–93, 1987.

Pederson RA: Early mammalian embryogenesis. In: *The Physiology of Reproduction,* Vol. 1, E Knobil and JD Neill, eds., Raven Press, New York, pp. 187–230, 1988.

Simpson ER, and MacDonald PC: Endocrine physiology of the placenta. Annu Rev Physiol 43:163–188, 1981.

Tulchinsky D, and Little AB (eds): *Maternal-Fetal Endocrinology,* Saunders, Philadelphia, 1994.

Veech LI: *Atlas of the Human Oocyte and Early Conceptus.* Williams & Wilkins, Baltimore, 1986.

Wassarman PM: Fertilization in mammals. Sci Am 259:78–84, 1988.

Weitauf HM: Biology of implantation. In: *The Physiology of Reproduction,* Vol. 1, E Knobil and JD Neill, eds., Raven Press, New York, pp. 231–262, 1988.

Yanagimachi R: Mammalian fertilization. In: *The Physiology of Reproduction,* Vol. 1, E Knobil and JD Neill, eds., Raven Press, New York, pp. 135–185, 1988.

12

Growth Regulation

PINCHAS COHEN
RON G. ROSENFELD

ENDOCRINE REGULATION OF GROWTH

Growth Hormone Secretion and Action

While multiple hormones influence somatic growth, the main regulator of postnatal growth is growth hormone (GH). GH is secreted in a pulsatile manner from the anterior pituitary primarily as a 22 kilodalton molecule (although other forms may be found). GH gene expression is regulated by the pituitary transcription factor Pit-1. Under normal waking conditions, GH levels are often low or undetectable, but several times during the day, and particularly at night during phase 3 of sleep, surges of GH secretion occur. GH secretion is mainly under hypothalamic control, which in turn is regulated by catecholaminergic neurotransmitters from higher cortical centers. The hypothalamic hormones, growth hormone releasing hormone (GHRH) and somatostatin, stimulate and inhibit GH secretion, respectively (see Chapter 6). Many other factors influence this interaction, notably, glucose, which inhibits and certain amino acids which stimulate GH secretion. Exogenous physiological and pharmacological factors are known to stimulate GH secretion. Indeed, some of these agents, which inlude clonidine, L-dopa, and the amino acid arginine, are used in GH stimulation tests. In plasma the majority of GH is bound to a carrier protein termed *GH-binding protein*, which appears to be identical to the extracellular domain of the GH receptor. GH binds to its receptor and mediates a cascade of cellular events. First dimerization of the GH receptor occurs, followed by the activation of a specific kinase, JAK-2. These events lead to the activation of insulin-like growth factor (IGF)-I gene transcription and modulation of other gene expression.

The Somatomedin Hypothesis

In the liver and other target cells, through interaction with its receptor, GH induces the production of somatomedins, or insulin-like growth factors (IGF-I and IGF-II). IGFs are found in plasma bound to a family of proteins called IGF-binding proteins (IGFBPs). The majority of IGFs are bound to IGFBP-3, which, together with a third protein known as the *acid labile subunit* (ALS), form a ternary complex in serum.

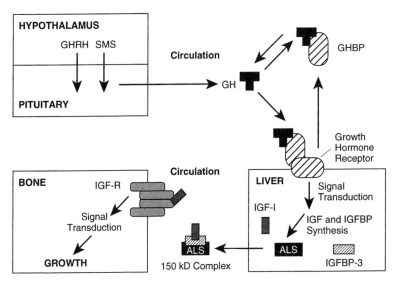

Fig. 12-1 Diagram showing regulation of growth. Growth hormone (GH) secretion is regulated by growth hormone-releasing hormone (GHRH) and somatostatin (SMS). In plasma, GH is bound to the GH-binding protein (GHBP). In the liver, GH-mediates secretion of insulin-like growth factor (IGF)-I, IGF-binding protein (IGFBP)-3, and the acid labile subunit (ALS). IGFs mediate growth by binding to the IGF receptor (IGF-R).

The somatomedins, particularly IGF-I (previously known as SM-C), are thought to interact with target organs, such as growing cartilage to induce growth; they may also feedback on the pituitary to inhibit GH secretion. Both IGFs and their main serum binding proteins (IGFBP-3 and ALS) are reduced in GH deficiency and elevated in conditions of GH excess. It is still uncertain, however, whether all of the anabolic actions of GH are mediated by the somatomedins. This cascade of growth control, known as the *somatomedin hypothesis*, is summarized in Fig. 12-1.

Additional Growth Regulators

Other endocrine factors which are involved in the regulation of growth include thyroid hormone, which is essential for normal linear skeletal growth, and glucocorticoids, which can stunt growth when present in excess. Sex steroids, through several direct and indirect mechanisms, can accelerate growth and normally facilitate the pubertal growth spurt. When present in excess at an earlier age, the resulting acceleration of growth is associated with premature closure of the epiphysis and reduced adult height. This effect has been demonstrated as being mediated by estrogen, regardless of the gender of the individual. Parathyroid hormone and vitamin D are essential for normal skeletal ossification, and the absence of or resistance to these agents is associated with abnormal growth patterns. Nonendocrine factors affecting growth are many and incompletely understood; by far the most important are genetic factors, which determine the growth rate, the age of puberty, and adult height. Psychological and social influences can prevent the genetic potential for growth from being fully expressed. Malnutrition, either through dietary deficiencies, malabsorption, or as a manifestation of chronic illness, can interrupt normal growth. Interestingly, GH levels are often elevated in these conditions, but somatomedin levels are depressed.

One critical growth period is the pubertal growth spurt, which is regulated by changes in GH secretion and by modulation of GH action, both related to the effects of sex steroids. Puberty is associated with an increase in the number and amplitude of nightly GH surges and a rise in serum IGF-I, both of which parallel the increase in the growth velocity of puberty (see Chapter 9). This peak height velocity occurs earlier in girls than in boys. Similarly, growth arrest due to epiphyseal fusion occurs earlier in girls. This, and the fact that puberty begins about 6–12 months earlier in girls, accounts for much of the difference in adult heights between men and women. It is interesting that peak height velocity does not parallel the levels of estrogen (which rise later in girls) or testosterone (which rises earlier in boys).

THE IGF AXIS

Insulin-Like Growth Factors (IGFs) and Their Role in Somatic Growth

IGF-I and IGF-II are two closely related peptide hormones of approximately 7 kD molecular weight. The IGFs were first identified in 1956 and were originally named *sulfation factors* or *somatomedins*. The IGFs belong to a family of peptide hormones that includes relaxin and insulin, and share a high degree of structural similarity with proinsulin. Like proinsulin, they are composed of A, B, and C domains but also include a D domain, which together form the mature IGF peptide. Both IGF-I and IGF-II are synthesized with an additional extension peptide known as the *E peptide*. In the liver and most other sites of IGF production this peptide is removed as part of the post-translational processing of IGF-I and IGF-II. Both IGF-I and IGF-II have a complex gene structure, with the IGF-I gene spanning 95 kb (containing 6 exons) at the long arm of chromosome 12 and the IGF-II gene being made up of 9 exons and having a total genomic size of 35 kb. Both genes are subject to multiple splicing and their mRNA species exist in several different sizes.

The IGFs are important metabolic and mitogenic factors involved in cell growth and metabolism. IGFs are produced in the liver, in bone cells, and in other tissues, at least partially under GH control. Circulating IGFs have direct (endocrine) effects on somatic growth and on the proliferation of many tissues and cell types, both *in vivo* and *in vitro*. However, the IGFs are also thought to be significant autocrine–paracrine factors involved in cellular proliferation. Locally produced IGFs have been demonstrated in bone, brain, prostate, muscle, mammary tissue, and other sites, where they are considered to be responsible for tissue growth and differentiation.

The cardinal role of IGF-I and IGF-II in somatic growth has been elegantly determined by gene-targeting models. Mice in which the IGF-II gene has been eliminated by targeted disruption are severely growth retarded *in utero,* being less than half the normal size at birth. After birth, however, IGF-II-lacking mice grow at a near-normal rate, although their size remains small relative to normal littermates. This illustrates the role of IGF-II as a fetal growth factor, with an unclear postnatal

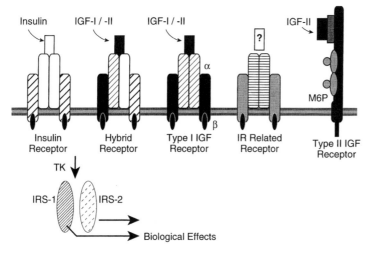

Fig. 12-2   Overview of the insulin-like growth factors (IGF) receptors. The IGFs bind to the type I IGF receptor and similar heterotetramer receptors containing α- and β-subunits and activate a tyrosine kinase (TK), which in turn phosphorylates the insulin-responsive substrates (IRS)-1 and -2. IR, insulin receptor; M6P, manmose-6-phosphate.

role. IGF-I gene–targeted mice are also very small (40% of normal) at birth but display dramatic growth arrest postnatally. It appears that IGF-I is critical for growth at all stages of development. In comparison, GH deficiency results in normal growth in utero and postnatal growth arrest. This indicates that IGF-I is not regulated by GH until the postnatal period and its expression appears to be under genetic, rather than endocrine, control in the prenatal period. Conversely, IGF-I transgenic mice are larger than normal, as are GH transgenic mice. IGF-II transgenic mice, however, do not exhibit rapid growth after birth. The IGFs and their cell surface receptors are depicted schematically in Fig. 12-2.

IGF Receptors (IGF-R)

The IGFs interact with specific receptors, designated as type I and type II IGF receptors, as well as with the insulin receptor. The type I IGF receptor binds IGF-I with high affinity, IGF-II with slightly lower affinity, and insulin with low affinity. The insulin receptor can bind IGF-I and IGF-II, but with much lower affinity than insulin. The type II IGF receptor binds only IGF-II with high affinity (in addition to being a mannose-6-phosphate receptor).

The type I IGF receptor and the closely related insulin receptor are heterotetramers composed of two pairs of α- and β-subunits, which result from post-translational processing of a single gene product which encodes for the entire receptor. The two α-subunits are linked by disulfide bonds and are primarily extracellular; they are involved in ligand binding. The β-subunits are connected to the α-subunit by disulfide bonds and function as intracellular tyrosine kinases. They undergo autophosphorylation after the interaction of the receptors with their ligands, followed by a conformational change. Subsequently, these kinases appear to phosphorylate cytoplasmic molecules known as the *insulin-responsive substrates* (IRS-1 and -2), which are involved in mediating many of the effects of insulin and IGFs. Gene targeting of the type I IGF receptor results in severe *in utero* growth retardation (to 30% of normal) similar to that seen in a double IGF-I + IGF-II gene targeting. These mice and multiple in vitro models demonstrate that the mitogenic effects of IGFs are mediated primarily through the type I IGF receptor. The metabolic effects of IGFs are probably mediated through the interactions of IGFs with the insulin receptor, as well as the type I IGF receptor, or may involve insulin–IGF hybrid receptors. Gene targeting of IRS-1 results in an impairment of both growth and carbohydrate metabolism, which is compatible with its role in mediating both insulin and IGF-I effects.

Recent evidence indicates that there is a class of receptors for the IGF family that has biological properties characterized between those of the insulin and the type I IGF receptors. Analysis of tissues where these receptors appear to be common (such as placenta), as well as transfection experiments, revealed that these receptors are composed of one insulin receptor α-β dimer and one type I IGF receptor α-β dimer. This receptor has been labeled the *hybrid receptor* and appears to have high affinity for insulin as well as for IGFs. The physiological role of the hybrid receptor remains elusive, but it may explain the potent insulin-like effects seen with intravenous administration of IGF-I to humans. While IGF-I binds to the insulin receptor with only 1–2% of the affinity of insulin, it mediates hypoglycemia *in vivo* with 7–10% the effectivity of insulin.

The cDNA for a newly described member of this family of receptors has recently been cloned and, due to the high homology it shares with the insulin receptor, has been designated the *insulin receptor-related receptor* (IRR). There are no known ligands for the IRR, and both IGFs and insulin bind to it very poorly. It appears, however, to be expressed in a specific manner in renal and neural tissues and may have a role in fetal development. In chimeric transfection experiments, it has been documented as a very potent mediator of cellular proliferation.

Although the type II IGF receptor is structurally distinct and binds primarily IGF-II, it also serves as a receptor for mannose-6-phosphate-containing ligands. It is not a member of the insulin receptor family but rather displays homology to certain cytokine receptors. The receptor is 270 kD in size and has 15 repeat extracellular domains. The type II IGF receptor has a very short intracellular domain and an unknown mechanism of signal transduction. It has been associated with changes in calcium influx and has been reported to mediate cell motility. Recently, it has been suggested that the type II IGF receptor serves as a targeting mechanism to mediate lysosomal destruction of excess IGF-II during fetal life. Gene targeting of the type II receptor results in mice that are larger than littermates, which is compatible with its negative growth-modulating effect.

Insulin-Like Growth Factor Binding Proteins (IGFBPs)

A recently recognized class of proteins with high affinities for the IGFs, the IGFBPs, have been shown to be involved in the modulation of the proliferative and mitogenic effects of IGFs on cells. These molecules appear to regulate the availability of free IGFs for interaction with the IGF receptors, as well as to directly interact with cell function. The human IGFBP family consists of six proteins. IGFBP-1 is a 25 kD

protein found in high concentrations in amniotic fluid and is also secreted by hepatocytes under negative control by insulin. IGFBP-2 has a molecular weight of 31
kD, is found in serum, cerebro-spinal fluid (CSF), and seminal plasma, is secreted
by many cell types, and is expressed in many fetal and adult tissues. IGFBP-3 is the
major binding protein in postnatal serum and is synthesized by hepatocytes and other
cells. In plasma, IGFBP-3 is found as part of a 150 kD complex, which also includes
the acid labile subunit and an IGF molecule, all of which are GH dependent. IGFBP-
4 is a 24 kD protein that has been identified in serum and in seminal plasma, as well
as in numerous cell types. IGFBP-5 is found in CSF and, in smaller amounts, in
serum; it is also observed in rapidly growing fetal tissues. IGFBP-6 is found in CSF
and is produced by transformed fibroblasts; IGFBP-6 has relative specificity for IGF-
II over IGF-I. A high degree of structural homology among the six cloned cDNAs
for the binding proteins and remarkable sequence conservation across species have
been demonstrated. IGFBPs are tightly regulated by various endocrine factors and
are uniquely expressed during ontogeny. The IGFBPs are schematically depicted in
Fig. 12-3.

The molecular mechanisms involved in the interaction of the IGFBPs with the
IGFs and their receptors remain enigmatic, but three possible models for their actions
have been suggested. The first mechanism suggested for IGFBP action is that these
molecules regulate the availability of free IGFs for interaction with the IGF receptors.
Indeed, the addition of IGFBPs to many *in vitro* cell culture systems results in the
inhibition of IGF actions within these experimental systems. Additionally, several
trophic hormones have been shown to suppress IGFBP production by their target
cells. These include TSH inhibiting IGFBP-2 in thyroid cells, FSH inhibiting IGFBP-
3 in Sertoli cells, and FSH inhibiting the elaboration of IGFBP-5 produced by granulosa cells. On the other hand, $1,25\text{-}(OH)_2$ vitamin D, which is inhibitory for bone
cells, stimulates the production of IGFBP-4. In these models, it has been assumed
that the suppression or stimulation of an inhibitory IGFBP stimulates or inhibits cell
growth, respectively.

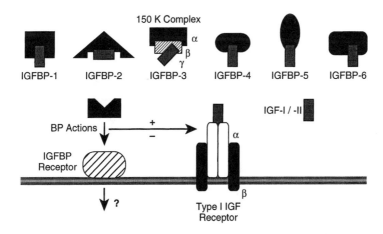

Fig. 12-3 The insulin-like growth factor binding proteins (IGFBP-1 through -6) are
shown at the cellular environment.

In other systems, however, IGFBPs have been demonstrated to enhance IGF action under circumstances that may involve cellular processing of the IGFBPs. These IGF-enhancing actions of IGFBPs have been demonstrated for IGFBP-1, IGFBP-3 and IGFBP-5. It appears that a processing step, such as an affinity change involving proteolysis or phosphorylation, may be required for the activation of this function.

Finally, as recently shown in several *in vitro* systems, IGFBPs have an IGF-independent mechanism of cell inhibition. The mechanism by which IGFBPs independently interact with cells has not yet been fully elucidated. IGFBP-1 and IGFBP-2 contain a unique recognition sequence, which may allow them to interact with a class of cell surface receptors known as *intergrin receptors*. IGFBP-3 does not contain such a sequence but has recently been shown to bind to specific receptors on the cell membrane. It appears that IGFBP-3 mediates an IGF-independent growth inhibitory effect via its own receptors.

IGFBP Proteases

Recently recognized as potential modulators of IGF action is a group of enzymes that are capable of cleaving IGFBPs. First identified in pregnancy serum, IGFBP-3 proteolytic activity is responsible for the disappearance of intact IGFBP-3 from the serum of pregnant individuals, with no change in IGFBP-3 immunoreactivity. IGFBP proteases have also been reported in the serum of severely ill patients in states of cachexia, in patients with growth hormone receptor deficiency, in prostate cancer patients, and in patients with other malignancies. Seminal plasma contains an IGFBP-3 protease which has been identified as prostate-specific antigen (PSA). This IGFBP protease belongs to the kallikrein family. CSF from patients with malignancies and urine from patients with renal failure also contain IGFBP proteases. IGFBP proteases have been described for all six IGFBPs and both acid and neutral activities have been detected. It has been speculated that the IGFBP proteases are important modulators of IGF bioavailability and bioactivity through their modification of the IGF carrier proteins. The proteolytic activity may play a role in regulating IGF availability at the tissue level by altering the affinity of the binding proteins for the

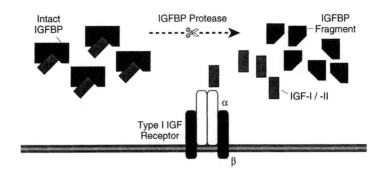

Fig. 12-4   Mechanism of action of insulin-like growth factor binding proteins (IGFBP) proteases. The proteases cleave IGFBPs and release free IGFs to interact with the IGF receptors.

growth factors, releasing free IGFs, and allowing increased receptor binding. A theoretical model of such interactions is illustrrated in Fig. 12-4.

ADDITIONAL GROWTH-REGULATING PEPTIDES

The FGF Family of Peptides and Receptors

Fibroblast growth factors (FGFs) constitute an important and rapidly expanding family of peptide cytokines which are important in the regulation of many tissues. Initially it was suggested that these factors may be specific for cells of stromal lineage, but it appears that many other cells respond to FGFs as well. There are at least 7 different FGFs (FGF-1 through -7) that have been identified, including the best characterized acidic FGF (aFGF or FGF-1), basic FGF (bFGF or FGF-2), and keratinocyte growth factor (KGF or FGF-7). The FGFs mediate their actions by binding to at least 3 receptors, FGFR1, FGFR2, and FGFR3, which have distinct tissue distributions. For the most part, FGFs appear to be autocrine–paracrine growth factors that participate in organ growth and differentiation as well as in carcinogenesis but not in somatic growth. An exception to this rule is the observation that the genetic form of dwarfism known as *achondroplasia* is caused by a mutation in the *FGFR3* gene, suggesting that normal FGFR3 signaling is essential for the normal growth of long bones. A human mutation in the *FGFR2* gene causes craniosynostosis, a disease characterized by abnormal closure of the bones in the skull but normal long bone growth. So far, targeted disruptions of murine *FGF*-2 and *FGF*-5 genes have been reported, and the phenotypes associated with them does not include growth abnormalities. Future studies will further define the role that different components of the FGF systems play in somatic and tissue growth.

The EGF System

Epidermal growth factors (EGFs) and their receptors are ubiquitous in many tissues and participate in developmental processes in mice such as precocious eyelid opening and tooth eruption. The mitogenic actions of EGF have been extensively explored in cell culture systems, and the receptor for EGF was characterized as a prototype model for signal transduction involving tyrosine kinases. The extensive *in vitro* data indicate multiple cellular functions of EGF. EGF has been identified in most body fluids of several mammalian species; however, neither EGF antibody administration to newborn animals nor gene targeting of EGF has caused major deleterious effects as might be expected from the *in vitro* studies. The EGF family of growth factors appears to be important in mammalian development and function, although the precise roles and significance are not yet clear. Members of the EGF family may have a role in embryogenesis and fetal growth since receptors have been identified in fetal tissues. It has been proposed that abnormal EGF–EGF receptor interactions may be instrumental in the development of cancer, but it appears that they are not involved in somatic growth.

Other Growth-Promoting Peptides

An ever-growing number of growth factors are being recognized and multiple hormones and peptides are being characterized as having growth-promoting activities in certain cell types. In general, these molecules appear to lack somatic growth-promoting effects but play important autocrine–paracrine as well as endocrine roles. Notable among these are groups of growth factors which have tissue specific effects. Endothelin, platelet-derived growth factor (PDGF), and vascular-epithelial growth factor (VEGF) regulate angiogenesis and other vascular processes in addition to modulating the function of numerous cultured cells. A variety of hematopoetic growth factors such as the granulocyte and macrophage colony-stimulating factors, (GCSF, MCSF), erythropoetin, and thrombopoeitin promote the growth of the different lineages of the hematopoetic cells. The growth of various cells of the immune system is stimulated by an array of cytokines including interleukins and interferons (see Chapter 4). The complex array of cells that comprise the nervous system is under the regulatory influence of specific growth factors such as nerve growth factor (NGF), the neurotrophins, and brain- and glial-derived neurotrophic factors (BDNF, GDNF). Other growth factors that have been attributed to specific tissues (such as hepatocyte growth factor [HGF]) are being recognized as having a general growth-promoting effect in numerous tissues. Additional organ-specific growth regulatory processes are being described in the gastrointestinal tract, kidneys, etc.

Growth Inhibitory Peptides

Of particular interest are a class of cytokines which can negatively modulate cellular growth. The transforming growth factor-β (TGF-β) can act both as an agent that mediates cellular growth and malignant transformation and as a growth inhibitory substance that has the potential of arresting the growth of normal and neoplastic cells. Tumor necrosis factors (TNF) and other compounds have been described to have similar effects. These molecules may regulate the entry of cells into programmed cell death (apoptosis). The growth inhibitory processes of TGF-β and other cytokines may prove to be of great importance in the development of cancer treatments.

DISORDERS OF GROWTH

Growth Curves and Growth Patterns

The growth of an individual is a dynamic function of height increment over time. Accordingly, to accurately evaluate growth, one must obtain serial and accurate measurements of a child's height and plot them on a growth curve. Growth curves are based on normative population data, and the ones commonly available reflect the North American means compiled by the National Center for Health Statistics of the U.S. Public Health Service. It is preferable to use those growth curves which show the standard deviations (SD) from the mean or Z score. These types of curves allow a more precise estimation of the degree of short stature, when it exists. However, it must be emphasized that such growth curves are cross-sectional rather than longitudinal, and they are particularly nonrepresentative during adolescence, when growth acceleration may normally occur at widely divergent ages. Consequently, it is preferable to perform consecutive measurements of height that are then used to calculate growth velocity and are plotted on a growth velocity chart.

Definition and Epidemiology of Short Stature

Since height is a normally distributed function, it is simple to provide a statistical definition of *normal height* as that falling between two standard deviations (Z scores) above and below the mean for each age-group. This will include those individuals between the 2nd and 98th percentiles but will exclude approximately 4% of the population, many of whom represent variants of the normal state. Prediction of adult height is helpful in determining the diagnosis and prognosis of short children. Factors influencing adult height include: mean parental height, the patient's current height and sexual maturation rating, the chronological age, and the bone age. Many methods have been developed for calculating estimated adult height. A simple method of predicting adult height is to determine the mean parental height and add or subtract (for males and females, respectively) 6.5 centimeters or 2.5 inches (genetic potential). The use of a bone age radiograph allows for a much more accurate prediction. The most commonly used method for the determination of bone age is the Greulich and Pyle (GP) method, which relies on an atlas of left hand and wrist radiographs of healthy children at various ages. Likewise, the simplest and most common method for adult height prediction is the Bayley-Pinneau (BP) method. This method assigns to each bone age a percentage of the adult height that has already been achieved, and the adult height can then be calculated by dividing the present height by the fraction given in the table. The confidence limits of the above predictions are usually within two inches of the actual achieved height.

Causes of Short Stature

A height greater than 2.5 standard deviations (SD) below the mean for age during adolescence can have many causes, including, as shown in Table 12-1, multiple endocrine and systemic disorders. Of these, the normal variants are by far the most common. When normal variant short stature is associated with delayed puberty, skeletal maturation, and growth, it is termed *normal variant constitutional delay* (NVCD). When it is found in individuals with normal progression through puberty and normal skeletal maturation, it is termed *normal variant short stature* (NVSS). NVSS is a growth pattern representing the genetic potential for growth of that individual and is not associated with any endocrine or systemic pathology of any kind.

NVSS is often, but not always, associated with a family history of short stature in at least one parent. These patients commonly have normal birth weights, and in the first few years of life, their growth pattern usually settles at or just below the third percentile and then follows this percentile with a pubertal growth spurt at a normal age. The adult height achieved by these patients is typically below the third percentile and is predictable by bone age measurements. Criteria for the diagnosis

Table 12-1   Causes of Short Stature

Normal variants
   Normal varient short stature (NVSS)
      familial
      sporadic
   Normal varient constitutional delay (NVCD)
Abnormal variants
   Chromosomal disorders
      Turner's syndrome
      Down's syndrome (trisomy 21)
      Other chromosomal conditions
   Dysmorphic syndromes
   Skeletal dysplasias (primary and secondary)
      Achondroplasia (FGFR3 deficiency)
   Systemic disease
      Malnutrition
      Eating disorders
      Malabsorption
      Other occult chronic disease
   Psychosocial dwarfism
   Endocrine disorders
      Growth hormone deficiency (GHD)
      Hypothyroidism
      Glucocorticoid excess
      Congenital abnormalities of bone metabolism

of NVSS include the exclusion of organic or emotional pathology by history, physical exam, and simple laboratory tests outlined later.

NVCD occurs in both sexes, although males are more likely to present to a physician with this condition. A family history of delayed maturation is usually present in at least one parent. The growth pattern typically observed in this condition consists of growth at or slightly below the fifth percentile throughout childhood, with further deviation from the normal growth curve at the time puberty should normally occur. Puberty develops late in these patients, but when it does, it is associated with a near normal growth spurt and the adult height achieved is entirely normal. The diagnosis of NVCD is made by exclusion. Systemic and endocrine causes must be ruled out by the appropriate tests before NVCD is diagnosed.

Other causes of growth failure may be the result of chromosomal disorders; among these, the most common are Down's syndrome (trisomy 21) and Turner's syndrome, which is caused by a monosomic 45,X state or other abnormalities of the X chromosome. Short stature is the most common finding in Turner's syndrome, with primary ovarian failure being the other important feature (see Chapter 8). Additional findings include typical dysmorphic features, and cardiac and renal anomalies. Nevertheless, short stature may be the only presenting sign in a girl with Turner's syndrome and even this finding may only manifest itself late in chidhood. It is therefore imperative to consider this condition in the evaluation of any short female.

Growth failure associated with systemic disease and malnutrition is usually characterized by normal or elevated growth hormone, but depressed somatomedin levels. During adolescence, these conditions are often associated with delayed puberty as well as a delayed bone age and must be differentiated from NVCD. In underdeveloped countries, overt protein calorie malnutrition is the most common form of malnutrition, but in developed countries, self-inflicted malnutrition (anorexia nervosa and bulimia) or malnutrition associated with chronic disease is also common. Specific mineral deficiencies, especially of iron and zinc, can also be associated with growth delay.

Gastrointestinal disease may also cause growth retardation, often with little or no appreciable symptoms, usually as a result of malabsorption. Insulin-dependent diabetes mellitus, when not adequately controlled, is associated with depressed somatomedin levels and poor growth. Adequate control of the diabetic state and regulation of the nutritional intake can restore growth rate in diabetics back to the normal range. Poverty, depression, and neglect can lead to a form of growth delay referred to as *psychosocial dwarfism*. Although nutritional factors may play a role in the etiology of this condition, defective GH secretion has also been documented.

Endocrine Causes of Short Stature

It is helpful to bear in mind that chronic illness frequently tends to retard weight to a greater degree than height, whereas endocrinopathies tend to do the opposite. Endocrine abnormalities account for only about 10% of the cases of short stature; it is, however, imperative to diagnose these children promptly, since adequate therapy can normalize the growth and correct the other abnormalities associated with these disorders.

Hypothyroidism in childhood is generally associated with poor linear growth. Indeed, poor growth and/or delayed puberty may be the presenting signs of chronic hypothyroidism. It is important to note that hypothyroidism may coexist with GH deficiency and that it may also cause false negative results on provocative GH testing. It is therefore imperative to consider hypothyroidism as the first step of short stature evaluation. Glucocorticoid excess, either iatrogenic or due to Cushing's disease or primary adrenal hypercortisolism, is associated with attenuation of the growth rate (see Chapter 14).

GH Deficiency

A classification of GH abnormalities is given in Table 12-2. Familial GH deficiency (GHD) displays a variety of hereditary patterns, ranging from autosomal recessive (type I), to autosomal dominant (type II) and X linked (type III). Congenital GHD may also be caused by a deficiency of the *Pit1* gene and be associated with additional pituitary hormone deficiencies. Congenital GHD as part of panhypopituitarism can also include abnormal secretion of corticotropin (ACTH), thyrotropin (TSH), vasopressin, or gonadotropins. Congenital GHD is usually recognized early in life, with the constellation of hypoglycemia, small phallus, relative adiposity, and subnormal length. Type I GHD is divided into two types. Type IA GHD has complete GHD secondary to gene deletion or point mutations and is associated with development of GH antibodies when treatment is initiated. Type IB patients have splice-site mutations in the GH gene, have very low levels of endogenous GH, and do not

Table 12-2   Classification of Growth Hormone Abnormalities

Growth hormone deficiency (GHD)
Familial
Type IA (complete, autosomal recessive)
Type IB (nearly complete, autosomal recessive)
Type II (autosomal dominant)
Type III (X-linked GHD)
Pit1 deficiency (panhypopituitarism)
Sporadic/idiopathic
Peri-sellar tumors
CNS irradiation and chemotherapy
Postinfectious/post-traumatic
Growth hormone insufficiency
Partial GH deficiency
Neurosecretory dysfunction
Biologically inactive GH
GH resistance
Genetic (Laron syndrome)
Acquired (liver diseases)

develop GH antibodies. Sporadic GHD can be either idiopathic (which is most common) or secondary to pathological processes at the level of the pituitary or hypothalamus. Space-occupying lesions of the brain are the second-most frequent cause of GHD. Infiltrative processes and post-traumatic and postinflammatory conditions can all cause GHD and growth failure.

A growing number of children presenting with GHD and growth failure in adolescence are survivors of malignancies who received central nervous system (CNS) irradiation and chemotherapy.

While some short children may successfully pass provocative GH testing, they may nevertheless respond to GH therapy. Integrated GH levels have been shown to be low in a subset of these children, and the term *neurosecretory dysfunction* (of GH secretion) has been used to describe their condition. Regardless of the etiology, children with characteristic GHD display short stature, decreased growth velocity, delayed bone age, and low serum IGF-I and IGFBP-3 levels. Failure to adequately secrete GH in two stimulation tests is currently required for the diagnosis of classical GHD, but abnormal serum levels of GH-dependent markers, such as IGF-I, IGFBP-3, and ALS, may soon replace it as the diagnostic tests of choice.

Biologically inactive GH is another potential cause of short stature. Patients with this disorder would be expected to have a normal response to exogenous GH, but no convincing cases of this condition have been documented.

Laron-type dwarfism (or GH receptor deficiency [GHRD]) is caused by mutations or deletions of the GH receptor gene and is associated with elevated GH levels and depressed IGFs and IGFBP-3 levels. This condition can now be diagnosed by assays of the plasma GH-binding protein.

Evaluation of Short Stature

Historical features that aid in the evaluation of short stature include the pregnancy and perinatal history, the growth pattern from infancy to adolescence (for both height and weight), the general medical history, and any symptoms of systemic disease. Particular emphasis should be placed on the review of systems for the presence of neurological, visual, or gastrointestinal disturbances and any symptom of hypothyroidism. The family history should be probed for parental heights and growth and maturation patterns, as well as for any short stature in the family background.

The complete physical exam should include accurate auxological measurements of height, weight, arm span, upper-to-lower segment ratios, and an exam of the thyroid and sexual maturity rating. Any abnormality in the physical exam, especially those suggesting systemic disease, or evidence of dysmorphism, should be carefully noted.

In the child with clinically significant short stature, where a systemic disease is not the obvious cause of the growth problem, laboratory evaluation is frequently necessary. The presence of occult chronic disease causing short stature can usually be ruled out with a complete blood count, erythrocyte sedimentation rate, a urinalysis, and a blood chemistry panel that includes serum electrolytes, blood urea nitrogen (BUN), creatinine, calcium, phosphorus, and alkaline phosphatase. Thyroid function can be evaluated with a TSH and free thyroxine (T_4). In girls, a karyotype is often necessary in order to rule out Turner's syndrome, even if no other stigmata exist. A bone age is essential for the diagnostic evaluation and prognosis of all short children. Measurements of serum IGF-I and IGFBP-3 levels can predict the presence of GHD. While IGF-I levels may be low in malnutrition, systemic disease, and hypothyroidism, IGFBP-3 levels are commonly normal in the absence of GHD. The clinical utility of serum growth factor levels is mainly as an adjunct or an alternative for GH stimulation tests. The serum levels of the third component of the IGF complex, the ALS, are also GH dependent and may soon become part of the diagnostic evaluation of GHD. Finally, in the adolescent male suspected to have NVCD and considered for therapy, a serum testosterone level is often helpful.

If the child being evaluated for possible GHD is growing poorly, (i.e., is shorter than 2.5 SD below the mean for age and has a low growth velocity and low serum IGF-I and IGFBP-3 concentrations), GH testing is often performed. In normal children random, unstimulated GH levels are typically under 5 ng/ml; therefore, if GHD is suspected, GH should be measured after appropriate pituitary stimulation. Pharmacological agents which stimulate GH include L-dopa, clonidine, propranolol, glucagon, insulin, and arginine. Levels of less than 10 ng/ml (by a polyclonal GH assay) are diagnostic of GHD, and levels greater than 10 are considered normal (although neurosecretory dysfunction of GH secretion remains a possibility). Priming the pituitary with estrogen will sometimes augment the GH response. An MRI of the head is recommended in patients with documented GHD to evaluate possible local pathology.

Management of Short Stature

After the evaluation of short stature has been completed, the physician must balance the needs of the child and the family, the severity of the growth retardation, and the expected adult height in deciding upon a therapeutic plan. It is important to appreciate both the potential of growth-enhancing therapy to affect these issues and the cost and possible complications of therapy. For some of the above mentioned disorders, particularly the normal variants, reassurance and support for the patient and the family, coupled with close follow-up, are all that may be required. On the other hand, for some of the pathological causes of short stature, as well as for selected severe cases of the normal variants, medical therapy is prudent. NVCD is associated with normal predicted adult height and, in the case of the socially well adjusted adolescent, reassurance is the most important aspect of therapy, since these patients will all enter puberty eventually and will experience a growth spurt at that time. Counseling the patient and the family should center around the expected growth that comes with puberty. If, however, the patient feels that he has become so stigmatized by the combination of short stature and sexual immaturity that reassurance alone is insufficient, a course of androgenic steroids (such as intramuscular testosterone) should be considered in male patients older than 14 years who have normal thyroid and GH evaluation and low testosterone levels.

Turner's syndrome patients have been shown to benefit from growth-enhancing therapy, and the currently recommended regimen for their management involves initiation of GH, regardless of the response to provocative testing. Combining GH with low-dose anabolic steroids such as oxandrolone enhances the early growth response. Other chromosomal and genetic syndromes do not respond to therapy quite as well and have not yet been adequately studied with regard to GH treatment.

Growth hormone deficiency of any type requires GH therapy, and, with the unlimited availability of recombinant GH, therapy can be provided to all these patients. If the diagnosis of GHD is made, GH therapy should be started immediately. Growth hormone is best given daily, at bedtime, as a subcutaneous injection at a dose ranging from 0.025 to 0.05 mg/Kg/day. The cost of GH may be a significant burden on the average family budget with current costs ranging between $10,000 and $50,000 a year. All of the above considerations should be weighed and discussed before embarking on therapy.

In patients with GH unresponsiveness, as is found in GHRD (Laron syndrome), therapy with IGF-I has been documented to be beneficial in improving growth velocity and adult height prediction.

In patients in whom a growth disorder is being managed around the time of puberty, closure of the epiphysis (either prematurely or at an appropriate age) may hamper the effects of the growth-promoting therapy employed. As the closure of the epiphysis has recently been demonstrated to be an estrogen effect, inhibition of puberty (with GnRH analogs) or estrogenic blockade (with estrogen receptor antagonists or aromatase inhibitors) can enhance and prolong the growth treatment and result in taller adult stature.

CONCLUSION

Over the last few years, the medical and scientific literature has witnessed an explosion of information regarding the physiology of growth and growth factors in general and the various components of the IGF axis in particular. Undoubtedly, the coming years will bring even more new information on the physiology and pathology of these key cellular regulators. Furthermore, these discoveries are likely to lead to the increasing use of diagnostic tests as well as to therapeutic applications of these agents.

SUGGESTED READING

Cohen P, and Rosenfeld RG: Growth problems in adolescence. In: *Textbook of Adolescent Medicine,* RE MacAnarney, ER Kreipe, DP Orr, and GD Comerci, eds., Saunders, Philadelphia, pp. 495–508, 1992.

Cohen P, and Rosenfeld RG: Physiological and clinical relevance of the IGFBPs. Curr Opin Pediatr 6:462–467, 1994.

Cohen P, and Rosenfeld RG: The IGF axis. In: *Human Growth Hormone, Basic and Scientific Aspects,* AL Rosenbloom ed., CRC Press, Boca Raton, FL, pp. 43–56, 1995.

Fraiser SD, and Lippe BM: The rational use of growth hormone during childhood. J Clin Endocrinol Metab 71:269–273, 1990.

Lamson G, Cohen P, Guidice LC, Fielder PJ, Oh Y, Hintz RL, and Rosenfeld RG: Proteolysis of IGFBP-3 may be a common regulatory mechanism of IGF action *in vivo.* Growth Reg 3:91–95, 1993.

Laurence DJ, and Gusterson BA: The epidermal growth factor. A review of structural and functional relationships in the normal organism and in cancer cells. Tumour Biol 11:229–261, 1990.

Ledoux D, Gannoun-Zaki L, and Barritault D: Interactions of FGFs with target cells. Prog Growth Factor Res 4:107–120, 1992.

LeRoith D, McGuinness M, Shemer J, Stannard B, Lanau F, Faria TN, Kato H, Werner H, Adamo M, and Roberts CT: Insulin-like growth factors. Biol-Signals 1:173–181, 1992.

Liu JP, Baker J, Perkins AS, Robertson EJ, and Efstratiadis A: Mice carrying null mutations of the genes encoding insulin-like growth factor I (Igf-1) and type 1 IGF receptor (Igf1r). Cell 75-59–72, 1993.

Phillips JA 3rd, and Cogan JD: Genetic basis of endocrine disease 6. Molecular basis of familial human growth hormone deficiency. J Clin Endocrinol Metab 78:11–16, 1994.

Wilson DM, Kraemer HC, Ritter PL, and Hammer LD: Growth curves and adult height estimation for adolescents. Am J Dis Child 141:565–570, 1987.

13

The Thyroid

JAMES E. GRIFFIN

Thyroid hormones are important for the normal growth and development of the maturing human. In the adult, thyroid hormones maintain metabolic stability by regulating oxygen requirements, body weight, and intermediary metabolism. Thyroid function is under hypothalamic–pituitary control, and thus like the gonads and adrenal cortex, it serves as a classical model of endocrine physiology. In addition, the physiological effects of thyroid hormones are regulated by complex extrathyroidal mechanisms resulting from the peripheral metabolism of the hormones. These mechanisms are not under hypothalamic–pituitary regulation.

SYNTHESIS AND SECRETION OF THYROID HORMONES

Iodide Kinetics

Adequate iodide intake is necessary for normal thyroid hormone synthesis since thyroid hormones are the only substances in the body that have iodine in their structure. The major sources of dietary iodide are iodated bread, iodized salt, and dairy products. Individuals may also be exposed to iodide in medications, disinfectants, and radiographic contrast agents. The minimum dietary requirement of iodide is about 75 μg/day. This amount or half of the recommended daily allowance of iodide (150 μg/day) could be obtained from 10 g of salt alone if iodized at the recommended World Health Organization rate of 1 part KI in 100,000 parts NaCl. In the U.S., iodized salt contains 1 part K1 in 10,000 parts NaCl.

Dietary iodine intake in the United States ranges from 200 to 500 μg/day. The higher number is used in Fig. 13-1 to depict normal iodine metabolism. Iodine is ingested in both inorganic and organically bound forms, and the organically bound form is converted to inorganic iodide. Iodide itself is efficiently absorbed; little is lost in the stool. Absorbed iodide is largely confined to the extracellular fluid, and its concentration in extracellular fluid is normally about 1 μg/dl and the total extracellular fluid pool about 250 μg. The major sites of removal of iodide from the extracellular fluid are the thyroid and the kidneys. The renal clearance of iodide—30–40 ml/minute—results from complete glomerular filtration and passive reabsorption. Iodide clearance in humans appears to be largely independent of serum

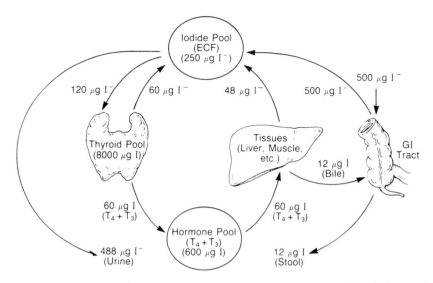

Fig. 13-1   Normal pathways of iodine metabolism in a state of iodine balance. T_4, thyroxine; T_3, 3,5,3'-triiodothyronine; ECF, extracellular fluid. Arrows indicate daily flux from one compartment to another. Numbers in parentheses indicate pool sizes. (Redrawn from Larsen PR, and Ingbar SH: In: *Williams Textbook of Endocrinology*, 8th ed., JD Wilson and DW Foster, eds., Saunders, Philadelphia, p. 361, 1992.)

iodide concentration. Total renal iodide excretion approximates intake. Iodide removed from the serum by the thyroid is returned to the circulation as iodothyronines (thyroid hormones), whose iodine is largely returned to the extracellular fluid after peripheral deiodination. The hormone pool of iodide includes that in the circulation as well as thyroid hormones in the tissues. The largest pool of iodine is in the thyroid, which contains about 8000 μg, virtually all in iodinated amino acids (Fig. 13-1). The thyroid iodine pool has a slow turnover of about 1% daily.

Iodide Transport and Organification

An overall scheme of thyroid hormone biosynthesis, secretion, and metabolism is depicted in Fig. 13-2. The thyroid concentrates inorganic iodide from the extracellular fluid by an active, saturable energy-dependent process. Under normal conditions the rate of inward clearance of iodide by the thyroid exceeds the rate of incorporation of iodide into amino acids (organification) and back diffusion so that the thyroid to plasma (T:P) ratios of iodide are greater than unity, usually about 20 to 40. Both thyroid-stimulating hormone (TSH) and an internal autoregulatory system influence this iodide transport mechanism. TSH stimulates iodide transport; in general, an increasing glandular content of organic iodine diminishes iodide transport and its response to TSH. Iodide transport is inhibited by certain anions, notably perchlorate and thiocyanate. Perchlorate both inhibits trapping of iodide by the gland and facilitates back diffusion out the gland. Thiocyanate mainly increases iodide efflux.

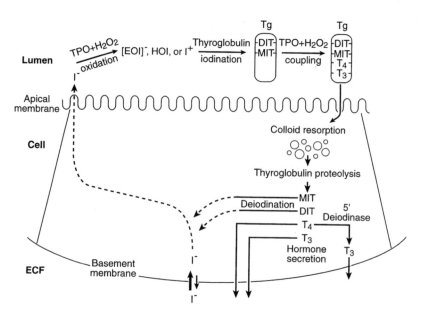

Fig. 13-2   Diagram of the steps of thyroid hormone biosynthesis and release. TPO, thyroid peroxidase; MIT, monoiodotyrosine; DIT, diiodotyrosine; T_4, thyroxine; T_3, 3,5,3′-triiodothyronine; I^+, activated iodide; ECF, extracellular fluid. (Redrawn from Taurog A: In: *Werner and Ingbar's The Thyroid*, 6th ed., LE Braverman and RD Utiger, eds., Lippincott, New York, p. 87, 1991.)

    The two principal thyroid hormones are thyroxine (3,5,3′,5′-tetraiodo-L-thyronine), or T_4, and 3,5,3′-triiodo-L-thyronine, or T_3 (see Fig. 13-3 for chemical structure). These hormones are produced from the precursor amino acids diiodotyrosine and monoiodotyrosine. These iodoamino acids are formed by the iodination of tyrosine residues within the matrix of thyroglobulin (Fig. 13-2), a large protein molecule unique to the thyroid. The iodination is catalyzed by thyroid peroxidase, a membrane-bound heme-protein enzyme. The source of the hydrogen peroxide is not known. The active iodinating agent is probably I^+ or hypoiodous acid (HOI) formed by peroxidase-catalyzed oxidation of I^-.

    The thyroid has a limited capacity to utilize excess iodides in hormonogenesis. Progressive increases in the concentration of iodide in extracellular fluid are associated with progressively decreasing values of the T:P ratio while the amount of iodide actively transported into the gland and organified rises progressively until reaching a maximum followed by a sudden decrease, termed the *acute Wolff–Chaikoff effect*. The increasing iodination occurs despite a falling percentage uptake of circulating iodide. The serum iodide level beyond which iodination decreases is 15–20 μg/dl. The inhibitory effect of excess iodide upon iodination in thyroid glands requires iodide organification. Inhibition is of brief duration and escape from or adaptation to iodide inhibition occurs after a few days.

    T_4 and T_3 form within thyroglobulin (Fig. 13-2) by a coupling reaction involving two diiodotyrosyl residues or a monoiodotyrosyl and diiodotyrosyl residue,

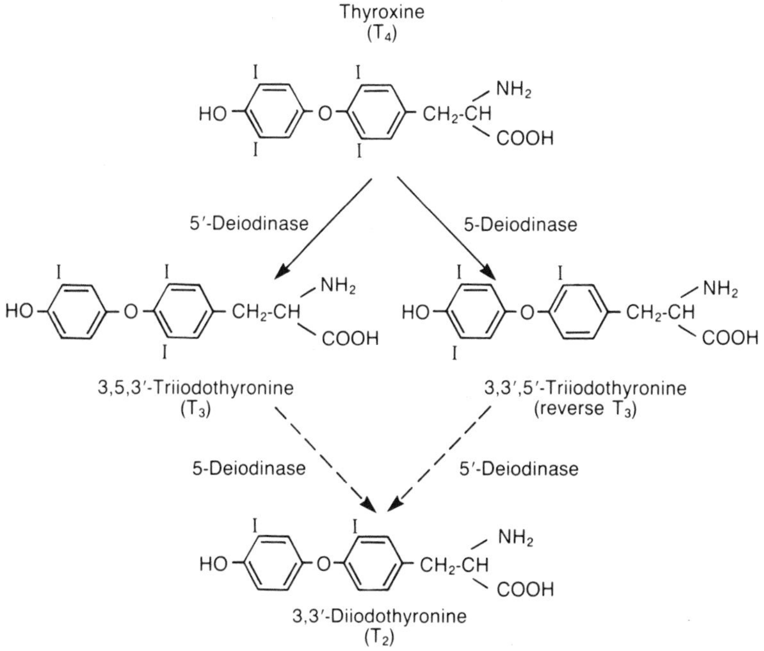

Fig. 13-3 Structures of the principal thyroid hormones, thyroxine, and 3,5,3'-triiodo-thyronine and part of the pathway of metabolism by deiodination. (Reprinted from Griffin JE: *Manual of Clinical Endocrinology and Metabolism,* McGraw-Hill, New York, p. 50, 1982.)

respectively. This coupling reaction occurs separately from iodination and also is catalyzed by thyroid peroxidase. The specific tertiary structure of thyroglobulin is thought to be important for efficient coupling since disruption of its native structure or substitution of other proteins for thyroglobulin results in very low levels of T_4 formation. Thyroglobulin is a large molecule with a molecular weight of 660,000. Almost all of the thyroglobulin in the normal thyroid gland is present as a soluble protein in the lumen of the thyroid follicle. Only three or four molecules of thyroxine are formed in each molecule of thyroglobulin, and certain tyrosines such as the one at position 5 from the amino-terminus appear to be favored for sites of thyroxine formation. The thyroid normally makes much more T_4 than T_3 if iodine intake is adequate, and the T_4 to T_3 ratio of normal thyroglobulin is 10 to 1.

Storage and Release of Thyroid Hormones

In contrast to most endocrine glands, which do not store appreciable amounts of hormone, the thyroid contains several weeks supply of thyroid hormones in the thyroglobulin pool. Thyroglobulin must be hydrolyzed to release T_4 and T_3. This cleavage of iodothyronines from thyroglobulin is accomplished by lysosomal proteases within the follicular cell (Fig. 13-2). In response to TSH stimulation, colloid droplets are formed at the apical surface of the follicular cells by endocytosis of adjacent luminal colloid. Lysosomes migrate from the basal end of the follicular cell to fuse with the colloid droplets and release their hydrolytic enzymes. Digestion of thyroglobulin apparently releases all iodinated amino acids in the free form. Iodotyrosines are largely prevented from being released into the circulation by the action of an intracellular deiodinase that is specific for iodotyrosines and not active on iodothyronines (Fig. 13-2). The iodide released from monoiodotyrosine and diiodotyrosine is then available for reutilization in the thyroid gland. The free T_4 and T_3 diffuse from the cell into the circulation.

Although the thyroid is capable of deiodinating T_4 to T_3, the contribution of this pathway of T_3 formation to overall T_3 production is considered small (see below).

Thyroglobulin itself is not normally released into the circulation in significant quantities. Some small quantities of thyroglobulin can be measured by sensitive immunoassays in peripheral blood of most normal subjects. It appears that this thyroglobulin reaches the blood via lymphatics. When the thyroid gland is damaged by disease processes, such as the inflammation associated with thyroiditis, significant quantities of thyroglobulin may leak into the circulation.

Release of thyroid hormones is inhibited by excess iodide. This effect of iodide is used in the treatment of severe hyperthyroidism. It may involve inhibition of adenylyl cyclase activity that increases as part of the normal response to thyroid stimulators.

TRANSPORT OF THYROID HORMONES AND TISSUE DELIVERY

Iodothyronines in the circulation are largely bound to thyroid hormone-binding globulin (TBG), transthyretin, and albumin. Under normal conditions at equilibrium, about 70% of both T_4 and T_3 is bound to TBG. About 10% of T_4 is bound to transthyretin and 15% to albumin. In contrast, very little T_3 is bound to transthyretin and about 25% is bound to albumin. About 3% of T_4 and 6% of T_3 are bound to lipoproteins. Only about 0.03% of T_4 and 0.3% of T_3 is "free" or dialyzable in *in vitro* studies. T_4 binds more tightly to serum-binding proteins than T_3, resulting in a lower metabolic clearance rate and longer serum half-life. The serum half-life of T_4 is about 7 days whereas the half-life of T_3 is less than 1 day.

Until recently it was thought that only the free fraction of thyroid hormones was available for entry into tissues. As discussed in Chapter 5, we now recognize that the capillary exchangeable or bioavailable thyroid hormone, as assessed by *in vivo* studies, is greater than the free fraction as assessed by *in vitro* analysis. Measured bioavailable T_3 in brain tissue (as an example of peripheral nonhepatic tissues) was found to equal 10% of the albumin-bound T_3, whereas bioavailable T_3 in liver tissue was equal to all of the albumin-bound fraction and more than half of the TBG-bound fraction. The bioavailable T_4 in brain is similar to T_3, whereas bioavailable T_4 in liver is primarily the albumin-bound fraction. Thus, like steroid hormones, the capillary exchangeable fraction of thyroid hormones more closely approximates the sum of the free plus the albumin-bound fraction than it does the free fraction.

THYROID HORMONE METABOLISM

Kinetics of Thyroid Hormone Production and Turnover

As mentioned above, the primary iodothyronine secreted by the thyroid gland is thyroxine. Most circulating T_3 is produced by monodeiodination of T_4 in peripheral tissues. About 80% of T_4 is monodeiodinated either in the 5' or 5 positions to form either T_3 or reverse T_3 (Fig. 13-3). About 40% of the 80 μg of T_4 secreted each day is peripherally metabolized via 5'-deiodination to produce about 80% of the 30 μg of T_3 produced each day (Fig. 13-4). Most of this conversion occurs in liver and kidney, and the T_3 that is formed is released into serum. The remaining 20% of T_3 production comes from direct secretion by the thyroid gland. In most systems, T_3 has about 10 times the potency of T_4 (see below). Since under physiological conditions most of the activity of T_4 can be accounted for by the T_3 formed from it, T_4 can be considered a prohormone.

The alternate monodeiodination product of T_4, formed by removal of an inner ring iodine, is 3,3',5'-triiodo-L-thyronine or reverse T_3 (rT_3) (Fig. 13-3). Nearly all

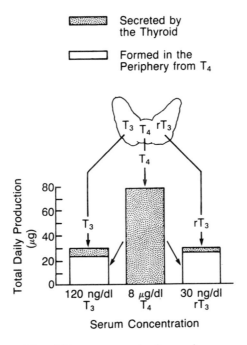

Fig. 13-4 Dynamics of thyroid hormone production and mean serum concentrations. The total daily production and distribution into the components of directed secretion by the thyroid and peripheral conversation of secreted thyroxine are shown. T_4, thyroxine; T_3, 3,5,3'-triiodothyronine; rT_3, reverse triiodothyronine or 3,3',5'-triiodothyronine. (Adapted with permission from Schimmel M, and Utiger RD: Thyroidal and peripheral production of thyroid hormones. Review of recent findings and their clinical implications. Ann Intern Med 87:760, 1977.)

rT_3 is produced extrathyroidally and a little more than a third of secreted T_4 is converted to this metabolite (Fig. 13-4). Serum rT_3 concentrations are lower than T_3 concentrations because of its more rapid metabolic clearance. Reverse T_3 has little or no thyroid hormone biological activity. Both T_3 and rT_3 are further deiodinated to 3,3'-diiodothyronine (Fig. 13-3) as well as other diiodothyronines and monoiodothyronines that are biologically inactive. T_4 and T_3 also form glucuronide conjugates that are excreted via the bile into the feces.

Enzymes Involved in Thyroid Hormone Metabolism

Several studies have led to the conclusion that there are two separate 5'-deiodinase enzymes (types I and II) and a distinct 5-deiodinase (type III) with different tissue localizations and different responses to changes in thyroid status and to drugs. The 5'-deiodinase I is primarily localized in the major sites of T_3 formation for delivery to the circulation, that is, the liver and kidney. Studies in rat tissues have shown that this reaction is inhibited by propylthiouracil, a drug known to impair T_4 to T_3 conversion in humans. In fasting animals, enzyme activity decreases in keeping with the observed lower T_3 levels in starved humans (see below). Since starvation is accompanied by increased rT_3 levels and decreased hepatic rT_3 deiodination to 3,3'-diiodothyronine, the activity of 5'-deiodinase in liver and kidney for rT_3 has properties similar to the enzyme metabolizing T_4 (Fig. 13-3). This suggests the presence of a single 5'-deiodinase in these tissues. Moreover, the activity of 5'-deiodinase in the liver and kidney increases in animals treated with excess thyroid hormone and decreases in thyroid hormone-deficient animals. These changes are consistent with the increased fractional rate of T_4 degradation in humans with clinical thyroid hormone excess and the decreased degradation of T_4 in subjects with thyroid hormone deficiency. Human and rat type I 5'-deiodinases have been cloned. Both are selenoproteins.

Indications that a second 5'-deiodinase enzyme exists came from studies of the effects of thyroid hormones on suppression of TSH secretion in the chronically hypothyroid animal. Infusion of T_4 or T_3 into such animals suppresses TSH levels. A 10-fold higher dose of T_4 than T_3 is required, however, and the resulting suppression persists for a longer time. During this time, serum T_3 levels fall below the level required for suppression when T_3 is infused. When labeled T_4 is given, 80–100% of the iodothyronine bound to the pituitary nuclear receptors is T_3. This T_3 does not originate from serum T_3 but from T_4 converted to T_3 within the pituitary. Propylthiouracil does not inhibit this 5'-deiodinase II enzyme. Studies in animals have shown that serum T_4 is a more important source, via intracellular T_3 production, of intracellular T_3 in the anterior pituitary than in the liver or kidney. This 5'-deiodinase II is enhanced in thyroid hormone deficiency and decreased in thyroid hormone excess, in contrast to the 5'-deiodinase I. The 5'-deiodinase II also differs from the 5'-deiodinase I in having a much higher affinity for the substrate T_4.

By studying the effects of thyroid hormone deficiency and T_4 treatment as well as substrate specificity and propylthiouracil sensitivity, it has been possible to demonstrate that both the 5'-deiodinase I characteristic of liver as well as the 5'-deiodinase II are present in the anterior pituitary, brain, and brown adipose tissue. The 5'-deiodinase II normally accounts for most of the locally produced T_3 in the pituitary, and it may also contribute significantly to peripheral T_3 levels in the thyroid-deficient state. The type III deiodinase is present in the central nervous system, placenta, and skin. It is also the most abundant deiodinase in the fetus and the major enzyme for formation of $r'T_3$. Preliminary cloning data suggests that it is also a selenoprotein.

Effects of Illness and Drugs on Thyroid Hormone Metabolism

The majority of patients hospitalized with nonthyroidal illness have serum T_3 levels below the normal range. These low values are not usually appreciated since serum T_3 levels are not measured. The routine screening tests (total T_4 and T_3 resin uptake, see below) are within the normal ranges in these patients. TSH levels and the TSH response to thyrotropin-releasing hormone (TRH, see below) are also usually normal. However, serum rT_3 levels are elevated, and the depressed T_3 and elevated rT_3 levels return to normal upon recovery from illness. These changes are due to a temporary decrease in extrathyroidal $5'$-deiodinase activity (Fig. 13-3). A decrease in $5'$-deiodinase would not only impair T_4 to T_3 conversion but would also impair further deiodination of rT_3, in effect, decreasing its clearance. A model for the low T_3 levels of illness is starvation. Total caloric deprivation results in a greater than 50% decrease in serum T_3 levels without major changes in T_4 or TSH. As in nonthyroidal illness, serum rT_3 levels increase about 50% with starvation.

A major question about the low T_3 state is whether patients with significantly reduced T_3 levels have thyroid hormone deficiency at the tissue level. Although the fact that TSH values are not usually increased would seem to argue against it, it could be postulated, however, that impaired hypothalamic TRH secretion might mask actual thyroid hormone deficiency. When fasted subjects with decreased serum T_3 levels are given exogenous T_3 replacement, muscle protein catabolism is enhanced, as evidenced by increased urinary urea, ammonia, and 3-methylhistidine excretion. This has been taken as evidence of a protein-sparing effect of the low T_3 levels in starvation. In clinical observations of patients with low T_3 levels associated with illness, basal metabolic rate and the cardiac measurement of thyroid status— pulse wave arrival time—are normal, again suggesting adequate thyroid effects at the tissue level. Dynamic tests of feedback at the hypothalamic–pituitary level, though normal in fasted subjects, may be subtly abnormal in some patients with illness and low T_3 levels. In general, however, there is little evidence for significant thyroid hormone deficiency at the tissue level, and the impaired T_4 to T_3 conversion is probably beneficial in sparing protein catabolism.

In the discussion of thyroid hormone metabolism above, the ability of propylthiouracil, a drug given to some patients with thyroid disease, to inhibit $5'$-deiodinase I activity was mentioned. Two iodine-containing drugs given to patients with nonthyroidal illness inhibit all deiodinases in all tissues and result in elevated T_4 levels (Fig. 13-5). They are iopanoic acid, a drug used for radiographic visualization of the gallbladder, and amiodarone, a drug used to treat cardiac arrhythmias. These compounds appear to mimic the structure of T_4 in the iodine substitutions on the benzene ring and the presence of an aliphatic side-chain (Fig. 13-5). The evidence for inhibition of $5'$-deiodinase II is that these drugs result in an elevation of TSH levels.

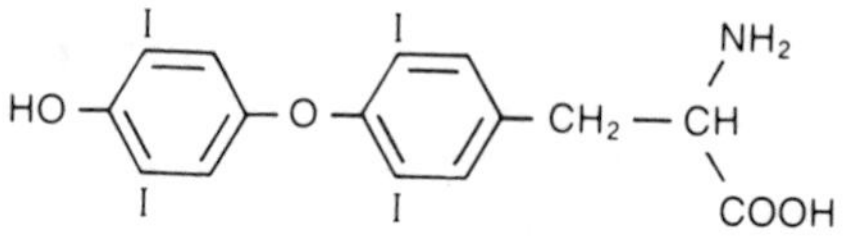

Thyroxine

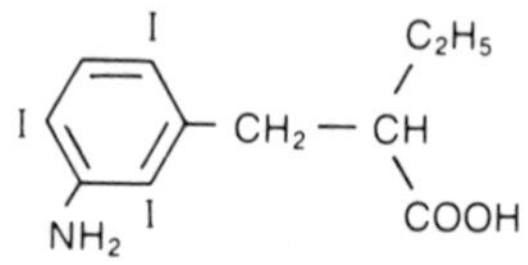

Iopanoic Acid

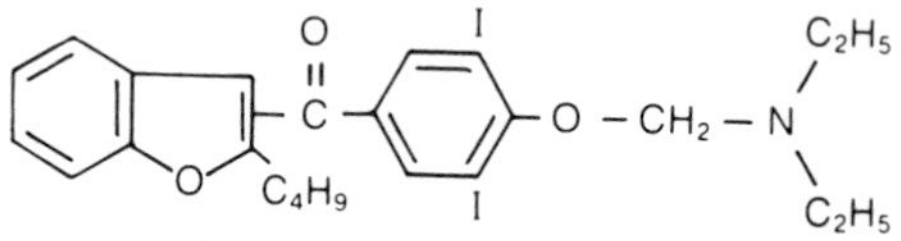

Amiodarone

Fig. 13-5 Structures of thyroxine, the oral cholecystographic agent iopanoic acid, and the antiarrhythmic agent amiodarone. The similarities of iodine substitutions on the benzene rings and aliphatic side chains are shown. (Reprinted from Griffin JE: Am J Med Sci 289:86, 1985, with permission.)

Upon discontinuing the amiodarone or waiting for the elimination of iopanoic acid, the alterations in thyroid function tests return to normal.

REGULATION OF THYROID FUNCTION

The Hypothalamic–Pituitary–Thyroid Axis

The function of the thyroid gland is regulated mainly by TSH, an anterior pituitary glycoprotein hormone with an α- and β-subunit and a molecular weight of about 28,000 (see Chapter 5). The α-subunit of TSH is identical to the α-subunit of gonadotropins, whereas the β-subunit is distinct and confers on the intact TSH molecule its biological activity. TSH stimulates the thyroid by interacting with specific cell surface, G protein–linked receptors on thyroid follicular cells to enhance the activity of adenylyl cyclase and thus stimulate the generation of cyclic AMP as a second messenger inside the cell. All of the intracellular effects of TSH stimulation can be mimicked by cyclic AMP in the absence of TSH.

The regulation of TSH secretion by the pituitary is primarily under the dual control of the hypothalamic tripeptide TRH and thyroid hormones (Fig. 13-6). TRH is derived from a pro-TRH peptide. Like other hypothalamic-releasing hormones, TRH reaches the anterior pituitary via the hypothalamic–pituitary portal circulation

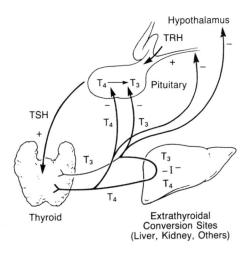

Fig. 13-6   The regulation of thyrotropin (TSH) secretion by the anterior pituitary. Positive effects of thyrotropin-releasing hormone (TRH) from the hypothalamus and negative effects of circulating 3,5,3'-triiodothyronine (T_3) and T_3 from intrapituitary and hypothalamic conversion of thyroxine (T_4).

(see Chapter 6). TRH interacts with specific receptors on pituitary thyrotrophs to release TSH and on mammotrophs to release prolactin. TRH belongs to the category of stimulating factors whose action starts with the hydrolysis of phosphatidylinositol 4,5-bisphosphate (PIP_2) (see Chapter 3). The hydrolysis of PIP_2 in the cell surface membrane generates two second messengers, inositol trisphosphate (IP_3) and 1,2-diacylglycerol. This hydrolysis is catalyzed by the membrane-bound enzyme phospholipase C and is associated with the rapid elevation of the concentration of free ionized calcium in the cell cytoplasm. The released calcium is thought to stimulate exocytosis. Together with these effects of IP_3 there is a parallel activation of protein kinase C by diacylglycerol that leads to phosphorylation of proteins involved in exocytosis and that mediates the sustained phase of secretion.

The release of TRH is controlled by central nervous system mechanisms (see Chapter 6). TRH is probably released in a pulsatile fashion, since pulsatile TSH secretion has been identified. The release of TSH is not absolutely dependent on TRH since low residual secretion of TSH persists even after the secretion of TRH has been stopped. Thus, the decrease in thyroid activity is more severe following hypophysectomy, which eliminates the source of TSH, than following hypothalamic lesions.

The ability of the thyrotroph to respond to TRH with increased TSH release is controlled by the feedback inhibition of thyroid hormones (Fig. 13-6). Either T_4 or T_3 is capable of inhibiting TSH secretion; however, as previously discussed, T_3 formed in the pituitary by 5'-deiodination of circulating T_4 appears to be a more important source of thyroid hormone to occupy nuclear receptors and mediate this feedback inhibition than is circulating T_3. Negative feedback by T_3, also derived from type II deiodinase, occurs at both the pituitary and the paraventricular nucleus of the hypothalamus. Negative feedback involves decreased transcription of TSH-α

and -β genes and also the pro-TRH gene. It is not known whether thyroid hormones also inhibit TRH release. Studies in which exogenous TRH was administered to normal subjects and patients with thyroid hormone deficiency or thyroid hormone excess indicate that feedback at the level of the pituitary is sufficient to explain the regulation of TSH secretion. The TSH response to administered TRH is enhanced in patients with thyroid hormone deficiency compared to normal subjects. In contrast, thyroid hormone excess diminishes or abolishes the TSH response to TRH. TRH receptors on thyrotrophs are increased in hypothyroidism and decreased in hyperthyroidism.

In addition to TRH and thyroid hormone, other substances of hypothalamic origin play some role in regulating TSH secretion. Somatostatin, another hypothalamic hormone (see Chapter 6), can inhibit TSH secretion. Since injection of antisomatostatin antiserum into normal animals increases TSH levels, somatostatin may be a tonic inhibitor of TSH secretion. The neurotransmitter dopamine may also be responsible for tonic inhibition of TSH release since dopamine antagonists cause elevation of serum TSH levels in normal subjects. It is not known by what mechanism dopamine inhibits TSH; it may act via effects on somatostatin. Finally, glucocorticoids, when present at supraphysiological levels, lead to partial inhibition of TSH secretion.

Thyroid Autoregulation

Although TSH is the primary regulator of thyroid function, the thyroid is also capable of some degree of autoregulation (as discussed above in relation to iodide transport and organification). The autoregulatory responses appear to operate in a manner to maintain thyroid hormone stores in the thyroid gland. Thus, in a state of iodide deficiency the efficiency of iodide transport is enhanced, with the opposite occurring in the presence of iodide excess. This response occurs without a detectable change in TSH levels. Proof of the autoregulatory nature of the inverse effect of iodide availability on iodide transport is found in the persistence of the phenomenon in the hypophysectomized animal. As previously mentioned, the mediator of the autoregulatory effect appears to be organic iodine rather than inorganic iodine in the gland.

EFFECTS OF THYROID HORMONES

Mechanism of Thyroid Hormone Action

Although there is some evidence to support a cell surface and mitochondrial site of action of thyroid hormones (see below), most of the characteristic biological effects of thyroid hormones are thought to be mediated by an interaction of T_3 with specific nuclear receptors. The mechanism of action of thyroid hormones is quite similar to that of steroid hormones in that the binding of the hormone is to a nuclear receptor with resultant alteration of transcription of specific messenger RNAs (see Chapter 3). In contrast to steroid hormone receptors that may not be firmly anchored in the nucleus before hormone binding (and thus are found in cytosol fractions following cell disruption), thyroid hormone receptors are thought to be tightly associated with acidic nonhistone nuclear proteins.

This nuclear thyroid hormone receptor has a low capacity (about 1 pmol/mg DNA) and a high affinity for T_3 (about 10^{-10} M). The affinity of the receptor for T_4 is an order of magnitude less. Initial studies of the nuclear receptor demonstrated a correspondence of receptor number and known responsiveness to thyroid hormones in a number of tissues. In the normal animal, about 85% of the total iodothyronine bound to liver and kidney nuclei is T_3 and the remaining 15% is T_4. Among the tissues examined for T_3 receptors, the anterior pituitary was notable for having the highest number of binding sites and greater than 50% of tissue T_3 bound to receptor sites. In contrast to the liver and kidney, in which most of the nuclear T_3 is derived from serum T_3, only about 40–50% of nuclear T_3 in the pituitary appears to originate from serum T_3. The remaining 50–60% of nuclear T_3 in the pituitary is derived from local conversion from T_4. As a consequence of this extra quantity of intracellular T_3, the nuclear receptors of the pituitary are about 80% saturated, compared to only 50% in the liver and kidney.

The number of thyroid hormone nuclear receptors appears to be altered by physiological and pathophysiological changes as well as by T_3 itself. In a cultured cell model of thyroid hormone action, incubation of the cells with T_3 resulted in a 50–60% loss of maximal binding capacity in about 10 hours. Fasting in rats decreases T_3 nuclear binding capacity by 30–50% in liver and kidney. Renal failure in rats was associated with a similar decrease in liver and kidney T_3 receptors. The change in amount of T_3 binding in each of these examples was not associated with a change in binding affinity.

All of the major effects of thyroid hormones appear to be mediated by the nuclear receptor. The production of many major classes of messenger RNA rises in response to thyroid hormones. Thyroid hormones stimulate the cell membrane enzyme Na^+, K^+-ATPase and thus increase oxygen consumption. Approximately 40% of oxygen consumption in many normal mammalian tissues seems to depend on this sodium pump. That this stimulation of ATPase activity is mediated through the nuclear receptor is indicated by the observation that the increased enzyme activity is due to an increased number of ''pump units'' rather than an alteration in preexisting enzyme molecules. Likewise, in studies of the effects of thyroid hormones on myocardial β-adrenergic receptors, the number but not the affinity of the receptors increased in response to thyroid hormones. Perhaps most important in elucidating the mechanism of thyroid hormone action has been the demonstration that T_3 enhances the concentration of specific messenger RNA for growth hormone in a cultured pituitary cell line and for α_2-microglobulin in rat liver. The concentration of both these proteins is known to increase in response to thyroid hormones.

The thyroid hormone receptor has been resistant to attempts to isolate it in a purified form using methods applied to other intracellular receptors. Thyroid hormone receptors were identified as being 50,000-Da nonhistone proteins. The actual description of the thyroid hormone receptor was achieved when two separate groups of investigators reported that c-erb-A, the product of the cellular homologue of the viral oncogene v-erb-A, has the ability to bind T_3 with high affinity and specificity. Since v-erb-A had been shown to be structurally related to steroid receptors, this confirmed that the thyroid receptor belonged to the same family of intracellular receptors (see Chapter 3). The most fascinating aspect of this story is that multiple thyroid hormone receptors exist. These different thyroid receptors (TR) are divided into α- and β-forms on the basis of sequence similarities and chromosomal location (Fig. 13-7). A TRβ2 cDNA has been found in rat, mouse, and chicken but not in frogs or humans. The TRβ gene is on human chromosome 3. TRβ2 differs from TRβ1 in that it is not widely distributed and thus far appears to be specific to pituitary and brain. TRα1 has been localized to human chromosome 17 and is widely distributed like TRβ1. Alternative splicing of the TRα gene transcripts yields a species called c-erb-Aα2, which is identical to TRα1 for the first 370 amino acids, including the DNA-binding domain, then diverges completely (Fig. 13-7). This latter c-erb-Aα does not bind T_3 and is thus not a TR. However, it is widely distributed and demonstrates regulation by T_3, as do the true TRs.

In addition to binding T_3, a true TR must also be able to bind to a thyroid hormone response element (TRE), a specific DNA sequence in the promoter region of thyroid-responsive genes. Generally, binding of T_3 and the TRE by TRs results in alteration (either activation or suppression) of gene transcription. TRβ1, TRβ2, and TRα1 all bind T_3 in a similar manner. The transcriptional activation of the rat growth hormone gene by T_3 is due to a TRE. The binding of TRs to TREs is independent of T_3. All three TRs are able to confer T_3 responsiveness when expressed in cells that normally do not respond to T_3. In contrast, c-erb-Aα2 does not produce a T_3 dependency of transcription of genes bearing TREs. Instead it appears to inhibit the action of TRs in a concentration-dependent manner. Although it may regulate thyroid hormone action, the exact *in vivo* functions of c-erb-Aα2 have not been defined. The v-erb-A protein has a similar dominant negative effect by means of its DNA-binding domain. In the absence of T_3, a TRE-bound TR may actually decrease basal T_3-responsive gene expression.

Based on evidence of heterodimer formation of TRs with the closely related retinoid X receptor (RXR), a dimerization domain has been identified in TRs. TRs

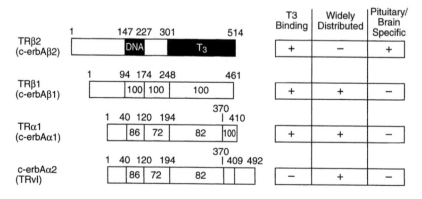

Fig. 13-7 Thyroid hormone receptors and the carboxy-terminal variant c-erb-Aα2. The sequences deduced from rat cDNAs are depicted with the amino acids numbered above. Putative DNA- and T_3-binding domains are indicated. Numbers in boxes indicate percent identity with TRβ2.

can form homodimers or heterodimers with each other or form heterodimers with one of the RXRs. When TRs bind T_3, they undergo a conformational change that can cause TR dimers to dissociate. The TR:RXR heterodimer may be the most important complex *in vivo* because 1) TR:RXR heterodimers bind with much higher affinity than TRs to each other; 2) binding of TR:RXR heterodimers to most TREs is more stable than binding of TR monomers or homodimers; and 3) transcriptional activation by TRs is enhanced by the presence of RXRs. Mutations in TRβ can cause thyroid hormone resistance.

A "mitochondrial pathway" of thyroid hormone action has also been proposed. In a number of rat tissues, high-affinity T_3 binding to inner mitochondrial membranes has been reported in association with enhanced oxidative phosphorylation by mitochondrial vesicles in response to T_3. Effects of T_3 on 2-deoxyglucose uptake in chick embryo heart cells and rat thymocytes have been reported. These effects are rapid, calcium-dependent, do not appear to require new protein synthesis, and may be mediated through a cell membrane action via stimulation of adenylyl cyclase. The overall importance of nonnuclear sites of thyroid hormone action remains to be determined.

Physiological Effects

Thyroid hormones have effects in almost all tissues of the body. In many respects, thyroid hormones may be viewed as tissue growth factors. Indeed, normal overall whole body growth does not occur in the absence of thyroid hormones despite adequate levels of growth hormone. In amphibians the effects of thyroid hormones on differentiation are dramatically evidenced by the failure of metamorphosis to occur in their absence.

Thyroid hormones have specific tissue effects that are perhaps best exemplified by the effects of thyroid hormone deficiency and excess (see below). These effects are mediated by the action of thyroid hormones on the concentration and activity of enzymes, the metabolism of substrates, vitamins, and minerals, and the function of other endocrine systems. The earliest recognized physiological effect of thyroid hormones is its ability to stimulate the basal metabolic rate or calorigenesis. Thyroid hormones stimulate oxygen consumption in the whole animal and in isolated tissues *in vitro*. Although direct mitochondrial effects of thyroid hormones have been postulated, as previously mentioned, the stimulation of synthesis of membrane Na^+,K^+-ATPase is thought to account for as much as half of the increased energy expenditure in going from the thyroid-deficient to the normal (euthyroid) state. This effect of thyroid hormones to enhance ATPase activity can be detected in many but not all tissues. Spleen and testes have very low concentrations of nuclear thyroid hormone receptors and fail to show increased calorigenesis in response to thyroid hormones. Thyroid hormones exert effects on thermogenesis and temperature regulation that are related to their effects on energy metabolism. Rats deficient in thyroid hormones are unable to survive in a cold environment.

Protein synthesis and degradation are stimulated by thyroid hormones. Stimulation of protein synthesis may be responsible for a portion of the calorigenic effect of thyroid hormones. The positive influence of thyroid hormones on normal body growth is derived largely from stimulation of protein synthesis. In the presence of a large excess of thyroid hormones there is accelerated protein catabolism, leading to increased nitrogen excretion.

Thyroid hormones affect most aspects of carbohydrate metabolism. They appear to enhance the actions of epinephrine in stimulating glycogenolysis and gluconeogenesis and to potentiate the actions of insulin on glycogen synthesis and glucose utilization. As with thyroid hormone effects on protein metabolism, a biphasic action of thyroid hormones on carbohydrate metabolism can be detected. Thus, low doses of thyroid hormones given to animals enhance glycogen synthesis in the presence of insulin, whereas large doses stimulate glycogenolysis. Thyroid hormones also enhance the rate of intestinal absorption of glucose and its rate of uptake by adipose tissue and muscle.

Thyroid hormones have multiple effects on lipid metabolism. Cholesterol synthesis and its metabolic conversions are depressed in thyroid hormone deficiency. However, since degradation is affected to a greater extent than synthesis, serum cholesterol levels increase in the thyroid-deficient state, and serum levels of cholesterol, phospholipids, and triglycerides decrease with thyroid hormone excess. One mechanism that may partially account for enhanced cholesterol metabolism in response to thyroid hormones is the ability of thyroid hormones to increase the number of low-density lipoprotein receptors on the cell surface (see Chapter 16). Fatty acid metabolism is affected by thyroid hormones that enhance lipolysis in adipose tissue.

Effects of Thyroid Hormone Deficiency

Severe thyroid hormone deficiency in infancy is termed *cretinism* and is characterized by mental as well as growth retardation. The developmental milestones of the infant such as sitting and walking are delayed. Linear growth impairment may lead to dwarfism, characterized by limbs that are disproportionately short compared with the trunk. When the onset of thyroid hormone deficiency is later in childhood mental retardation is less prominent, and impairment of linear growth is the major feature. The result is a child who appears younger than his or her chronological age. Epiphyseal development is delayed so that bone age is less than chronological age.

The onset of thyroid hormone deficiency in the adult is usually insidious; the symptoms and signs gradually appear over months or years. The early symptoms are nonspecific. Tiredness and lethargy are common, and constipation may develop or worsen. Eventually there is an overall slowing of mental function and motor activity. Although some weight gain usually occurs, appetite is usually decreased so that gross obesity is rare. Cold intolerance may be the first manifestation of thyroid hormone deficiency, the individual complaining of feeling cold in a room in which others are comfortable. Women may experience abnormalities of menstrual function with increased menstrual flow being more common than cessation of menses. Decreased clearance of adrenal androgens may facilitate extraglandular estrogen formation leading to anovulatory cycles with infertility. Hair loss may occur, as may brittle nails and dry, coarse skin. The voice may become hoarse. When thyroid hormone deficiency is long-standing and severe, there is an accumulation of mucopolysaccharides in subcutaneous tissues and other organs termed *myxedema*. Dermal infiltration results in thickened features, periorbital edema, and swelling of the hands and feet that does not indent with pressure. Stiffness and aching of muscles may be followed by muscle swelling as an early manifestation. Delayed muscle contraction and relaxation lead to slow movement and delayed tendon reflexes. Both stroke volume and heart rate are reduced so that cardiac output decreases. The heart may enlarge and a pericardial effusion develops. Pleural and peritoneal fluid rich in protein and mucopolysaccharides may accumulate. The slowing of mental function is characterized by impaired memory, slow speech, decreased initiative, and eventually somnolence. Mild hypothermia may lead to more severe hypothermia if there is environmental exposure. Eventually, coma may develop in conjunction with hypoventilation.

Effects of Thyroid Hormone Excess

The earliest manifestations of thyroid hormone excess are usually nervousness, irritability or emotional instability, a feeling that the heart is pounding, fatigue, and heat intolerance. As with thyroid hormone deficiency, this latter manifestation may be manifested as altered comfort in a room in which others are comfortable and lead to resetting the thermostat or requiring lighter clothing. Increased sweating is commonly noticed. Although fatigue invariably occurs, the nervousness and irritability may give the impression of increased energy. Weight loss despite normal or increased food intake is one of the most common manifestations. The increased food intake occasionally can be so great as to overcome the hypermetabolic state and result in weight gain. Interestingly, most patients relate that their increased caloric intake is predominantly in the form of carbohydrates. Women have decreased or absent menstrual flow. The number of bowel movements per day often increases, but true watery diarrhea is rare. Physical findings may include warm, moist skin of a velvety texture often compared to that of a newborn, a change in the fingernails, termed *onycholysis*, involving separation of the nail from the nail bed, and proximal muscle weakness often making it difficult for the patient to rise from a sitting or squatting position. In contrast to thyroid hormone deficiency, the hair has a very fine texture, but hair loss may similarly occur. On extending the fingers of an outstretched hand a fine tremor may be seen. The eyelids may retract, leading to the impression that the patient is staring. Tachycardia that persists during sleep is characteristic, and atrial arrhythmias and congestive heart failure may develop.

CLINICAL ASSESSMENT OF THYROID STATUS

To assess thyroid status one must search the medical history for symptoms of thyroid hormone deficiency or excess and examine the patient for thyroid enlargement as well as other physical findings typical of thyroid disease (see above). Because thyroid disease is relatively common and the clinical manifestations may be subtle, laboratory tests are widely used to screen for the presence of thyroid dysfunction and to confirm the suspected diagnosis.

As discussed above, iodothyronines in the circulation are largely protein bound, primarily to TBG. Since most clinical assays measure total T_4 or T_3 in serum, and since bioavailable thyroid hormone is only a fraction of the total, and since abnormalities in thyroid binding occur, the interpretation of the total hormone concentration requires some assessment of the level of TBG or available binding sites in the serum. The total concentrations of T_4 and T_3 are usually measured by immunoassay. Although the amount of TBG can be measured by immunoassay, this is not routinely done. Instead, an *in vitro* test to assess total available binding sites for thyroid hormones present in the serum is performed. This test is termed the *T_3-resin uptake* or *T_3-uptake ratio* and is performed by incubating a serum sample with radioactive T_3. After equilibration a resin or other substance is added to the mixture to bind the radioactive T_3 not bound to serum proteins. The percentage of the total radioactive T_3 bound to the resin is related in a reciprocal manner to the thyroid hormone-binding capacity in the sample. As shown in Fig. 13-8, there are two possible interpretations of both an increased and a decreased T_3-resin uptake. An increased T_3-resin uptake (i.e., more T_3 bound to the resin and less to the serum-binding proteins) could be due to greater saturation of a normal amount of TBG secondary to increased thyroid hormone production (i.e., thyroid hormone excess). Alternatively, because of a decreased concentration of TBG with normal thyroid hormone production, there could be fewer available binding sites, resulting in more T_3 being bound to the resin (Fig. 13-8). Conversely, a decreased T_3-resin uptake (i.e., less T_3 bound to the resin and more to the serum-binding proteins) could be due to decreased saturation of a normal amount of TBG because of decreased thyroid hormone production (i.e., thyroid hormone deficiency). Alternatively, because of an increased concentration of TBG with normal thyroid hormone production, there could be more available binding sites, resulting in less T_3 being bound to the resin (Fig. 13-8). The value of the T_3-resin uptake is either reported as a percentage (Fig. 13-8) or compared to the T_3-resin uptake of a normal serum pool and reported as a T_3-uptake ratio, with the normal being unity. It is used to correct the total T_4 or total T_3 measurement to result in a calculated unitless number, the free T_4 index and free T_3 index. The free T_4 index is the product of the serum total T_4 and the T_3-resin uptake. This calculation is usually made only for the total T_4, but it can be done for the total T_3 as well.

The effect of pregnancy and other conditions that alter the level of TBG (Table 13-1) is to increase or decrease total T_4 and T_3 concentrations without changing the level of free, unbound hormone in the steady state. For example, if TBG increases in concentration, the amount of free hormone decreases temporarily until compensatory mechanisms result in a new steady state with increased total hormone concentration but normal free hormone. The correction of the increased total T_4 by the decreased T_3-resin uptake results in a normal free T_4 index. Thus, changes in the T_3-resin uptake are reciprocal to the changes in the number of available binding sites on TBG and the total T_4 when the primary abnormality is a change in binding proteins. When there is a primary alteration in thyroid hormone production, the change in the T_3-resin uptake is in the same direction as the change in the total T_4 level. Within the usual variation of TBG levels, these free hormone indices correlate fairly well with free hormone concentrations measured by equilibrium dialysis. Such a correlation has not been reported for tissue available hormone (see Chapter 5). The free T_4 concentration (and free T_4 index) does correlate well with clinical thyroid hormone deficiency and excess. However, in using the free hormone index to estimate tissue-available hormone, one must be aware that in extreme alterations in TBG levels the T_3-uptake ratio tends not to completely estimate the binding abnormality. For example, with near absence of TBG the T_3-uptake ratio is not as high as it ought

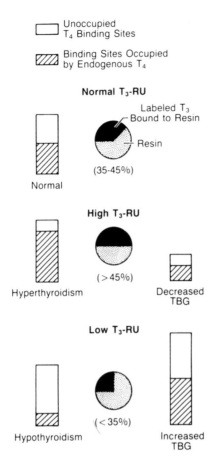

Fig. 13-8 The possible interpretations of results of a T₃-resin uptake test. A high T₃-resin uptake (T₃-RU) could indicate hyperthyroidism with saturation of binding sites on a normal amount of thyroid-binding globulin (TBG) or a low serum TBG concentration. A low T₃-resin uptake could indicate hypothyroidism with decreased occupancy of binding sites on a normal amount of TBG or a high serum TBG concentration. (Redrawn with permission from Spaulding SW, and Utiger RD: The thyroid: Physiology, hyperthyroidism, hypothyroidism, and the painful thyroid. In: *Endocrinology and Metabolism*, P Felig, JD Baxter, and LA Frohman, eds., McGraw-Hill, New York, p. 295, 1981.)

to be, so that the decreased total T_4 is not brought up to a normal free T_4 index. In such circumstances additional tests may be required to confirm the normal thyroid status of the individual (see below).

In normal individuals there are variations in the concentration of thyroid hormones with age. In the fetus the T_3 level is decreased as it is in euthyroid individuals who are fasted or who have acute or chronic illnesses (see above). Shortly after birth the T_3 level rises. The normal range (see Chapter 5) for total T_4 in adults varies somewhat with the specific assay but is usually about 5–12 μg/dl. The normal range of total T_3 in adults is about 70–200 ng/dl. The mean T_3 level declines slightly with

Table 13-1 Conditions and Hormones That Alter the Level of Thyroid Hormone-Binding Globulin (TBG)

Increased TBG	Decreased TBG
Pregnancy	Androgens and anabolic steroids
Newborn state	Large dose of glucocorticoids
Oral contraceptives and other sources of estrogens	Chronic liver disease
Infectious and chronic active hepatitis	Severe systemic illness
Genetically determined	Active acromegaly
	Kidney disease with proteinuria
	Genetically determined

advanced age, but the lower limit of normal does not fall until after age 70 in men and age 80 in women. The T_4 level does not decline with age.

Effect of Nonthyroidal Illness on Thyroid Function Tests

Abnormal results on thyroid function tests are frequently encountered in the evaluation of patients without obvious thyroid disease. Some of these abnormal results are caused by altered levels of TBG (see above). In the remainder it is necessary to decide whether subtle thyroid hormone deficiency or excess exists or whether the abnormal values are somehow spurious. In general, one uses the additional or confirmatory tests described below to clarify the diagnosis. The T_4 and free T_4 index may be either decreased or increased in subjects who are actually euthyroid but suffer from acute or chronic nonthyroid illness. As discussed in the section on thyroid hormone metabolism, the most common effect of nonthyroidal illness on thyroid function tests is to lower T_3 levels.

In addition to the low T_3 levels, some patients with nonthyroidal illness have a decreased total T_4 and free T_4 index. These are often the more critically ill patients. Studies have shown that the decreased level of T_4 is due to a circulating inhibitor of thyroid hormone binding to serum-binding proteins and that this inhibitor likely comes from tissue breakdown associated with severe illness. Since the inhibitor is not detected in the T_3-resin uptake test, the decreased total T_4 is not adjusted in calculations of the free T_4 index. Free T_4 concentrations are not decreased, and the T_4 production rate is normal. Again, these alterations of thyroid function tests revert to normal on recovery from the serious illness.

Nonthyroidal illness may also be associated with temporary alterations in TSH levels in hospitalized patients.

DERANGED THYROID PHYSIOLOGY

Deranged thyroid physiology may involve altered thyroid growth, in either a diffuse or a nodular pattern, while thyroid hormone production remains normal. Although

his simple increase in thyroid size, termed euthyroid goiter, is common in the popu-
ation, the physiological derangement(s) leading to its development in areas of iodide
sufficiency are poorly understood. Thus, to illustrate the effects of thyroid hormones
ind pathophysiological mechanisms of thyroid hormone secretion, only hypothyroidism and hyperthyroidism will be considered here.

Hypothyroidism

Hypothyroidism is a clinical state resulting from decreased production of thyroid
hormones. It may begin *in utero*, or later in life, resulting in a different set of clinical
features as we saw in the discussion of thyroid hormone deficiency.

Hypothyroidism is relatively common, occurring in about 2% of adult women
n a clinically overt form. It is much less common in men. Spontaneous hypothyroidism may be due to disorders of the thyroid (primary hypothyroidism), of the
anterior pituitary or hypothalamus (secondary hypothyroidism), or, rarely, of pe-
ipheral target tissues (severe generalized thyroid hormone resistance). Primary hypothyroidism accounts for over 95% of patients with hypothyroidism; it may be
associated either with a decrease in thyroid tissue or with thyroid enlargement. Both
he nongoitrous and goitrous forms are commonly the result of an autoimmune thyroiditis leading to destruction of thyroid parenchyma. In those who have goiters,
ymphocytic infiltration of the thyroid is seen and the condition is termed *Hashimoto's disease*. This autoimmune destruction of the thyroid may be part of a polyglandular endocrine deficiency syndrome in which the adrenals, parathyroids, gonads, pancreatic islets, and stomach parietal cells are variably involved, but the
hypothyroidism is usually an isolated finding. The most common cause of primary
hypothyroidism in adults is destruction of the thyroid tissue by prior surgery or
radioactive iodine treatment for hyperthyroidism.

In areas in which iodine is not sufficient, goitrous hypothyroidism may be due
o endemic iodine deficiency. This is rare in North America but quite common in
many other parts of the world. The incidence of endemic goiter has been greatly
educed in many areas by the introduction of iodized salt. Defects in thyroid hormone
biosynthesis are rare inherited causes of goitrous hypothyroidism. They are usually
transmitted as autosomal recessive conditions.

Secondary hypothyroidism is due to decreased TSH secretion. Although an
solated abnormality of TSH release has been described, most secondary hypothyroidism is due to pituitary insufficiency involving multiple anterior pituitary hormones (see Chapter 6). The pituitary may be affected directly or indirectly as a result
of hypothalamic disease that inhibits the delivery of releasing factors.

As with almost all hormones, resistance to the action of thyroid hormones has
been described as a disease mechanism. In more severely affected individuals the
esistance is present in all target tissues. This leads to increased thyroid hormone
production in the absence of symptoms of thyroid hormone excess and occasionally
n the face of symptoms of thyroid hormone deficiency (delayed bone age, epiphyseal
stippling, and decreased hydroxyproline excretion). In such patients the serum TSH
s normal and responsive to TRH. The defect in most instances is due to a TRβ
mutation.

Most patients with clinical hypothyroidism have a free T_4 index below the lower range of normal. Since the normal range is relatively broad, some hypothyroid patients may have T$_4$ levels that are within the low normal range. Because most patients with hypothyroidism have primary disease of the thyroid gland rather than secondary hypothyroidism resulting from hypothalamic or pituitary disease, the serum TSH is a useful test to confirm the presence of overt primary hypothyroidism and to identify the presence of subtle hypothyroidism or decreased thyroid reserve when the serum-free T$_4$ index is normal (Fig. 13-9). The upper limit of normal for TSH is usually <7 μU/ml. Most subjects with primary hypothyroidism have basal TSH values >20 μU/ml (Fig. 13-9). Although the TSH response to TRH stimulation is enhanced in primary hypothyroidism, such testing is not necessary for the detection of

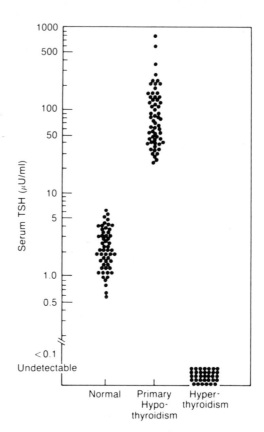

Fig. 13-9   Serum thyrotropin (TSH) levels in normal subjects, in patients with primary hypothyroidism, and in patients with hyperthyroidism. Data from hypothyroid subjects are from a standard TSH radioimmunoassay. (Replotted from Hershman JM, and Pittman JA: Utility of the radioimmunoassay of serum thyrotropin in man. Ann Intern Med 74:481, 1971.) The data from normal and hyperthyroid subjects are from a sensitive immunoenzymometric TSH assay. (Replotted from Spencer CA, et al: Thyrotropin secretion in thyrotoxic and thyroxine-treated patients: Assessment by a sensitive immunoenzymometric assay. J Clin Endocrinol Metab 63:349, 1986.) The TSH values by the sensitive assay in hyperthyroid patients are below those in normal subjects and are undetectable.

hypothyroidism because the basal TSH is elevated. If the T_4 level is clearly low and the TSH level is not high, hypothalamic or pituitary dysfunction should be suspected (see Chapter 6).

The thyroid radioactive iodine uptake (RAIU) is a direct test of thyroid function; a tracer dose of ^{123}I is given orally, and the percentage accumulation in the thyroid is measured, usually at 24 hours. The RAIU varies directly with the functional activity of the thyroid. However, this test is not useful in suspected hypothyroidism since many normal subjects have RAIUs in the range observed in patients with hypothyroidism.

Hyperthyroidism

Hyperthyroidism is the clinical state resulting from increased circulating levels of available thyroid hormones. The manifestations are, in general, an exaggeration of the normal physiological actions of thyroid hormones. In fact, except for the specific form of eye changes termed *ophthalmopathy*, certain skin changes (which may or may not be present depending on the cause of the hyperthyroidism), and goiter, all of the clinical manifestations can be related to excess thyroid hormone. In contrast to hypothyroidism, which tends to be insidious in onset and slowly progressive in course, hyperthyroidism usually results in recognized symptoms and may (in the case of the most common cause, Graves' disease) undergo periods of exacerbation and remission. The clinical manifestations common to all forms of hyperthyroidism were described in the section on thyroid hormone excess.

As is true for hypothyroidism, hyperthyroidism is relatively common, occurring in 2% of women but only one-tenth as many men. The predominant cause is diffuse toxic goiter, usually termed *Graves' disease.* The mechanism of the autonomous thyroid hormone secretion in Graves' disease is stimulation of the gland by an IgG immunoglobulin that interacts with the TSH receptor. In contrast to most antireceptor antibodies, this antibody does not inhibit but profoundly activates the receptor. Most patients with Graves' disease have a palpably enlarged thyroid gland. A specific infiltrative ophthalmopathy is seen only in Graves' disease. The eyes become protuberant due to infiltration of the extraocular tissues with mucopolysaccharides. Trapping of extraocular muscles in the confined space of the orbit may result in paralysis of eye movement and double vision. In some patients the ophthalmopathy may be the most severe manifestation of Graves' disease. The specific cause of the infiltrative ophthalmopathy of Graves' disease is unknown.

Other causes of hyperthyroidism such as nontoxic multinodular goiters in elderly subjects or inflammation of the thyroid gland (thyroiditis) account for only 10–15% of patients. Excess dietary or medication iodide may lead to a low RAIU form of hyperthyroidism, and ingested thyroid hormones may mimic endogenous hyperthyroidism.

The appropriate initial tests in patients with suspected hyperthyroidism is the total T_4 and T_3-resin uptake, to determine the free T_4 index. Since the average increase in circulating T_4 in patients with the common form of hyperthyroidism is approximately 2-fold, whereas T_3 levels increase 3-fold or 4-fold, the measurement of the total serum T_3 concentration is useful in patients in whom the free T_4 index is not clearly elevated. Until recently, most TSH immunoassays were unable to distinguish normal from low values. Thus, the response of serum TSH to the intravenous injection of TRH was the dynamic endocrine test of choice to evaluate equivocal hyperthyroidism. Recall that the level of feedback inhibition of thyroid hormones on the hypothalamic–pituitary system is at the level of the pituitary and hypothalamus. Thus, if the pituitary is exposed to supraphysiological levels of thyroid hormones from any source (endogenous or exogenous), the response of TSH to an intravenous bolus of TRH (usually 500 μg) is blunted or absent (Fig. 13-10). Except for some older but otherwise normal men there is usually at least a 2 μU/ml increment in serum TSH in the 20–30 minutes following TRH injection. A normal TSH response to TRH excludes hyperthyroidism.

Sensitive immunometric serum TSH assays have been developed and are now generally available. With these immunometric assays TSH levels can be detected in all normal subjects, and abnormally low values in patients can be identified (Fig. 13-9). These sensitive TSH assays have nearly eliminated the need for TRH testing to confirm the presence of hyperthyroidism.

Although the thyroid RAIU is of limited value in the evaluation of suspected hypothyroidism, it has a role in the assessment of hyperthyroidism. Elevation of the

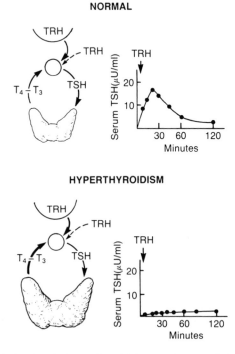

Fig. 13-10 Hypothalamic–pituitary–thyroid interactions and the serum thyrotropin (TSH) response to thyrotropin-releasing hormone (TRH) injection in normal subjects and patients with hyperthyroidism. The increased thyroid hormone levels in the hyperthyroid patients inhibit the TSH response to TRH. T_4, thyroxine; T_3, 3,5,3'-triiodothyronine. (Adapted from Kaplan MM, and Utiger RD: Diagnosis of hyperthyroidism. Clin Endocrinol Metab 7:97, 1978.)

RAIU above the normal range indicates thyroid hyperfunction. However, some patients with endogenous thyroid hormone excess have RAIUs in the normal or low range.

SUGGESTED READING

Brent GA: The molecular basis of thyroid hormone action. N Engl J Med 331:847–853, 1994.

Danforth E Jr: Effects of fasting and altered nutrition on thyroid hormone metabolism in man. In: *Thyroid Hormone Metabolism,* G Hennemann, ed., Marcel Dekker, New York, pp. 335–358, 1986.

Engler D, and Burger AG: The deiodination of the iodothyronines and of their derivatives in man. Endocr Rev 5:151–184, 1984.

Gershengorn MC: Mechanism of thyrotropin releasing hormone stimulation of pituitary hormone secretion. Annu Rev Physiol 48:515–526, 1986.

Griffin JE: The dilemma of abnormal thyroid function tests—is thyroid disease present or not? Am J Med Sci 289:76–88, 1985.

Griffin JE: Hypothyroidism in the elderly. Am J Med Sci 299:334–345, 1990.

Hamburger JI: The various presentations of thyroiditis. Diagnostic considerations. Ann Intern Med 104:219–224, 1986.

Kolesnick RN, and Gershengorn MC: Thyrotropin-releasing hormone and the pituitary. New insights into the mechanism of stimulated secretion and clinical usage. Am J Med 79:729–739, 1985.

Larsen PR: Thyroid-pituitary interaction. Feedback regulation of thyrotropin secretion by thyroid hormones. N Engl J Med 306:23–32, 1982.

Larsen PR, and Ingbar SH: The thyroid gland. In: *Williams Textbook of Endocrinology,* 8th ed., JD Wilson and DW Foster, eds., Saunders, Philadelphia, pp. 357–487, 1992.

Larsen PR, Silva JE, and Kaplan MM: Relationships between circulating and intracellular thyroid hormones: Physiological and clinical implications. Endocr Rev 2:87–102, 1981.

Lazar MA: Thyroid hormone receptors: Multiple forms, multiple possibilities. Endocr Rev 14:184–193, 1993.

Nicoloff JT, and Spencer CA: The use and misuse of the sensitive thyrotropin assays. J Clin Endocrinol Metab 71:553–558, 1990.

Oppenheimer JH, Schwartz HL, Mariash CN, Kinlaw WB, Wong NCW, and Freake HC: Advances in our understanding of thyroid hormone action at the cellular levels. Endocrine Rev 8:288–308, 1987.

Pisarev MA: Thyroid autoregulation. J Endocrinol Invest 8:475–484, 1985.

Samuels HH, Forman BM, Horowitz ZD, and Ye ZS: Regulation of gene expression by thyroid hormone. J Clin Invest 81:957–967, 1988.

Silva JE, and Larsen PR: Regulation of thyroid hormone expression at the prereceptor and receptor levels. In: *Thyroid Hormone Metabolism,* G Hennemann, ed., Marcel Dekker, New York, pp. 441–500, 1986.

Taurog A: Hormone synthesis: Thyroid iodine metabolism. In: *Werner and Ingbar's The Thyroid: A Fundamental and Clinical Text,* 6th ed., LE Braverman and RD Utiger, eds., Lippincott, New York, pp. 51–97, 1991.

Wartofsky L, and Burman KD: Alterations in thyroid function in patients with systemic illness: The "Euthyroid Sick Syndrome." Endocr Rev 3:164–217, 1982.

Weetman AP, and McGregor AM: Autoimmune thyroid disease: Developments in our understanding. Endocr Rev 5:309–355, 1984.

14

The Adrenal Glands

NORMAN M. KAPLAN

The adrenal glands, weighing only 6–10 g, are essential for life. In their absence, death occurs in a few days.

The overall function of the adrenals is to protect the organism against acute and chronic stress, a concept popularized by Walter Cannon in 1929 as the "fight-or-flight" response for the medulla and by Hans Selye in 1936 as the "alarm" reaction for the cortex. The steroid hormones of the cortex and the catecholamines of the medulla probably developed as protection against immediate stress or injury and more prolonged deprivation of food and water. In acute stress, catecholamines mobilize glucose and fatty acids for energy and prepare the heart, lungs, and muscles for action. Glucocorticoids protect against overreactions from the body's responses to stress, which themselves threaten homeostasis. In the more chronic stress of food and fluid deprivation, the cortical steroid hormones stimulate gluconeogenesis to maintain the supply of glucose and increase sodium reabsorption to maintain body fluid content.

Diseases associated with adrenal hypofunction are relatively rare; those associated with adrenal hyperfunction are slightly more common. Synthetic hormones similar to their natural secretions are widely used to treat numerous diseases.

ANATOMY

The adrenals are located on top of the kidneys (ad-renal) (Fig. 14-1). In keeping with their essential function, they are well supplied with arterial blood from branches of the aorta, the renal arteries, and the phrenic arteries. Per gram of tissue, they have about the highest rate of blood flow in the body. Arterial blood enters from the outer cortex, flows through fenestrated capillaries between the cords of cells, and drains inwardly into venules in the medulla. On the right, the adrenal vein directly enters the inferior vena cava; on the left, it usually drains into the left renal vein. Awareness of this arrangement is important when attempts are made to obtain blood from or inject radiographic dye into the adrenals in the diagnosis of various primary adrenal diseases.

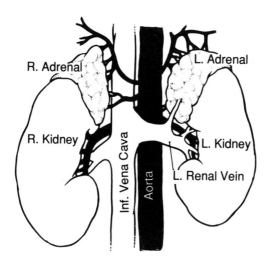

Fig. 14-1   Gross anatomy of the adrenal glands.

Anatomical Zonation

The cortex is derived from mesodermal tissue and comprises 90% of the gland. It is composed of three zones below its fibrous capsule: the outer zona glomerulosa, the middle zona fasciculata, and the inner zona reticularis (Fig. 14-2). The major secretions of the three zones differ as do the mechanisms primarily responsible for the modulation of their activity (Table 14-1).

The medulla, which is derived from neuroectodermal cells from the neural crest, synthesizes and secretes catecholamines. Whereas norepinephrine is the primary neurotransmitter secreted from sympathetic ganglia and neurons, the adrenal medulla

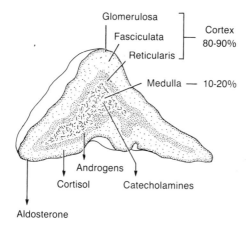

Fig. 14-2   Secretion of adrenocortical hormones by the different zones of the adrenal cortex. (From Guyton AC: *Textbook of Medical Physiology,* Saunders, Philadelphia, p. 909, 1986.)

Table 14-1   Functional Zonation of the Adrenal Cortex

Zone	Major secretion	Major control mechanism
Z. glomerulosa	Aldosterone (mineralocorticoid)	Renin–angiotensin
Z. fasciculata	Cortisol (glucocorticoid)	ACTH
Z. reticularis	Dehydroepiandrosterone (androgen)	ACTH

secretes mainly epinephrine, reflecting the presence of the N-methyltransferase enzyme that converts norepinephrine into epinephrine only within the adrenal medulla.

In the fetus, the adrenal glands are much larger relative to body size than in the adult. In absolute size, they are almost as large at term as the fetal kidneys. The fetal gland has an outer subcapsular zone, the precursor of the three zones of the postnatal cortex, and a much larger inner zone, comprising about 80% of the fetal cortex. The cells of the inner fetal cortex are very active but, since the enzyme 3β-hydroxysteroid dehydrogenase needed to convert the early precursors into either glucocorticoids or mineralocorticoids is not active, they produce instead large quantities of dehydroepiandrosterone, which serves as a precursor for the synthesis of estrogens by the placenta (see Chapter 11). After birth, the fetal zone rapidly involutes and the outer subcapsular zone differentiates into the three zones of the postnatal gland.

Functional Zonation

The arrangement of blood flow from the outer cortex into inner medulla may be involved in the differentiation of secretions from the various anatomical zones. Subcapsular cells are capable of generating all three cortical zones. One theory for their differentiation is that the youngest cells are beneath the capsule, forming the zona glomerulosa, and over time the cells migrate downward. The inner zona reticularis cells, being the oldest, have lower rates of cell division and more pigmentation and other signs of a progressive loss of functional activity.

The arrangement of blood flow from the outer cortex into inner medulla also appears to serve the function of maintaining high levels of synthesis and secretion of the major catecholamine required for the response to acute stress, epinephrine. The enzyme responsible for the conversion of norepinephrine to epinephrine, N-methyltransferase, is specifically induced by cortisol, the major steroid secreted by the cortex. Therefore, after the immediate response to stress by release of stored epinephrine, the secretion of cortisol will ensure a continued supply of the hormone.

ADRENAL CORTEX

The three major secretory products of the adrenal cortex will be considered separately, but first the biosynthetic pathways for all steroidogenesis will be examined.

Steroidogenesis

All adrenal steroids are derived from cholesterol by various enzymatically mediated modifications of its structure (Fig. 14-3). Cholesterol can by synthesized within the adrenal from acetyl CoA, but most is taken up from the blood by specific plasma membrane receptors that bind low-density lipoproteins (LDL) that are rich in cholesterol. The LDL–cholesterol is transferred into the cell by endocytosis, the cholesterol split from the lipoprotein, esterified, and stored in cytoplasmic vacuoles. Stimulation by corticotropin (ACTH) activates the cholesterol esterase, which releases free cholesterol from its storage depots, providing the substrate for steroid synthesis.

The synthesis of the various adrenal hormones involves cytochrome P-450 enzymes, which are mixed oxygenases that catalyze steroid hydroxylations. The rate-limiting step in steroid biosynthesis is the first step, the action of the cytochrome P-450 side-chain cleavage enzyme, 20,22-desmolase, needed to convert cholesterol

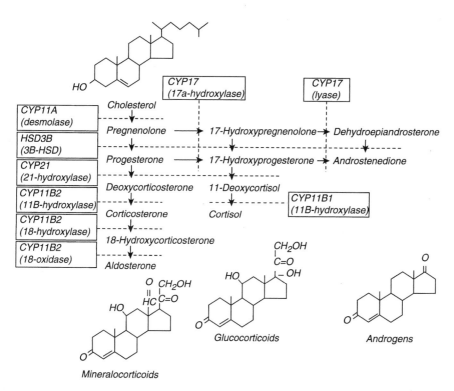

Fig. 14-3 Pathways of adrenal steroidogenesis. Gene products mediating each biosynthetic step are boxed with the corresponding enzymatic activity listed in parentheses. Note that one enzyme may have more than one activity. The planar structures of cholesterol, the mineralocorticoid aldosterone, the glucocorticoid cortisol, and the androgen androstenedione, are shown. 3β-HSD, 3β hydroxysteroid dehydrogenase. 18-hydroxylase and 18-oxidase are also called corticosterone methyloxidase I and II or aldosterone synthase. (From White PC, and Speiser PW: Steroid 11β-hydroxylase deficiency and related disorders. Endocrinol Metab Clin North Am 23:325–339, 1994.)

into pregnenolone. All stimuli of steroidogenesis, including ACTH, increase the interaction of cholesterol with the side-chain cleavage enzyme.

Once pregnenolone is available, it is rapidly removed from mitochondria and sequentially modified by dehydrogenases and hydroxylating enzymes within the endoplasmic reticulum and mitochondria to form the three major types of adrenal hormones.

An Example of Aberrant Steroidogenesis

As seen in Fig. 14-3, human adrenal cells have two different 11β-hydroxylase isozymes, CYP11B1 (also called *P450c11* or *11β-hydroxylase*) and CYP11B2 (also called *P450c18* or *aldosterone synthase*), that are responsible for cortisol and aldosterone biosynthesis, respectively. CYP11B2 first 18-hydroxylates and then 18-oxidizes corticosterone to aldosterone; these activities are also referred to as *corticosterone methyloxidase I and II.*

These two enzymes are encoded by two genes on the long arm of chromosome 8 (8 q21-q22). A form of familial hypertension has been shown to be caused by an unequal crossing over of the 5' end of CYP11B1 (11-OHase), including regulatory sequences, with the 3' end of CYP11B2 (aldosterone synthase) yielding a chimeric hybrid gene (Fig. 14-4) that is regulated like the 11β-hydroxylase located in the zona fasciculata but which leads to increased secretion of 18-hydroxylated derivatives with mineralocorticoid activity. Thus the crossover empowers the zona fasciculata to produce mineralocorticoids whose origin is normally only from the zona glomerulosa. Though rare, this syndrome provides a fascinating picture of normal physiology gone slightly awry.

Adrenal Steroid Secretion

As noted, the major hormones secreted from the adrenal are cortisol, aldosterone, and dehydroepiandrosterone (DHEA) (Table 14-2). Small amounts of these hormones are stored in the adrenal. When more is needed, the biosynthetic pathway must be activated, particularly at the rate-limiting 20,22-desmolase reaction. Most of the DHEA is sulfated and plasma levels of DHEA sulfate are much higher than that of the free hormone.

Syndromes of Congenital Enzymatic Deficiency

Some precursors of the major end products are also secreted from the adrenal gland. They tend to be less active and present in smaller concentrations. However, congenital deficiencies of various adrenal enzymes may block the normal biosynthetic path and cause the progressive build-up of whichever precursors can be formed. This build-up tends to be progressive since, when cortisol synthesis is severely limited by the enzyme deficiency, there is insufficient circulating cortisol to inhibit the release of more ACTH. The high levels of ACTH continue to stimulate the gland to synthesize as much as possible in the presence of an enzymatic deficiency. The resulting hyperplasia of the gland gives these relatively rare congenital syndromes their name: *congenital adrenal hyperplasia.*

The most common form of adrenal enzymatic defect, responsible for about 95% of all cases, involves the 21-hydroxylase enzyme. This enzyme is needed to convert

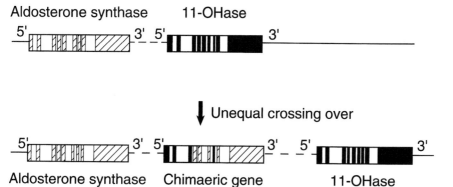

Fig. 14-4 Diagram depicting unequal crossing over between aldosterone synthase and 11-OHase genes. Each gene is represented by a wide bar, with the location of exons indicated by either black (11-OHase) or stippled (aldosterone synthase) bands. One of the two products of unequal crossing over will have a chimeric gene fusing sequences of the normal AldoS and 11-OHase genes. In this example, unequal crossing over is depicted as occurring in the intron between exons 3 and 4. (From Lifton RP, Dluhy RG, Powers M, Rich GM, Cook S, Ulick S, and Lalouel JM: A glucocorticoid-remediable aldosterone synthase gene causes glucocorticoid-remediable aldosteronism and human hypertension. Nature 355:262–265, 1992.)

both 17-hydroxyprogesterone to 11-deoxycortisol, the precursor of cortisol, and progesterone to 11-deoxycorticosterone, the precursor of aldosterone (Fig. 14-5). Since cortisol is deficient, increasing amounts of ACTH stimulate the gland to produce more and more of the precursors formed prior to the 21-hydroxylation step, precursors which can be converted into androgens. The excess androgens lead to virilization. This is usually obvious at birth in girls, who display enlargement of the clitoris and fusion of the labia. The enzymatic deficiency is variable. In some girls, the defect is less complete, leading only to excessive hair growth with menstrual irregularities in adult life. In others, the defect may be so severe as to cause misassignment

Table 14-2 Secretion Rates and Plasma Concentrations of Major Adrenocortical Hormones

Class	Steroid	Secretion rate (mg/day)	Plasma concentration (ng/ml)[a]
Glucocorticoid	Cortisol	8–25	40–180
	Corticosterone	1–4	2–6
Mineralocorticoid	Aldosterone	0.05–0.20	0.05–0.20
	Deoxycorticosterone	0.1–0.6	0.05–0.20
Androgen[b]	DHEA	7–15	2–8
	DHEA-S		400–1200
	Androstenedione	2–3	1–2

[a]Plasma concentrations vary with time of day and, in women, with phase of menstrual cycle.
[b]Secretion rates vary with gender.

NORMAL STEROID SYNTHESIS

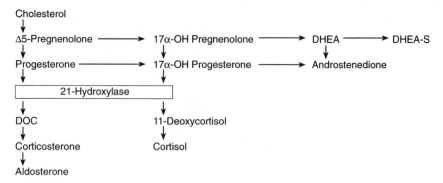

21-HYDROXYLASE DEFICIENCY

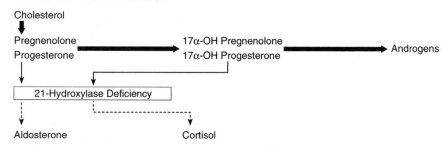

Fig. 14-5   Schematic representation of normal adrenal steroid biosynthesis and of the 21-hydroxylase deficiency from congenital adrenal hyperplasia. Wide arrows denote pathways with increased synthesis proximal to the block. Dashed arrows denote pathways with decreased or absent steroid synthesis. The box shows the deficient enzyme. (From Baxter JD, and Tyrrell JB: In: *Endocrinology and Metabolism,* 2nd ed., P Felig, JD Baxter, AE Broadus, and LA Frohman, eds., McGraw-Hill, New York, p. 625, 1987.)

of sex at birth, since the external genitalia may be completely virilized, with a penile urethra and a ''scrotum'' (see also Chapter 8).

In affected males, the virilization usually does not become clinically obvious until after age 2 and sometimes not until age 10. The manifestations include a rapid spurt in body height along with early maturation of the genitalia.

Much less common are those congenital adrenal enzyme deficiencies involving other enzymes in the biosynthetic pathway. Their manifestations vary. If the defect is early, all steroids may be deficient. If the defect involves the 17-hydroxylase enzyme, mineralocorticoid levels may be high. If it involves the 18-hydroxylase enzyme, aldosterone may be deficient with cortisol normal. With knowledge of the normal pathway, the clinical expression of each enzyme deficiency can be easily determined.

Glucocorticoids

Cortisol is the primary glucocorticoid for humans and most mammals. Corticosterone serves this function in some rodents.

Regulation of Glucocorticoid Secretion

As described in Chapter 6, the hypothalamic–pituitary axis controls adrenal cortical function through the release of ACTH (Fig. 14-6). After it binds to its adrenal plasma membrane receptor, ACTH activates adenylyl cyclase, increasing cyclic AMP and thereby activating protein kinases that convert cholesterol esters to free cholesterol. In concert with this and other effects, ACTH stimulates the rate-limiting 20,22-desmolase reaction, increasing the conversion of cholesterol to Δ^5-pregnenolone and starting the biosynthetic cascade to cortisol (Fig. 14-3).

Plasma levels of cortisol rise within minutes of the intravenous infusion of ACTH, increasing 2 to 5 times by 30 minutes. With chronic ACTH stimulation, the mass of adrenal tissue increases, so that cortisol secretion may rise to 200–250 mg/day, some 10 to 20 times the basal levels.

Secretion. Under ordinary circumstances, the daily secretion of cortisol is both episodic and variable (Fig. 14-7). The episodes follow by 15–30 minutes those of plasma ACTH, and there is a major burst of activity in the early morning hours before awakening. Thereafter, the release of ACTH from the pituitary and of cortisol from the adrenal generally occur only for brief bursts, 7–15 episodes per day. After each episode, plasma cortisol rises enough to suppress further ACTH release. Circulating levels of cortisol then progressively fall, and after variable periods reach the set point of hypothalamic–pituitary negative feedback control so that more corticotropin-releasing factor (CRF) and ACTH are then released.

The normal fine tuning of hypothalamic–pituitary control of cortisol secretion may be interrupted by administration of supraphysiologic doses of glucocorticoids, usually to control serious inflammatory or immune reactions, in keeping with their actions as described in the next section. Various steroids are available, derivatives

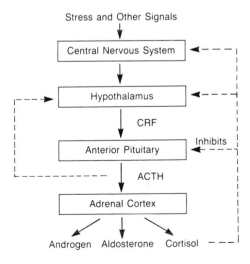

Fig. 14-6 Mechanism for regulation of glucocorticoid secretion. Solid arrows indicate stimulation; dashed arrows indicate inhibition. Glucocorticoids are known to inhibit both pituitary ACTH and hypothalamic corticotropin-releasing factor (CRF) secretion. ACTH also exerts a short negative feedback effect on CRF release.

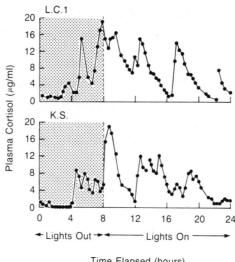

Fig. 14-7   A typical pattern of cortisol secretion during the 24-hour day. Note the oscillations in secretion as well as a daily secretory surge an hour or so after waking in the morning. (From Weitzman ED, Fukushima D, Noseire C, Roffwarg H, Gallagher TF, and Hellman L: Twenty-four hour pattern of the episodic secretion of cortisol in normal subjects. J Clin Endocrinol Metab 33:14, 1971.)

of cortisol with substitutions that enhance the glucocorticoid and reduce the mineralocorticoid effect (Table 14-3).

Exogenous glucocorticoids are initially administered in doses far beyond the normal physiologic levels, usually enough to provide the equivalent of 200–300 mg of cortisol per day, the maximal secretory capacity of the adrenal. When that amount is given, or any amount beyond the equivalent of the 20–30 mg of cortisol per day

Table 14-3   Relative Glucocorticoid and Mineralocorticoid Potency of Natural Adrenal Steroids and Some Derivatives

	Glucocorticoid	Mineralocorticoid
Cortisol	1.0	1.0
Prednisone (1,2 double bond)	4	<0.1
6α-Methylprednisone	5	<0.1
9α-Fluoro-16α-hydroxyprednisolone (triamcinolone)	5	<0.1
9α-Fluoro-16α-methylprednisolone (dexamethasone)	30	<0.1
Aldosterone	0.25	500
Deoxycorticosterone	0.01	30
9α-Fluorocortisol	10	500

that is normally secreted, the hypothalamic–pituitary axis remains suppressed. With no stimulation by ACTH, the adrenal atrophies both in function and in structure.

As long as the patient receives exogenous steroid, suppression of the hypothalamic–pituitary–adrenal system is irrelevant. However, when the exogenous steroid is discontinued, the release of ACTH for a short time, and the release of cortisol for a longer time, may remain suppressed (Fig. 14-8). The levels of ACTH return rather quickly, and, since the adrenal remains unresponsive, they rise above normal. The levels of cortisol take considerably longer to return to the normal range. Obviously, after prolonged adrenal suppression the patient remains susceptible to the development of adrenal insufficiency if stressed when the adrenal is not capable of normal responsiveness.

Metabolism. Once in the circulation, cortisol is largely bound to a specific glucocorticoid-binding α_2-globulin, transcortin. The normal level of transcortin, 3–4 mg/dl, is saturated at a level of cortisol of about 28 μg/dl. Of that which is not bound to transcortin, 15–20% is bound less tightly to albumin, leaving only about 5% of circulating cortisol as unbound.

The hepatic synthesis of transcortin is increased by estrogen, in a manner similar to that of thyroid-binding globulin (TBG) and other hormone carrier proteins. With more transcortin, more cortisol is bound. Therefore, women taking exogenous estrogen or with high endogenous levels during pregnancy have high total plasma cortisol levels but, since their free cortisol levels are normal, they have no manifestations of cortisol excess. On the other hand, transcortin levels are lower in diseases with reduced blood protein concentrations, for example, cirrhosis of the liver; here again, free cortisol levels are normal.

The half-life of exogenous cortisol is about 70–90 minutes. Cortisol can be interconverted to its 11-keto analogue cortisone, which was the first glucocorticoid isolated and made available for therapeutic use. Most cortisol, sometimes referred to as compound F, and most cortisone, compound E, are metabolized in the liver to

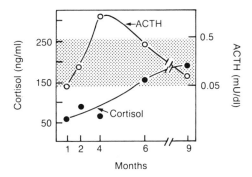

Fig. 14-8 Recovery of the hypothalamic–pituitary–adrenocortical axis after exposure to a long-term elevated concentration of glucocorticoids. All measurements were made at 0600 hours and closed circles represent median values from 14 patients. The dotted area represents the range for normal subjects. (Based on original data of Graber AL, Ney RL, Nicholson WE, Island DP, and Little GW: J Clin Endocrinol Metab 25:11, 1965.)

tetrahydro forms and then conjugated to glucuronides. The glucuronides are more water soluble and are the major forms excreted into the urine. The measurement of these metabolites in the urine by colorimetric assays, as 17-hydroxycorticoids, has been the usual way to estimate cortisol production. Now plasma and urine levels of cortisol are usually measured by radioimmunoassays.

The 24-hour urinary excretion of unmetabolized cortisol is in many ways the most accurate measure now available. Since only free cortisol is excreted into the urine, the total urinary free cortisol in normal people will be below 100 μg/day. However, if cortisol secretion rises and the binding capacity of transcortin is exceeded, at a level of plasma cortisol above 25–30 μg/dl, a progressively larger share will remain free and therefore available for filtration and excretion into the urine.

Molecular Mechanisms of Action

Adrenal steroids enter target cells by passive diffusion and are bound to a specific glucocorticoid receptor, a single-chain polypeptide (about 94,000 Da) that binds glucocorticoids with high affinity in a specific, saturable manner. The cytosolic steroid receptor is complexed with other proteins including a heat-shock protein. After the steroid is bound, the hormone–receptor complex translocates to the nucleus and binds to DNA (Fig. 14-9). By interaction with specific regulatory DNA sequences, called *glucocorticoid response elements*, a change occurs in the transcriptional rate of specific genes, altering production of messenger RNA and changing the cell's content of specific proteins.

Mineralocorticoids preferentially bind to receptors found primarily in their target tissues—kidney, colon, and salivary glands—that have a high affinity for aldosterone. These are called *type I receptors* whereas the glucocorticoid receptors in these tissues are called *type II*. The type I receptors have specificity for mineralocorticoids *in vivo* but not *in vitro*, where they bind glucocorticoids just as avidly as mineralocorticoids. *In vivo,* mineralocorticoid target tissues have an 11β-hydroxysteroid dehydrogenase enzyme that converts the natural glucocorticoid cortisol (which has a high affinity for the type I receptor) into cortisone, which has a low affinity. Therefore, despite the much higher concentration of cortisol in the glomerular filtrate, its conversion to cortisone prevents its binding to the type I receptors in the kidney and these receptors bind aldosterone passing through the tubule. A few patients lack the 11β-hydroxysteroid dehydrogenase enzyme and the normal levels of cortisol are taken up by the type I receptors and exert a massive mineralocorticoid effect; this syndrome is called *apparent mineralocorticoid excess*. The glycyrrhizic acid present in licorice inhibits the 11β-hydroxysteroid dehydrogenase enzyme. People who consume large amounts of licorice, now added as a flavor to some chewing tobaccos, may manifest the same syndrome with all of the features of mineralocorticoid excess.

Functions

Cortisol, the primary glucocorticoid, is essential for life. Exactly why this is so is uncertain. In a simplistic way, cortisol is necessary to maintain critical processes at times of prolonged stress and to contain the reactions to inflammation. Most of its effects are *permissive,* or not directly responsible for the initiation of metabolic or circulatory processes but necessary for their full expression.

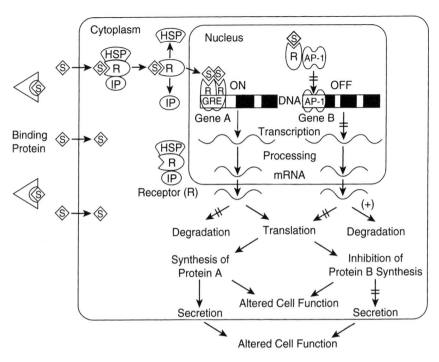

Fig. 14-9 Mechanisms of glucocorticoid action. Steroid hormone (S) circulates as a free molecule or as a complex with plasma-binding protein. After the steroid enters the cell, it binds to receptors (R) that reside in the cytosol complexed to heat-shock protein (HSP) and immunophilin (IP). Binding of the ligand to the complex causes dissociation of HSP and IP. The receptor–ligand translocates into the nucleus where it binds at or near the 5'-flanking DNA sequences of certain genes (glucocorticoid-responsive elements [GRE]). Receptor binding to the regulatory sequences of the responsive genes increases or decreases their expression. In the first instance (ON), glucocorticoids increase the transcription or stability or both of messenger RNA, which is translated on ribosomes to the designated protein. In the second instance (OFF), glucocorticoids repress (cross-hatched arrows) certain genes at the transcriptional level by interacting with and preventing the binding of nuclear factors required for activation of the gene (for example, activator protein [AP]-1 nuclear factor). In other instances, glucocorticoids exert their effects post-transcriptionally by either increasing the degradation of messenger RNA or by inhibiting the synthesis or secretion of the protein. (From Boumpas DT, Chrousos GP, Wilder RL, Cupps TR, and Balow JE: Glucocorticoid therapy for immune-mediated diseases: Basic and clinical correlates. Ann Intern Med 119:1198–1208, 1993.)

Intermediary Metabolism. Figure 14-10 demonstrates the balance between the anabolic and catabolic effects of cortisol, with a predominance of the catabolic. The major catabolic effect is to facilitate the conversion of protein in muscles and connective tissue into glucose and glycogen. Gluconeogenesis involves both the increased degradation of protein already formed and the decreased synthesis of new protein. Gluconeogenesis from amino acids is critical during prolonged fasting, since circulating glucose and stores of liver glycogen will be used up in less than 24 hours.

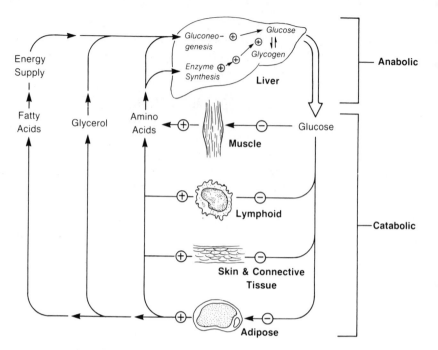

Fig. 14-10 Glucocorticoid influences on peripheral tissues. +, stimulation, and −, inhibition. (From Baxter JD, and Tyrrell JB: In: *Endocrinology and Metabolism,* 2nd ed., P Felig, JD Baxter, AE Broadus, and LA Frohman, eds., McGraw-Hill, New York, p. 545, 1987.)

Carbohydrate Metabolism. Cortisol not only increases the supply of glucose through gluconeogenesis but also decreases the utilization of glucose by cells. The former likely reflects the activation of DNA transcription in liver cells to increase various enzymes involved in the conversion of amino acids into glucose and glycogen. The mechanism for the decreased utilization of glucose by cells involves direct inhibition of glucose transport into the cells.

As a result of these effects, a deficiency of cortisol increases the susceptibility to and decreases the recovery from hypoglycemia. An excess of cortisol raises the blood sugar and decreases the sensitivity to insulin.

Protein Metabolism. Cortisol reduces the utilization of amino acids for the formation of protein everywhere except in the liver. Extrahepatic protein stores are reduced and amino acid levels in the blood increase. Extrahepatic utilization is decreased because amino acids are not transported into muscle cells, thereby reducing protein synthesis. Breakdown of cellular proteins continues, so that plasma amino acids are further increased.

In the liver, the higher levels of blood amino acids are transported more avidly into hepatic cells where they are utilized for gluconeogenesis, glycogen formation, and protein synthesis.

A deficiency of cortisol is not followed by any measurable increase in protein synthesis. An excess of cortisol, however, is associated with progressive loss of protein, atrophy and weakness of muscles, thinning of skin, and loss of bone matrix and mass. Bone formation is reduced, and less calcium is absorbed and more excreted into the urine.

Fat Metabolism. Cortisol also increases the mobilization of fatty acids and glycerol from adipose tissue, increasing their concentration in the blood and making more glycerol available for gluconeogenesis. Not only is fat broken down but less is formed, perhaps because less glucose is transported into fat cells.

The actions of cortisol on fat metabolism are not all lipolytic: it also stimulates appetite and the laying down of additional fat in the central or truncal areas. With an excess of cortisol, the extremities lose fat and muscle whereas the trunk and face become fatter. The mechanism for this redistribution of body fat is unknown.

These various effects of cortisol on intermediary metabolism contribute to the maintenance of blood glucose during food deprivation and the mobilization of extra glucose during stress. But these protective effects come at a cost of decreased protein and fat.

Circulatory. Cortisol is needed to maintain normal vascular integrity and responsiveness, and the volume of body fluids. In the absence of cortisol, abnormal vasodilation occurs, so that even without an external loss of fluid, the filling of the vascular bed is reduced and the blood pressure falls.

In addition, normal renal function requires cortisol. In its absence, glomerular filtration falls and water cannot be excreted as rapidly. Cortisol also has a mineralocorticoid effect that is responsible for part of normal sodium retention and potassium excretion. Although aldosterone is some 300–600 times more potent a mineralocorticoid than cortisol, under normal circumstances 200 times more cortisol than aldosterone is secreted. Therefore, a substantial part of overall mineralocorticoid activity is derived from cortisol. Of course, when body fluids or blood pressure are reduced, the renin–angiotensin mechanism (described in the next section) is activated to stimulate aldosterone, and not cortisol. Thus, when extra mineralocorticoids are needed, aldosterone and not cortisol serves that function.

Inflammatory and Immune Responses. The reactions to various injuries and foreign substances involve multiple anti-inflammatory and immune responses. Cortisol, overall, blocks these responses. The effects may be salutary, as when swelling and the degree of tissue damage are reduced or resolution of the inflammatory responses and wound healing are speeded. On the other hand, the blockade of inflammatory and immune responses may prove harmful if necessary reactions to trauma or foreign substances are thereby impaired. Latent infections such as tuberculosis may be thereby reactivated.

The mechanisms for these actions of cortisol include stabilization of lysosomal membranes so that proteolytic enzymes are not released, decrease in permeability of capillaries so that less plasma and fewer cells enter inflamed areas, depression of phagocytosis by white blood cells, and suppression of thymus-derived lymphocytes. The concept that much of what cortisol does is to protect against the body's normal reactions to stress suggests that some of the enzymes induced by cortisol may de-toxify mediators released during the body's response to stress.

Cortisol suppresses synthesis of several key proteins which accentuate the inflammatory process, including interleukin-1, and stimulates synthesis of others that are anti-inflammatory, including lipocortins that inhibit the generation of proinflam-matory eicosanoids.

Most of these anti-inflammatory effects are seen only with large amounts of cortisol. With large doses of cortisol derivatives, these glucocorticoid effects are widely utilized in clinical medicine in the treatment of many diseases in which the inflammatory reactions are harmful or there is a need to block immune responses. The treatment of rheumatoid arthritis is an example of the former and the prevention of rejection of transplanted organs is an example of the latter use of glucocorticoids.

Central Nervous System. In addition to the physiological negative feedback control of hypothalamic–pituitary release of ACTH, cortisol modulates perception and emo-tion. This is usually recognized only in disease: with cortisol deficiency, the senses of taste, hearing, and smell are accentuated; with cortisol excess, initial euphoria but subsequent depression is common and the threshold for seizure activity may be lowered.

Developmental Effects. Cortisol is important for the maturation of various fetal organs, again in a permissive manner. It is involved in the maturation of intestinal enzymes and various aspects of pulmonary function, including the synthesis of sur-factant, a phospholipid needed for maintaining alveolar surface tension. In children, glucocorticoids inhibit linear skeletal growth by direct effects on bone and connec-tive tissue.

Features of Cortisol Deficiency

The classic features of cortisol deficiency, as listed in Table 14-4, are seen with *primary* hypofunction of the adrenal glands (Addison's disease). This was once most commonly caused by destruction of the glands by tuberculosis, but now the process is usually of unknown cause or is related to antibodies against the person's own adrenal tissue, that is, to autoimmune disease.

Hypofunction of the adrenals can also be *secondary* to a loss of hypothalamic–pituitary function resulting in a deficiency of ACTH. The features of secondary adrenal insufficiency are usually intermingled with manifestations of deficiencies of other glands that are also under the control of the hypothalamic–pituitary system (Table 14-5). Although most instances of pituitary insufficiency involve all of its hormones, a few cases of isolated ACTH deficiency have been reported. The defi-ciency of ACTH, isolated or as part of panhypopituitarism, is rarely complete, so that cortisol deficiency is not usually as marked as in primary adrenal insufficiency. Moreover, the adrenal synthesis of the mineralocorticoid aldosterone, under the con-trol of renin–angiotensin, is not affected by loss of ACTH, so that manifestations of mineralocorticoid deficiency usually do not arise.

The two forms of adrenal insufficiency, both quite rare, can be most easily diagnosed either by measurement of endogenous ACTH levels in the blood or by the response to exogenous ACTH. When the adrenal is destroyed, that is, as in

Table 14-4 Major Functions of Cortisol and Their Clinical Expression

Effect	Clinical expression	
	Cortisol deficiency	Cortisol excess
Carbohydrate metabolism		
Increased gluconeogenesis, decreased glucose utilization	Hypoglycemia	Hyperglycemia
Decreased sensitivity to insulin	Insulin sensitivity	Insulin resistance
Protein metabolism		
Decreased extrahepatic amino acid utilization	Hypoglycemia	Decreased protein structure of bone, skin, muscle
Increased gluconeogenesis		Poor wound healing
Fat metabolism		
Increased lipolysis, decreased lipogenesis	Weight loss	Hyperlipemia
Distribution of fat		Redistribution of body fat, truncal obesity
Circulatory		
Maintain ECF volume	Vasodilation, hypotension	Hypertension
Maintain capillary integrity		
Mineralocorticoid		
Sodium retention	Hypovolemia	Hypervolemia
	Hyponatremia	Hypernatremia
Potassium excretion	Hyperkalemia	Hypokalemia
Inflammatory and immune responses		
Stabilize lysosomes	Propensity toward autoimmune disease	Decreased inflammatory response
Suppress synthesis of antibodies		Increased susceptibility to infection
Decrease capillary permeability		Decreased fibrous tissue formation
Decrease phagocytosis		
Hematopoietic		
Stimulate red cell production	Anemia	Erythrocytosis
Lympholysis	Lymphocytosis	Lymphopenia
Inhibit neutrophil accumulation at inflammatory sites		Leukocytosis
Central nervous system	Anorexia	Euphoria
	Fatigue	Depression
Hypothalamic–pituitary feedback control of ACTH	Increased ACTH secretion	Decreased ACTH secretion
	Pigmentation	If excess is 2° to hypothalamic–pituitary drive, ACTH increased

Table 14-5   Distinctions between Primary and Secondary Adrenal Insufficiency

	Primary	Secondary[a]
Site	Adrenal	Hypothalamic–pituitary
ACTH secretion	Increased	Decreased
Pigmentation	Increased	Decreased
Headaches, visual loss	Rare	Frequent
Body weight	Decreased	Variable
Other pituitary hormones		
Growth hormone	No change	Decreased
Gonadotropin	No change	Decreased
Adrenal hormones		
Aldosterone	Deficient	Normal
Androgens	Deficient	Variable
Cortisol	Markedly decreased	Moderately reduced
Response to exogenous ACTH	None	Sluggish

[a]Rare examples of isolated ACTH deficiency have been reported with normal levels of other pituitary hormones.

primary hypofunction, and cortisol is deficient, the plasma level of ACTH increases markedly since the pituitary is no longer inhibited by the negative feedback control of circulating cortisol. Additional exogenous ACTH can do no more to stimulate damaged glands. On the other hand, when the problem resides in the hypothalamus or pituitary and the adrenal becomes secondarily suppressed, plasma ACTH levels are low. The adrenals can respond to exogenous ACTH, although after prolonged suppression of their structure and function by the deficiency of ACTH stimulation their response may be sluggish.

Features of Cortisol Excess

Cortisol excess is usually exogenous, induced for management of various diseases. The much less common disorders causing endogenous cortisol excess may arise within the adrenal glands or, more commonly, from overproduction of ACTH either by the pituitary or by ectopic ACTH-producing tumors, with the adrenals secondarily stimulated (Table 14-6). The manifestations of cortisol excess (Cushing's syndrome), as delineated in Table 14-4, are similar in both conditions. They may be difficult to differentiate since their clinical features are almost exclusively related to the excess levels of circulating cortisol.

However, the manuevers shown in Table 14-6 will usually differentiate the two, along with direct visualization of the sites by CT scans. If the problem arises primarily in the adrenal gland, usually from a tumor, pituitary ACTH secretion will be suppressed by the high levels of circulating cortisol. With hypersecretion from the pituitary or an ectopic tumor, levels of ACTH obviously should be elevated. Before measurements of plasma ACTH were easily obtained, the distinction was usually made by differential suppression of the pituitary and adrenals with exogenous glucocorticoids. This approach utilizes a number of principles of endocrine physiology, adapted to clinical circumstances. These principles are applied, first, to document cortisol excess and, second, to differentiate the source.

Table 14-6 Differential of Adrenal Hyperfunction

	Primary (30%)	Secondary (70%)	
Site	Adrenal	Hypothalamic–pituitary	Ectopic tumors
Adrenal pathology	Tumor	Bilateral hyperplasia	Bilateral hyperplasia
ACTH secretion	Decreased	Increased	Markedly increased
Suppression with exogenous glucocorticoid			
Low dose	None	Minimal	None
High dose	None	Marked	None

The suppression tests are based upon the ability of exogenous cortisol to exert negative feedback control on hypothalamic–pituitary release of ACTH. Those that are normal but are suspected of having cortisol excess because of suggestive features will be normally suppressed by amounts of glucocorticoid just beyond the normal, physiological level. Since exogenous cortisol would only take the place of endogenous cortisol, suppression as measured by levels of cortisol in the blood could not be demonstrated with cortisol itself. Recall, however, that potent glucocorticoid derivatives, with 20–30 times more suppressive activity per milligram than cortisol are available (Table 14-3). The one usually used is dexamethasone. A highly reliable screening test utilizes 1 mg of dexamethasone (equivalent to about 30 mg of cortisol) taken at bedtime to measure plasma cortisol the next morning. Normal suppression is to below 5 μg/dl; patients with adrenocortical hyperfunction will not suppress to that level.

To elucidate further the nature of adrenal hyperfunction, more prolonged suppression tests can be done. One-half milligram of dexamethasone is given 4 times a day, for a total of 2 mg, which is equivalent to 60 mg of cortisol a day. That is enough to completely suppress the normal hypothalamic–pituitary axis, so that ACTH will not be released and the adrenal will make little cortisol. Cortisol measured in a urine or blood sample should be very low (Fig. 14-11). Those with cortisol excess, whatever the cause, will not be significantly suppressed.

The second differential suppression test builds on the same principles of pituitary–adrenal control. If the problem is caused by hyperfunction of the hypothalamic–pituitary axis with increased secretion of ACTH, near-normal amounts of cortisol will not go above the set point of negative feedback suppression, since by definition that set point has been raised to an abnormally high level in hyperfunctioning pituitary ACTH-secreting cells or in hypothalamic cells secreting excess CRF. However, the set point is suppressible if enough exogenous glucocorticoid is given. By trial and error, that dose was found to be 2.0 mg of dexamethasone 4 times a day for 2 days, a total of 8 mg of dexamethasone a day, which is equivalent to 240 mg of cortisol, or approximately 10 times the normal secretion. With that level, most hypothalamic–pituitary hyperfunctioning states will be suppressed, so that urine cortisol levels will fall by 40% or more.

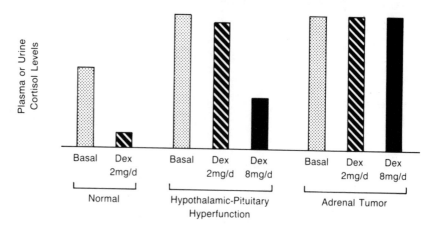

Fig. 14-11 Expected changes in plasma or urine cortisol values after 2 days of low (2 mg/day) and high (8 mg/day) dexamethasone in divided doses in normals and patients with Cushing's syndrome caused by hypothalamic–pituitary hyperfunction or an adrenal tumor.

On the other hand, if the cortisol excess arises from an autonomous adrenal tumor or from an ectopic ACTH-producing tumor, no amount of exogenous gluco-corticoid will usually stop the secretion of cortisol. The high levels of cortisol are suppressed slightly or not at all by the equivalent of 240 mg of cortisol.

Mineralocorticoids

Aldosterone is the primary mineralocorticoid. It arises from the outer zona glomeru-losa and is under the control of angiotensin primarily, and of serum potassium and ACTH to a lesser degree. In its absence, fluid and electrolyte status is altered, al-though if enough cortisol is present, cortisol itself may exert a mineralocorticoid effect sufficient to prevent progressive depletion of body fluids.

Regulation of Mineralocorticoid Secretion

Since the major function of aldosterone is to control body fluid volume by increasing the reabsorption of sodium by the kidneys, it is appropriate that the major stimulus for aldosterone synthesis and secretion arises in the kidneys (Fig. 14-12). The juxta-glomerular apparatus consists of modified myoepithelial cells in the renal afferent arterioles, contiguous to the macula densa of the distal tubule. These cells synthesize and secrete the proteolytic enzyme renin, which splits off the 10 amino acid peptide, angiotension I, from its protein substrate, angiotensinogen. Angiotensin I is inactive but is rapidly converted into the very active 8 amino acid, angiotensin II, by the action of a converting enzyme that is found throughout the body, arising from the plasma membrane of vascular endothelial cells.

Angiotensin II exerts two major actions, one as a direct arteriolar vasoconstric-tor and the other as a stimulus to aldosterone secretion. These two actions, in concert,

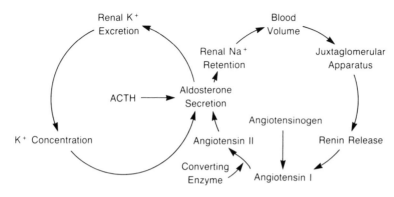

Fig. 14-12 Regulation of aldosterone secretion by the renin–angiotensin system, the potassium concentration, and corticotropin (ACTH). (From Griffin JE: *Manual of Clinical Endocrinology and Metabolism*, McGraw-Hill, New York, p. 78, 1982.)

maintain the volume and pressure of the arterial circulation. They provide the major support to the circulation in times of fluid loss or falling blood pressure.

Stimuli to Renin Release. The major stimulus to the release of renin is a decrease in the perfusion pressure of blood traversing the renal afferent arterioles which is sensed by the juxtaglomerular apparatus functioning as a baroreceptor. When a decrease in either systemic blood volume or blood pressure leads to a fall in renal perfusion pressure, renin is secreted.

Other factors also stimulate renin secretion, including the concentration of sodium and chloride in the macula densa, the concentration of plasma electrolytes, and the concentration of circulating angiotensin II, a short feedback control loop. Some of these stimuli work through the stimulation of renal sympathetic nerves that innervate the juxtaglomerular apparatus.

Other Stimuli for Aldosterone. All of the actions of renin are mediated through its generation of angiotensin II, which serves as the primary stimulus to aldosterone synthesis. The stimulation occurs at various sites in the biosynthetic pathway. Plasma potassium concentration and ACTH also stimulate aldosterone. The potassium effect appears appropriate since the effects of aldosterone include a major increase in the renal excretion of potassium. If levels of plasma potassium are suddenly raised, as after a large meal of potassium-rich foods, the secretion of aldosterone is increased, acting on the kidney to excrete the excess potassium load. As described below, however, the effects of aldosterone to increase renal sodium retention and potassium excretion require an hour or more; thus, there must be more rapid ways to remove high levels of potassium entering the circulation, since such levels could lead to instant and serious trouble. The rapid transfer of potassium from extracellular fluid into cells is accomplished by a combination of insulin and epinephrine effects on potassium transport across cell membranes.

The effect of ACTH is powerful but short-lived. Although the physiological role of ACTH on aldosterone secretion is limited, when ACTH is deficient, the zona glomerulosa may atrophy and be less able to respond to other stimuli.

Secretion. Aldosterone is so powerful a mineralocorticoid that very little is needed for its primary effect. Only 0.05–0.20 mg (50–200 μg) is produced each day. Since there is no specific aldosterone carrier protein in the blood, very little circulating aldosterone is bound. Plasma concentrations are around 10 ng/dl. Secretion can be increased 2- to 6-fold by sodium depletion and even more so by persistent shrinkage of effective arterial blood volume as in cirrhosis of the liver with ascites and marked edema.

Metabolism. Aldosterone is metabolized mainly in the liver into a tetrahydroaldosterone derivative that is excreted in the urine as a glucuronide. Some is excreted as a glucuronide without being metabolized, and this acid-labile fraction, about 10% of total secretion, is what is usually measured. Therefore, a normal urine aldosterone level is 5–20 μg/day.

Functions

Aldosterone's primary action is to increase sodium reabsorption in the distal tubules. To do so, it is first bound to cytosolic receptors, then the hormone–receptor complex is transferred to the nucleus where it induces specific portions of DNA to form messenger RNA, which in turn increases the formation of enzymes that are involved in sodium transport across cell membranes. These processes take 30–60 minutes, thus the effect of aldosterone is delayed by that length of time.

As aldosterone increases active sodium reabsorption, an electrochemical gradient is established that facilitates the passive transfer of potassium from tubular cells into the urine. Therefore, potassium is excreted, not as a direct exchange for sodium, but in a manner that depends directly upon the active reabsorption of sodium. If almost all sodium is reabsorbed more proximally, as in the presence of severe volume depletion, little will reach the distal reabsorptive site. Therefore, despite high levels of aldosterone, potassium excretion will be minimal in the absence of sodium delivery to the distal tubule. On the other hand, a high sodium intake will increase potassium excretion. This is particularly true if the individual is receiving a diuretic that blocks part of the reabsorption of sodium higher in the tubule, causing even more sodium to reach the distal reabsorptive site.

Sodium Concentration. As sodium is reabsorbed in the kidney, an equivalent amount of water is also reabsorbed, thus preserving the normal concentration of sodium in the extracellular fluid. If not enough water is reabsorbed so that the concentration of sodium rises, the increase in osmotic pressure will activate the release of antidiuretic hormone, effectively increasing renal water reabsorption and increasing water intake via thirst.

Escape. Under usual circumstances of mild sodium depletion leading to increased aldosterone secretion, the extracellular fluid volume re-expands as more sodium is reabsorbed. The degree of expansion should be adequate to replenish the shrinkage that initiated the process so that it is then turned off. Under some circumstances, such as persistent leak of effective blood volume into the abdomen in a patient with cirrhosis of the liver and ascites, the stimulus to aldosterone never turns off. The extra sodium that is reabsorbed does not replenish the shrunken effective fluid volume because of the continuous leak. Therefore, very high levels of aldosterone, that is, secondary hyperaldosteronism, are seen in chronic edematous states.

When there is no continuous leak but aldosterone levels remain high, as with the administration of exogenous aldosterone or an aldosterone-secreting tumor, extracellular volume expands by a few liters but then the expansion stops. This *escape* phenomenon likely results from a confluence of effects: proximal reabsorption of sodium is decreased, probably because of a natriuretic factor that arises in the hypothalamus; glomerular filtration is increased, probably because of the actions of the atrial natriuretic factor (see Chapter 7); and renin–angiotensin levels are suppressed by the rise in systemic blood pressure, renal perfusion pressure, and sodium content in the macula densa.

Although sodium reabsorption escapes, potassium excretion continues since more sodium is delivered to the distal tubule.

Features of Primary Aldosterone Excess

Autonomous benign tumors of the adrenal gland that secrete large amounts of aldosterone have been recognized since 1954 as a rare but interesting form of curable hypertension. The clinical hallmarks are hypokalemia and hypertension. The blood pressure rises because plasma volume expands due to the aldosterone-mediated reabsorption of sodium. Even though "escape" occurs, the extra few liters of body fluid volume are enough to keep the blood pressure high. Similarly, as noted above, the increase in potassium excretion continues so that hypokalemia, with its various manifestations, is typical (Fig. 14-13).

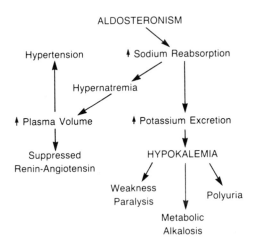

Fig. 14-13   The pathophysiology of primary aldosteronism. (From Kaplan NM: In: *Clinical Hypertension,* 5th ed., NM Kaplan, ed., Williams & Wilkins, Baltimore, 1990.)

Features of Secondary Aldosterone Excess

More common than aldosterone-secreting tumors are conditions in which effective arterial blood volume is shrunken, calling forth high levels of renin–angiotensin that continue to stimulate the adrenal synthesis of aldosterone, to the point that bilateral adrenal hyperplasia develops (Table 14-7). These include chronic edematous diseases such as severe heart failure, cirrhosis of the liver with ascites, and the nephrotic syndrome. Despite the high levels of renin–angiotensin and the secondarily increased aldosterone excess, fluid volume remains shrunken and the blood pressure is low-normal.

It is possible to block aldosterone's effects in the kidney by use of spironolactone, a nonmineralocorticoid steroidal compound that competes directly with aldosterone for the mineralocorticoid receptors in the renal tubule. Since it is more weakly bound than aldosterone, amounts of 50–500 mg/day are needed to competitively block the 0.5–10 mg of aldosterone that may be secreted in secondary aldosteronism.

Extrarenal Effects

Aldosterone also increases the reabsorption of sodium in other places where sodium may be excreted—saliva, sweat, and stool. When physical activity is first begun in a hot climate, a great amount of sodium may be lost in the large volumes of sweat that are formed. The resulting shrinkage of fluid volume activates the renin–angiotensin–aldosterone mechanism that causes more and more sodium to be retained, which is the process of acclimitization to a hot environment. After a few days, sweat will be sodium-free. Thus, there is no need to increase sodium intake beyond the usual amounts obtained in the diet when undergoing continued exercise in hot climates. Water intake should be adequate to replenish the volume of sweat.

Androgens

Androgens, the third secretion of the adrenals, unlike the other two, are not critical for life, but in women they may contribute to sexual responsiveness. Libido, as well as axillary and pubic hair growth in both sexes, depends largely upon the androgens

Table 14-7 The Two Major Forms of Aldosterone Excess

	Primary	Secondary
Pathology	Adrenal tumor	Adrenal hyperplasia
Mechanism	Autonomous hypersecretion	Driven hypersecretion (renin–angiotensin)
Effective arterial blood volume	Expanded	Shrunken
Total body fluid volume	Slightly expanded	Markedly expanded (edema)
Renin–angiotensin	Suppressed	Activated
Clinical expression	Hypertension	Low-normal blood pressure
	Potassium wastage	Potassium wastage
		Edema

testosterone and dihydrotestosterone. In the male, these come almost entirely from the testes. In the female, the smaller amount of testosterone comes largely from peripheral conversion of androstenedione of ovarian and adrenal origin. In the post-menopausal women whose ovaries are no longer active, adrenal androgens also serve as the major source of estrogens, derived by peripheral conversion of adrenal androstenedione.

The clinical situations in which adrenal androgens are most prominent are the congenital adrenal hyperplasias that shuttle large amounts of precursors into the androgen pathway. As noted previously in this chapter and in Chapter 8, the presence and degree of androgen excess or deficiency depend upon the site of enzymatic defect.

Adrenal androgens are formed primarily in the zona reticularis and are largely under the control of ACTH and, perhaps, of prolactin.

ADRENAL MEDULLA

Catecholamines are the hormones secreted by the sympathetic nervous system. In the peripheral nerves, norepinephrine is the locally produced and secreted neurotransmitter; in the adrenal medulla, epinephrine is the major secretion. It is useful to regard norepinephrine as a local hormone that is involved in neurotransmission and that spills over into the general circulation only after rather intense activation of the sympathetic nervous system. In contrast, epinephrine is released from the adrenal medulla in response to stress and circulates throughout the body to prepare the organism for "flight-or-fight." The effects of adrenergic nervous activity are mediated via α- and β-receptors on effector tissues that often have antagonistic actions, such as α-vasoconstriction versus β-vasodilation.

Synthesis and Release of Epinephrine

The adrenal medulla is a modified sympathetic ganglion with postganglionic cells, but no axons, that is specialized to secrete catecholamines from chromaffin (take up chromium) cells. One preganglionic fiber from the splanchnic nerve innervates a number of medullary cells, so that a few impulses can cause a massive discharge of catecholamines.

Epinephrine is synthesized in the adrenal medulla (and probably in certain brain neurons) from norepinephrine by action of the enzyme phenylethanolamine-N-methyltransferase (PNMT) (Fig. 14-14). The synthesis of PNMT is induced by cortisol. Since the adrenal medulla is cradled by the adrenal cortex, it is bathed with very high concentrations of cortisol, so that PNMT is available. About 80% of adrenal medullary catecholamine secretion is epinephrine; the other 20% is norepinephrine.

The synthesis of catecholamines is regulated by changes in the levels of the rate-limiting enzyme, tyrosine hydroxylase. Acutely, when neural activity causes a release of cytoplasmic catecholamines, the enzyme is released from end-product inhibition. With chronic stimulation more of the enzyme is synthesized.

The release of catecholamines occurs by "stimulus–secretion coupling": preganglionic nerve impulses release acetylcholine which increases membrane per-

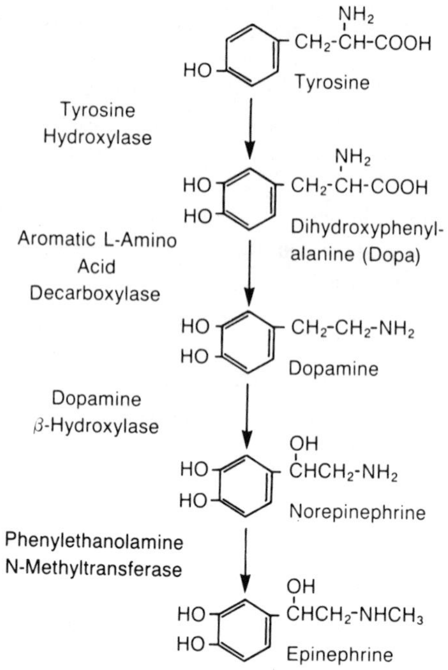

Fig. 14-14 Catecholamine biosynthesis.

meability, thus depolarizing the cell membrane, and increases calcium uptake. This induces reverse pinocytosis of the secretory granules or exoctyosis.

Accurate assays for the very small quantities of catecholamines in the blood are now available. Increased levels under various clinical circumstances reflect the involvement of epinephrine in cardiovascular, metabolic, and other responses to stress. Repetitive sympathetic activation may eventuate in persistently high blood pressure and potentiate atherosclerosis.

Fate of Secreted Catecholamines

Catecholamines are rapidly removed from the synapses and circulation by a number of processes:

1. Reuptake by catechol-secreting cells for reuse or metabolism.
2. Uptake by receptors on effector cells.
3. Metabolism by inactivating mechanisms, mainly in the liver, involving the catecholamine-O-transferase and monamine oxidase enzymes. The products are mainly excreted through the kidney as metanephrines or vanilmandelic acid (VMA) (Fig. 14-15).

Action of Epinephrine

Alpha- and β-receptors for epinephrine and norepinephrine are present in many tissues and the responses are varied. Both epinephrine and norepinephrine can in-

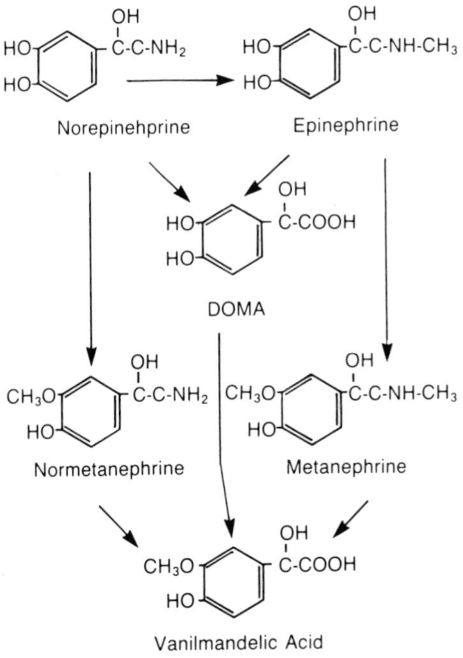

Fig. 14-15 Pathways of catecholamine metabolism.

teract with both α- and β-adrenergic receptors, so that the response of a given system to adrenergic stimulation is largely determined by the type of receptors that populates that system (Table 14-8).

Since the adrenal medulla normally accounts for only about 10% of sympathetic nervous activation, we can live quite well without it. The major actions of epinephrine are as follows:

Table 14-8 Adrenergic Receptors and Function

α-Receptor	β-Receptor
Vasoconstriction	Vasodilatation (β_2)
Iris dilatation	Cardioacceleration (β_1)
Intestinal relaxation	Increased myocardial strength (β_1)
Intestinal sphincter contraction	Intestinal relaxation (β_2)
	Uterus relaxation (β_2)
Pilomotor contraction	Bronchodilatation (β_2)
Bladder sphincter contraction	Calorigenesis (β_2)
	Glycogenolysis (β_2)
	Lipolysis (β_1)
	Bladder wall relaxation (β_2)

From Guyton AC: *Textbook of Medical Physiology*, Saunders, Philadelphia, p. 671, 1991.

Arousal:

alerting, pupillary dilation, piloerection, sweating, bronchial dilation, tachycardia, inhibition of smooth muscle in the gastrointestinal tract, constriction of sphincters, relaxation of uterine muscles.

Metabolic:

In keeping with the need for ready fuel, epinephrine's actions provide more glucose and free fatty acids. This occurs through the release of glucose from glycogen (glycogenolysis) by activation of phosphorylase in the liver and muscle via the Cori cycle (see Chapter 16). In muscle, without glucose-6-phosphatase the glucose is metabolized to lactate. Epinephrine also breaks down adipose cell lipids (lipolysis) by activating the lipase that converts triglycerides into free fatty acids and glycerol. As a result of these calorigenic effects, the metabolic rate increases by 20–30%.

Cardiovascular:

In addition to direct stimulation of the heart rate and vasoconstriction, repeated bursts of epinephrine have been hypothesized to result in permanent hypertension via a number of possible effects. In addition, there is some evidence that epinephrine, by its neuronal reuptake, may be responsible for prolonged pressor effects, far beyond the short duration of its expected action. The fact that hypertension usually disappears after removal of a pheochromocytoma, however, argues against this hypothesis.

Both agonists and antagonists of α- and β-adrenergic receptors are widely used in clinical medicine. Examples include β-agonists such as isoproterenol to dilate bronchi for the treatment of asthma and β-antagonists such as propranolol to decrease cardiac output for the treatment of hypertension. Epinephrine is also used for allergic reactions and norepinephrine is used as a pressor agent.

Features of Catecholamine Excess

Pheochromocytomas are usually benign tumors arising from the adrenal medulla, with clinical features that reflect the physiology of epinephrine and norepinephrine gone awry. Thus, most patients have paroxysms of hypertension, tachycardia, sweating, tremor, palpitations, and nervousness. Most lose weight and are hyperglycemic. The diagnosis is made by measuring either plasma for free catecholamines or urine for metanephrine and by direct visualization of the typically large adrenal tumors, usually by computerized tomographic scan.

ASSESSMENT OF ADRENAL FUNCTION

The various forms of adrenal hypo- and hyperfunction are relatively rare, but many more patients than the few who turn out to have these diseases may be suspected of having them, in large part because of the multiple and diffuse effects of adrenal hormones. Thus, among the large number of women with obesity of predominantly truncal distribution and primary hypertension, some will be suspected of having cortisol excess (Cushing's syndrome), since truncal obesity and hypertension are typical features of the syndrome. Although only a few will have the syndrome, many may need an assessment to rule it out. Therefore, screening tests to assess adrenal function are rather frequently needed to exclude adrenal disorders. If they are positive, more definitive tests may be needed to establish the presence of the disorder and to determine its precise form (Table 14-9). In addition to these, anatomic localization and determination of the type of adrenal pathology may be obtained by various X-ray and isotopic scanning procedures, in particular, computerized tomography and magnetic resonance imaging.

As a general approach to screening for either hypofunction or hyperfunction, simple measurements of plasma or urine concentrations of adrenal hormones under basal conditions are usually not adequate in themselves, mainly because of the wide range of values seen in normal people. In screening for hypofunction, however, the presence of low levels of the adrenal hormone along with high levels of the physiological stimulating hormone, that is, ACTH for glucocorticoids and renin–angiotensin for mineralocorticoids, may be useful. Better results are obtained by measuring plasma levels after an appropriate stimulus, that is, ACTH for glucocorticoids and acute volume depletion by a diuretic for mineralocorticoids. For more definitive evidence, the responses of either plasma or urine levels of the hormones are measured after repeated administration of the appropriate stimulus.

To screen for hyperfunction, attempts at rapid suppression of the adrenal hormones are made. For glucocorticoid excess, the response of plasma cortisol to a single dose of the exogenous glucocorticoid dexamethasone is usually adequate. For

Table 14-9   Assessments of Adrenal Function

	Hypofunction		Hyperfunction	
	Screening	Definitive	Screening	Definitive
Cortical				
Glucocorticoid	Low cortisol, high ACTH Response to single ACTH injection	Response to repeated ACTH injections	High cortisol Response to single dose of glucocorticoid	Response to repeated doses of glucocorticoid
Mineralocorticoid	Low aldosterone, high renin Response to acute volume depletion	Response to chronic volume depletion	High aldosterone, low renin Response to acute volume expansion	Response to chronic volume expansion
Androgen	Low plasma levels	—	High plasma levels	—
Medullary	—	—	High catecholamines in single plasma or urine samples	Response to sympathetic nervous system suppressants

mineralocorticoid excess, the response of plasma aldosterone to acute volume expansion by a short infusion of normal saline is measured. For more definitive testing, the responses of either plasma or urine levels of the hormones after more prolonged attempts at suppression are used.

There are few indications to test for deficiencies of adrenal androgens and even fewer for catecholamine deficiency. Excesses of androgens are seen with congenital adrenal enzymatic defects. Catecholamine-secreting tumors (pheochromocytomas) are detected by measuring plasma and urine levels of the catecholamines or their metabolic products.

In summary, the assessment of adrenal function is similar to that of other endocrine disorders: hypofunction is best diagnosed by attempting to stimulate with the appropriate physiological trophic agent and hyperfunction, by suppressing with the appropriate physiological maneuver.

SUGGESTED READING

Baxter JD, and Tyrrell JB: The adrenal cortex. In: *Endocrinology and Metabolism,* 2nd ed., P. Felig, JD Baxter, AE Broadus, and LA Frohman, eds., McGraw-Hill, New York, pp. 511–650, 1987.

Betito K, Mitchell JB, Bhatnagar S, Boksa P, and Meaney MJ: Regulation of the adrenomedullary catecholaminergic system after mild, acute stress. Am J Physiol 267:R212–R220, 1994.

Boumpas DT, Chrousos GP, Wilder RL, Cupps TR, and Balow JE: Glucocorticoid therapy for immune-mediated diseases: basic and clinical correlates. Ann Intern Med 119:1198–1208, 1993.

Dluhy RG, and Lifton RP: Glucocorticoid-remediable aldosteronism. Endocrinol Metab Clin North Am 23:285–297, 1994.

Ebeling P, and Koivisto VA: Physiological importance of dehydroepiandrosterone. Lancet 343:1479–1481, 1994.

Genuth SM: The adrenal glands. In: *Physiology,* RM Berne and MN Levy, eds., Mosby, St. Louis, pp. 950–982, 1988.

Gomez MT, Magiakou MA, Mastorakos G, and Chrousos GP: The pituitary corticotroph is not the rate-limiting step in the postoperative recovery of the hypothalamic–pituitary–adrenal axis in patients with Cushing syndrome. J Clin Endocrinol Metab 77:173–177, 1993.

Goulding NJ, Godolphin JL, Sharland PR, Peers SH, Sampson M, Maddison PJ, and Flower RJ: Anti-inflammatory lipocortin 1 production by peripheral blood leucocytes in response to hydrocortisone. Lancet 335:1416–1418, 1990.

Griffin JE: *Manual of Clinical Endocrinology and Metabolism,* McGraw-Hill, New York, pp. 75–96, 1982.

Grinspoon SK, and Biller BMK: Laboratory assessment of adrenal insufficiency. J Clin Endocrinol Metabol 79:923–931, 1994.

Guyton AC: The adrenocortical hormones. In: *Textbook of Medical Physiology,* 8th ed., Guyton AC, ed., Saunders, Philadelphia, pp. 667–678, 842–854, 1991.

Kaplan JR, Pettersson K, Manuck SB, and Olsson G: Role of sympathoadrenal medullary activation in the initiation and progression of atherosclerosis. Circulation 84(suppl VI):VI-23–VI-32, 1991.

Kaplan NM: Adrenal diseases: Pheochromocytoma; primary aldosteronism; Cushings syndrome and congenital adrenal hyperplasia. In: *Clinical Hypertension,* 6th ed., Williams & Wilkins, Baltimore, pp. 367–388, 389–408, 409–422, 1994.

Kater CE, and Biglieri EG: Disorders of steroid 17α-hydroxylase deficiency. Endocrinol Metab Clin North Am 23:341–357, 1994.

Landsberg L, and Young JB: Catecholamines and the adrenal medulla. In: *Williams Textbook of Endocrinology,* 8th ed., JD Wilson and DW Foster, eds., Saunders, Philadelphia, pp. 621–705, 1992.

Orth DN, Kovacs WJ, and DeBold CR: Disorders of the adrenal cortex. In: *Williams Textbook of Endocrinology,* 8th ed., JD Wilson and DW Foster, eds., Saunders, Philadelphia, pp. 489–619, 1992.

Walker RB, and Edwards CRW: Licorice-induced hypertension and syndromes of apparent mineralocorticoid excess. Endocrinol Metab Clin North Am 23:359–377, 1994.

White PC: Disorders of aldosterone biosynthesis and action. N Engl J Med 331:250–258, 1994.

White PC, and Speiser PW: Steroid 11β-hydroxylase deficiency and related disorders. Endocrinol Metab Clin North Am 23:325–339, 1994.

15

Calcium Homeostasis

NEIL A. BRESLAU

IMPORTANCE OF MAINTAINING CALCIUM HOMEOSTASIS

The physiological importance of calcium falls into two broad categories. The calcium salts in bone provide the structural integrity of the skeleton. The calcium ion in cellular and extracellular fluids is essential to the normal function of a number of biochemical processes including neuromuscular excitability, blood coagulation, hormonal secretion, and enzymatic regulation. The biochemical role of calcium requires that its extracellular and cellular concentrations be maintained within a very narrow range, and this is achieved by an elaborate system of controls.

The control of cellular calcium homeostasis is as carefully maintained as its concentration in extracellular fluids. The concentration of calcium ions in the cytoplasm is approximately one one-thousandth that in the extracellular fluids. Most of the intracellular calcium is contained in the mitochondria and microsomes. Three "pump–leak" transport systems located in the plasma, mitochondrial, and microsomal membranes control the concentration of calcium in the cytosol. Calcium leaks passively into the cytosol by diffusion across these three membranes, but each pump actively shifts calcium away from the cytosolic pool.

Several examples may serve to illustrate the importance of these three calcium transport systems in integrating cellular calcium metabolism. Calcium ion is the coupling factor linking excitation and contraction in skeletal and cardiac muscle. The sarcoplasmic reticulum, composed primarily of microsomes, is the main storehouse of intracellular calcium in muscle. Depolarization of the plasma membrane permits entry of a small amount of extracellular calcium into the muscle cell, which then triggers release of large quantities of stored calcium from the sarcoplasmic reticulum. The abrupt increase in cytosolic calcium interacts with troponin, a specific calcium-binding protein that undergoes conformational change culminating in the actin–myosin interaction that causes muscle contraction. Then the microsomes rapidly reaccumulate the cytosolic calcium, permitting relaxation.

Another important role of the intracellular calcium ion is in stimulus–secretion coupling. In secretory cells, it is frequently the mitochondria that act as the major intracellular storehouse of calcium. Once released from the mitochondria by an appropriate stimulus, cytosolic calcium links secretory vesicles to the plasma membrane, resulting in exocytosis of preformed cell products. Calcium has been implicated as an important coupling factor in neurotransmitter release, exocrine secretion (e.g., amylase), and endocrine secretion (e.g., insulin). In a number of these systems, cyclic adenosine monophosphate (cyclic AMP) operates as a second or intracellular messenger that modulates mitochondrial calcium transport, with the change in cytosolic calcium ion actually mediating cellular secretion.

Intracellular calcium ion is also important in the control of key enzymes regulating intermediary metabolism. In the kidney, increased intracellular calcium enhances gluconeogenesis by activating phospho*enol*pyruvate carboxykinase and inhibiting pyruvate kinase. In mammalian skeletal muscle, which often works under hypoxic conditions and depends on glycogenolysis for its immediate energy needs, the major effect of calcium is upon the action of phosphorylase *b* kinase, a key enzyme in the glycogenolytic sequence. Thus, calcium, which is the coupling factor between excitation and contraction in muscle, may also play a role in providing the energy for contraction.

In humans, maintenance of extracellular calcium ion concentration within narrow limits is essential for a number of vital functions including mineralization, blood coagulation, and membrane function. Proper mineralization of bone requires maintenance of a normal calcium $\times$ phosphate ion product. The calcium ion serves as a cofactor for clotting factors VII, IX, and X. Ionic calcium also plays an important role in the permeability and excitability of plasma membranes. When the extracellular fluid concentration of calcium ion falls below normal, the nervous system becomes progressively more excitable because of increased permeability of the neuronal membrane to sodium. Nerve fibers become hyperexcitable, spontaneously depolarized, and initiate nerve impulses to peripheral skeletal muscles, thus eliciting tetanic contraction. This effect is the basis for the muscle spasms and marked hyperreflexia seen in hypocalcemic tetany. The traditional positive Chvostek sign—a twitching of the facial muscles in response to tapping over the facial nerve at the angle of the mandible—is one manifestation of hypocalcemic tetany. In severe hypocalcemia, increased central nervous system (CNS) irritability may lead to seizures. An increase in serum ionized calcium depresses central and peripheral neural excitability, resulting in mental sluggishness and hyporeflexia.

There are three definable fractions of calcium in serum: ionized calcium (50%), protein-bound calcium (40%), and calcium that is complexed to other constituents in serum, mostly citrate and phosphate forming soluble complexes (10%). Both complexed and ionized calcium are ultrafilterable, so that about 60% of the total calcium in serum crosses semipermeable membranes. Nearly 90% of the protein-bound calcium is bound to albumin and the remainder to globulins. The binding of calcium to albumin is pH dependent. Acute acidosis decreases binding and increases ionized calcium, whereas acute alkalosis increases binding with a consequent decrease in ionized calcium. Since it is the ionized fraction of calcium that is physiologically important, it is not surprising that patients who hyperventilate and develop an acute respiratory alkalosis may develop circumoral numbness, carpal spasm, and even seizures due to increased neural excitability.

REGULATION OF CALCIUM METABOLISM IN HUMAN BEINGS

Total serum calcium concentration in humans is maintained between 8.5 and 10.5 mg/dl by the flux of calcium between blood and three organs: bone, kidney, and intestine (Fig. 15-1). Calcium is constantly entering and leaving the blood by exchange through these organs. The movement of calcium in this system is controlled by three main hormones: parathyroid hormone (PTH), calcitonin, and the most important metabolite of vitamin D: 1,25-dihydroxyvitamin D [1,25-$(OH)_2$D]. PTH increases blood calcium concentration by increasing resorption of bone, by promoting reabsorption of calcium from the glomerular filtrate, and by increasing the rate of formation of 1,25-$(OH)_2$D in the kidney. This vitamin D metabolite accelerates the rate of absorption of calcium from the intestine. Moreover, it directly stimulates bone resorption and plays a permissive role in the action of PTH on bone. Calcitonin inhibits bone resorption but is not believed to play as important a role in the day-to-day regulation of calcium homeostasis as the previous two hormones.

The daily net changes in calcium transfer for an adult man in the steady state are approximately as follows. About 1000 mg of calcium is ingested per day, but only one-third of this amount is absorbed from the intestine (360 mg) and enters the extracellular fluid (ECF). Part of the calcium in the ECF is lost through enteric secretion (190 mg) back into the gut. The net absorption is thus 170 mg of calcium, approximately the same as that lost per day via urinary excretion. Bone is the principal reservoir of calcium in the body, containing 1000 g. Each day about 550 mg of calcium is exchanged between bone and body fluid. In the steady adult state, the transfer rates for calcium entering and leaving bone are the same. Obviously, in a growing child, calcium retention would occur in proportion to bone growth. The increased calcium entering bone would be balanced by greater intestinal absorption and less urinary excretion of calcium, thereby maintaining a constant serum calcium concentration.

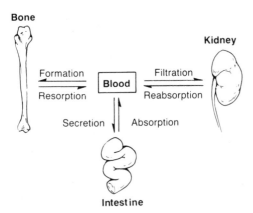

Fig. 15-1   Calcium homeostasis is regulated by the transfer of calcium between the blood and three major target organs.

BIOCHEMISTRY AND PHYSIOLOGY OF THE MAJOR CALCIUM-REGULATING HORMONES

Parathyroid Hormone (PTH)

Parathyroid hormone is produced in the parathyroid glands, of which there are generally two pairs situated behind the thyroid gland. Each parathyroid gland measures 5 × 4 × 2 mm and weighs 30–50 mg. Supernumerary parathyroids (up to eight glands) are present in 2–5% of the population. The chief cell, the principal parenchymal parathyroid cell, is responsible for the synthesis and secretion of PTH.

PTH is a single-chain polypeptide composed of 84 amino acids with a molecular weight of 9500. The structure of human PTH and its biosynthetic sequence are now known. The gene for PTH actually encodes a precursor termed *prepro-PTH*, which is a polypeptide of 115 amino acids with a molecular weight of 13,000. This prehormone is short-lived, being rapidly cleaved to a smaller peptide, pro-PTH. Pro-PTH is composed of 90 amino acids and has a molecular weight of 10,200. A second enzymatic cleavage of the 6 amino acid extension at the amino terminus of the molecule results in the final secretory product: 1–84 PTH. The function of the amino-terminal leader sequences in the precursor hormones may be to transport the polypeptide through the membrane channels of the endoplasmic reticulum. In the Golgi apparatus, PTH is packaged into secretory granules for storage and subsequent secretion. The precursor forms are not secreted and are biologically inactive.

The ability of the parathyroid cell to sense minute fluctuations in the extracellular ionized calcium concentration is essential for maintaining mineral ion homeostasis. Until recently, the mechanism(s) through which the parathyroid cell and other cells recognize and respond to changes in serum calcium has remained unclear. An extracellular calcium-sensing receptor has now been cloned and characterized. It is present in the cell membrane of parathyroid cells, renal tubule cells, and in the C cells of the thyroid gland. As shown in Fig. 15-2, the protein is predicted to have three major structural domains. The first is a 613 amino acid, putatively extracellular amino-terminus which has several regions rich in acidic amino acids (glutamine and aspartic acids) that may potentially be involved in binding ionized calcium. The second comprises seven membrane-spanning segments that are characteristic of the superfamily of G protein–coupled receptors. The last structural domain is a 222 amino acid, presumably cytoplasmic carboxy-terminus. In response to an increase in extracellular ionized calcium, the receptor becomes activated and via a G protein–coupled mechanism, activates phospholipase C. This leads to the hydrolysis of polyphosphoinositides, accumulation of inositol triphosphate, and release of ionized calcium from intracellular stores. Moreover, the calcium-sensing receptor inhibits adenylyl cyclase via an inhibitory G protein, thereby reducing intracellular cyclic AMP. Both the increase in intracellular calcium and the decrease in cyclic AMP act to suppress PTH secretion. The calcium-sensing receptor also recognizes other divalent cations such as magnesium, and even trivalent cations such as gadolinium.

The major regulator of both the synthesis and secretion of PTH is the concentration of ionized calcium in the blood. There is an inverse relationship between PTH secretion and serum calcium. As serum calcium decreases acutely, PTH secretion may rise up to five times the normal secretion rate. If the hypocalcemia is chronic,

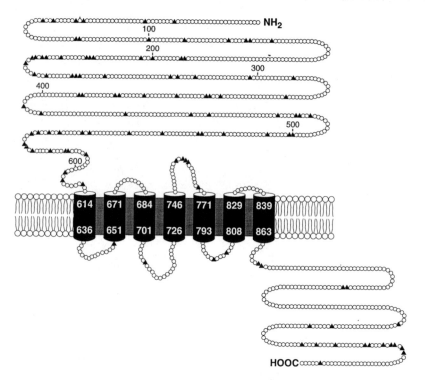

Fig. 15-2 Schematic diagram illustrating the principal structural features of the predicted parathyroid extracellular calcium-sensing receptor. Solid triangles indicate acidic amino acids. (Adapted from Brown EM, Pollak M, Hebert SC: Physiology and cell biology update: Sensing of extracellular Ca^{2+} by parathyroid and kidney cells: Cloning and characterization of an extracellular Ca^{2+}-sensing receptor. Am J Kidney Dis 25:506–513, 1995).

gross parathyroid hyperplasia occurs and PTH secretion may approach 50 times the basal rate. A rising serum calcium will suppress PTH secretion, but there is a calcium-independent, nonsuppressible component of PTH secretion, which represents about 15% of the normal secretion rate. Within the physiological range, the effect of magnesium on PTH secretion is in the same direction as that of calcium, but much weaker. Severe hypomagnesemia is associated with an impairment in PTH secretion. β-Adrenergic receptors have been identified in parathyroid cells, and isoproterenol and epinephrine have been shown to stimulate PTH secretion. The influence of catecholamines on PTH secretion may provide a partial explanation for the hypercalcemia noted in some patients with pheochromocytoma (see Chapter 14). There are receptors for 1,25-$(OH)_2$D in the parathyroid glands and this metabolite has been shown to suppress PTH secretion. There are also H_2 histamine receptors in the parathyroid glands, and under certain circumstances, the H_2 receptor blocker cimetidine has been shown to reduce PTH secretion. Glucocorticoids directly stimulate PTH secretion. Inorganic phosphorus has no direct influence on PTH secretion. Rather, it is the phosphorus-induced reduction in the concentration of serum ionized calcium that stimulates PTH secretion.

Intact, biologically active 1–84 PTH (molecular weight 9600) is the major secretory product of the parathyroid glands. Its half-life is several minutes and it constitutes only about 10% of the circulating parathyroid fragments. However, it accurately reflects parathyroid function and its clearance is largely independent of kidney function. This peptide may be measured by a double antibody sandwich technique known as the *immunoradiometric assay* (IRMA) (see Chapter 5). Measurement of intact PTH by the IRMA technique has become the assay of choice for assessing parathyroid function. Intact PTH is metabolized by the kidney and liver to amino- and carboxy-terminal fragments. The amino-terminal 1–34 PTH (molecular weight 3000–4000) possesses biological activity, but also has a short circulating half-life measured in minutes. It, too, constitutes 10% of circulating parathyroid fragments. Like intact PTH, the amino-terminal fragment is metabolized by its target tissues: kidney and bone. Carboxy-terminal fragments (molecular weight 6000–7000), which are inactive, derive not only from hepatic and renal cleavage of the intact hormone, but also to some extent from the parathyroid glands themselves. The secretion of these inactive fragments is particularly evident at elevated ambient calcium concentrations. The carboxy-terminal peptides have a half-life measured in hours and constitute about 80% of the circulating PTH fragments. Carboxy-terminal fragments are not metabolized by target organs and are eliminated from the circulation solely by glomerular filtration. Therefore, patients with renal insufficiency may develop a very high concentration of these inactive fragments that does not accurately reflect the parathyroid activity. In the presence of renal disease, the intact or amino-terminal PTH assays provide a more accurate assessment of parathyroid status.

By interacting with its three major target organs (bone, kidney, and intestine), PTH is primarily responsible for the regulation of calcium homeostasis in humans. In response to systemic needs, bone mineral is made available by the combined processes of osteocytic osteolysis and osteoclastic bone resorption. Osteocytic osteolysis refers to the rapid mobilization of bone mineral by lacunar osteocytes in haversian bone. This type of mineral mobilization may occur within minutes, involves activation of existing bone cells, and may proceed without actual resorption or destruction of bone. There is evidence that the rapid rise in serum calcium in response to PTH through this mechanism requires the presence of $1,25\text{-}(OH)_2D$. A more delayed (hours to days) response of bone to PTH is osteoclastic resorption. Following activation by PTH, precursor cells differentiate into multinucleated osteoclasts that resorb a localized area of mineralized bone. This process results in the destruction and removal of both bone mineral and bone matrix. Osteoclastic resorption is a delayed response that occurs only after exposure of bone to prolonged stimulation by PTH. Over a period of many months or years, prolonged excess of PTH leads to increased bone resorption and development of large cavities filled with multinucleated osteoclasts (classic osteitis fibrosa cystica of severe hyperparathyroidism). If PTH is given intermittently, however, it increases osteoblastic activity and promotes bone growth through enhancement of bone remodeling in which a longer osteoblastic phase follows short-term induction of new osteoclasts. PTH is believed to influence bone cell activity by stimulating membrane adenylyl cyclase to form cyclic AMP and by promoting calcium influx into bone cells.

In the kidney, PTH decreases proximal tubular phosphate reabsorption and increases calcium reabsorption in the distal nephron. Both effects are important in the minute-to-minute regulation of the serum concentrations of phosphorus and calcium. Another very important effect of PTH in the kidney is to stimulate 1,25-$(OH)_2D$ synthesis, which exerts control over the intestinal absorption of calcium and phosphorus. In the proximal tubule, in addition to inhibiting phosphate reabsorption, PTH also inhibits reabsorption of sodium and bicarbonate. Thus, it has diuretic effects and may provoke a mild hyperchloremic acidosis. PTH decreases renal clearance of uric acid and hyperparathyroidism may be associated with hyperuricemia and gout. Even though PTH stimulates distal tubular calcium reabsorption, the urinary calcium excretion is usually greater than normal in hyperparathyroidism because of the increase in filtered calcium load (owing to the bone and intestinal effects of PTH). Hypoparathyroid individuals with a normal serum calcium concentration of 9–10 mg/dl will excrete 3-fold more calcium in the urine than normal subjects because they lose the distal tubular effect of PTH. The renal tubular actions of PTH are believed to be mediated by cyclic AMP. Exposure of the kidney to PTH results in a marked increase of urinary cyclic AMP excretion.

Calciferol (Vitamin D)

The biochemical activation of 7-dehydrocholesterol to the active metabolite 1,25-$(OH)_2D$ requires three separate organ systems (skin, liver, and kidney) plus interaction with the environment, so there are many potential loci for disease (Fig. 15-3). Vitamin D_3 (cholecalciferol) is produced in the skin from 7-dehydrocholesterol, an inert precursor whose availability in skin decreases with age. A "previtamin D_3" is synthesized photochemically in the high dermis and lower epidermis when the 7-dehydrocholesterol is exposed to photolytic wavelengths of 250–310 nm (mid-ultraviolet). Melanin absorbs the same wavelengths and interferes with calciferol synthesis. Overproduction is also prevented by a photochemical equilibrium that favors synthesis of lumisterol and tachysterol (inert metabolites) during prolonged sun exposure. Intensity and duration of solar 286–310 nm radiation is strongly dependent on the angle of the sun (time of day, season of year), clouds, air pollution, and habitat (window glass and most electrical fixtures absorb all radiation less than 315–350 nm). Clothing and social customs further modify mid-ultraviolet exposure. Calciferol deficiency was common in temperate latitudes before diets were fortified and still occurs in elderly or housebound individuals.

Over a period of several days, the previtamin D_3 formed in the skin undergoes a temperature-dependent isomerization to vitamin D_3. The vitamin D–binding protein in blood has a 1000-fold higher affinity for vitamin D_3 than previtamin D_3, so that vitamin D_3 is preferentially transported from the skin into the circulation. The vitamin D_3 removed into the circulation is replaced by slow thermal conversion of previtamin D_3 to vitamin D_3. This series of steps appears to constitute an important mechanism whereby relatively brief exposure to ultraviolet radiation may lead to a continuous supply of vitamin D for days to weeks.

The terms *parent vitamin D* and *calciferol* are synonyms that actually refer to two sterols. Cholecalciferol (vitamin D_3) is the form of the vitamin found in animal tissues, fish liver oils, and irradiated milk. Ergocalciferol (vitamin D_2) is the form of vitamin D found in plants and irradiated yeast and bread. Most foodstuffs, from either plant or animal sources, contain only inactive vitamin D precursors and require ultraviolet irradiation for the conversion of these precursors to the calciferols. In

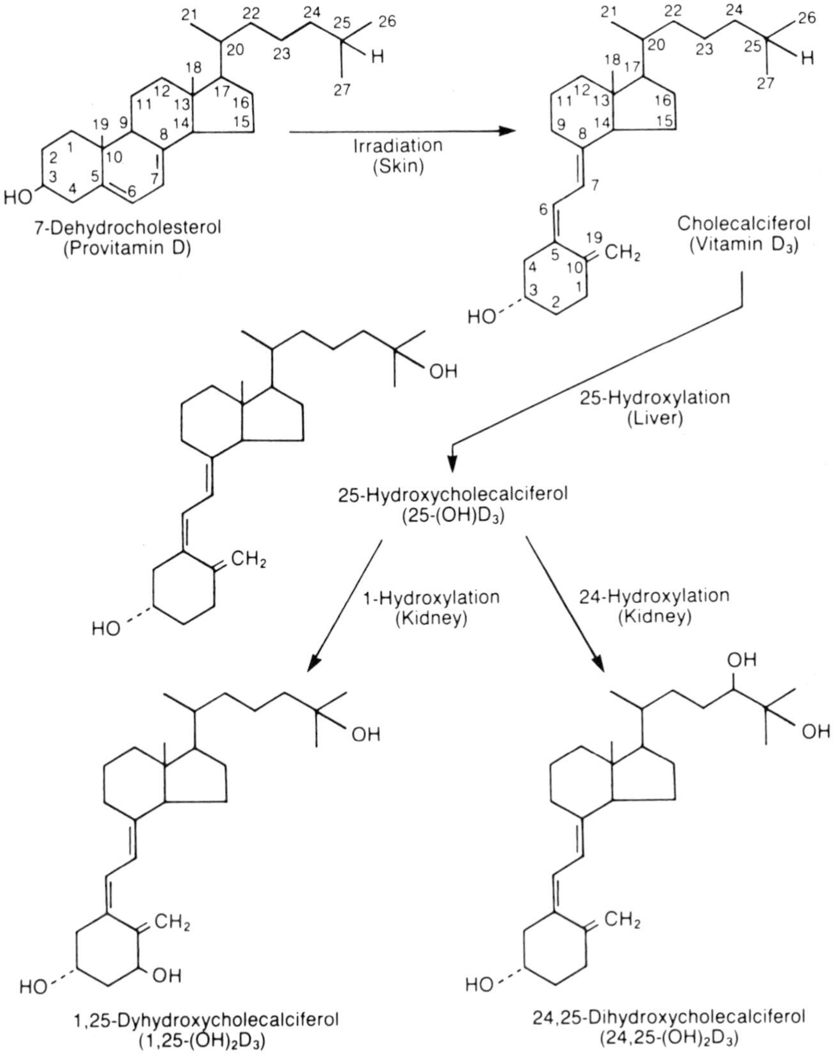

Fig. 15-3 Chemical structure and site of formation of the major vitamin D metabolites.

addition to the photochemical synthesis of vitamin D_3 in the skin, humans may supplement their vitamin D stores by dietary intake of vitamin D_2 or D_3. In humans, the storage, transport, metabolism, and potency of vitamin D_2 and vitamin D_3 are identical, and the net biological activity of vitamin D *in vivo* results from the combined effects of the hydroxylated derivatives of vitamin D_2 and vitamin D_3. The adult requirement for calciferol is considered to be 400 USP units daily (1 mg of calciferol is equivalent to 40,000 USP or international units). In the United States, vitamin D additives are present in a number of foods including milk, milk products, and cereals.

Cholecalciferol, a biologically inactive prohormone, is transported to the liver bound to vitamin D-binding protein. There it is hydroxylated at C-25 to form 25-

(OH)D, a partially active product (Fig. 15-3). 25-(OH)D constitutes the major circulating form of vitamin D in humans and is a good index of the vitamin D status of the individual. The typical concentration of 25-(OH)D in serum is 7–42 ng/ml with a circulating half-life of 15 days. The circulating pool of this metabolite is in equilibrium with a storage pool of 25-(OH)D in muscle and fat. 25-Hydroxylation of vitamin D takes place in the hepatic microsomes. Clinical studies have shown that 25-hydroxylation of vitamin D is inhibited by vitamin D and 1,25-(OH)$_2$D. Whether 25-(OH)D plays a physiologic role in regulating its own synthesis has not been established. Inhibition of 25-hydroxylation by vitamin D seems to be a relatively ineffective control mechanism in view of the fact that marked increases in circulating 25-(OH)D are responsible for the abnormal mineral metabolism in vitamin D intoxication. In contrast, feedback regulation of 25-(OH)D production by 1,25-(OH)$_2$D appears to be effective. 1,25-(OH)$_2$D completely inhibits the increase in serum 25-(OH)D produced by vitamin D challenge in normal subjects. The mechanism by which hepatic production of 25-(OH)D is modulated by 1,25-(OH)$_2$D is not known. Systemic calcium and phosphorus status do not influence hepatic 25-hydroxylase activity, and the importance of this enzyme as a biological control site is questionable. Anticonvulsants induce hepatic microsomal mixed oxidase enzymes that convert vitamin D metabolites to more polar hydroxylated, biologically inactive products, which are excreted in the bile. This mechanism may partially explain the reduced circulating 25-(OH)D levels and/or clinical osteomalacia (impaired mineralization of bone) in patients receiving prolonged high-dose anticonvulsant therapy.

25-(OH)D is transported to the kidney, where it is hydroxylated either at C-1 to produce 1,25-(OH)$_2$D, the most potent vitamin D metabolite, or at C-24 to produce 24,25-(OH)$_2$D, a less active compound of unclear biological significance. Although there is evidence of regulation of both cholecalciferol production in the skin and the initial hydroxylation of vitamin D in the liver, the hydroxylase reactions in the kidney are the key points of biological regulation of the metabolic activation of vitamin D. PTH is the principal regulator of the renal synthesis of 1,25-(OH)$_2$D. Serum 1,25-(OH)$_2$D is increased in primary hyperparathyroidism and is reduced in hypoparathyroidism. Circulating 1,25-(OH)$_2$D is increased when parathyroid extract is given to normal subjects and to patients with hypoparathyroidism. In normal adults, the renal production of 1,25-(OH)$_2$D is tightly regulated, and serum 1,25-(OH)$_2$D changes very little in response to vitamin D challenge. This occurs because an increase in circulating 1,25-(OH)$_2$D raises intestinal calcium absorption and hence the serum-ionized calcium concentration. The rise in serum-ionized calcium concentration suppresses PTH secretion and thereby inhibits the renal production of 1,25-(OH)$_2$D. Moreover, as already noted, receptors for this vitamin D metabolite have been demonstrated in parathyroid cells and there is considerable evidence for a direct feedback inhibition of PTH synthesis and secretion by 1,25-(OH)$_2$D.

PTH may control 1-hydroxylation by promoting urinary phosphate excretion, which in turn lowers the serum phosphate level. In animals and humans it has been shown that 1-hydroxylation can be stimulated even in the absence of the parathyroid glands by decreasing serum phosphate concentration. The action of PTH on the renal tubular cell, which is mediated by cyclic AMP, probably also involves decreased intracellular phosphate concentrations owing to inhibition of phosphate reabsorption.

The importance of these biochemical stimuli is underscored by the observation that urinary cyclic AMP and phosphate excretion change very little and serum 1,25-$(OH)_2D$ remains low in response to parathyroid extract in patients with a PTH-resistant state known as *pseudohypoparathyroidism*. These same patients have a brisk rise in serum 1,25-$(OH)_2D$ associated with a marked phosphaturia in response to dibutyryl-cyclic AMP infusions.

Another major regulator of the renal production of 1,25-$(OH)_2D$ is the vitamin D status of an individual. In the absence of PTH, as in hypoparathyroidism, serum 1,25-$(OH)_2D$ varies with the concentration of serum 25-$(OH)D$, suggesting that there is substrate dependency. There is also evidence that 1,25-$(OH)_2D$ itself and an elevated serum calcium level inhibit synthesis of 1,25-$(OH)_2D$. Moreover, 1,25-$(OH)_2D$ induces renal 25-hydroxyvitamin D 24-hydroxylase activity. Thus, in the vitamin D–replete state, the production of the relatively inactive metabolite 24,25-$(OH)_2D$ is enhanced.

The normal serum level of 1,25-$(OH)_2D$ is 20–50 pg/ml and its circulating half-life is about 3 hours. As noted, the renal production of this potent vitamin D metabolite is carefully regulated. However, under certain circumstances, extrarenal production of 1,25-$(OH)_2D$ may occur without appropriate regulation, resulting in disturbances in calcium metabolism. A classic example is a granulomatous disorder known as sarcoidosis, in which sarcoid lymph node tissue produces 1,25-$(OH)_2D$. Patients with this disorder may have elevated calcium levels in the blood and urine, sometimes resulting in kidney stones. Defective regulation of circulating 1,25-$(OH)_2D$ in sarcoid can be shown by the lack of change in serum 1,25-$(OH)_2D$ in response to increased calcium intake as compared with the reduction of circulating 1,25-$(OH)_2D$ that occurs in normal subjects. Other granulomatous conditions in which hypercalcemia associated with abnormal elevation of 1,25-$(OH)_2D$ has been reported include tuberculosis, disseminated candidiasis, silicon-induced granuloma, and lymphoma. The human placenta has also been demonstrated to produce 1,25-$(OH)_2D$, which may be important in meeting the increased calcium demands of pregnancy.

Both 1,25-$(OH)_2D$ and 24,25-$(OH)_2D$ may undergo an additional hydroxylation to form 1,24,25-trihydroxyvitamin D [1,24,25-$(OH)_3D$]. *In vivo,* the synthesis of 1,24,25-$(OH)_3D$ from 1,25-$(OH)_2D$ is probably the more important pathway, since 1,25-$(OH)_2D$ induces the 24-hydroxylase that can use 1,25-$(OH)_2D$ as substrate. 1,24,25-$(OH)_3D$ has only one-third the activity of 1,25-$(OH)_2D$ in the intestine and it is only marginally active in bone resorption. Many investigators favor the view that the hydroxylation of the vitamin D nucleus at C-24 represents a step in the direction of inactivation of the hormone. Since 24-hydroxylase activity has been detected in the intestine and bone as well as in the kidney, it is conceivable that this may represent an initial inactivation step when 1,25-$(OH)_2D$ reaches its target organs. Upon oxidative cleavage of its side chain, 1,25-$(OH)_2D$ forms a C-23 carboxylic acid, calcitroic acid, which is biologically inactive. Production of calcitroic acid in target tissues may be another important mechanism for inactivation of the biologically potent 1,25-$(OH)_2D$. Side-chain oxidation appears to proceed more rapidly in the presence of hydroxyl groups at C-24 and C-25, a finding compatible with the view that 24-hydroxylation may represent an initial inactivation step. In addition, vitamin D metabolites undergo an enterohepatic circulation in which they are concentrated in bile, secreted into the intestine, and then reabsorbed. Consequently, certain gastrointestinal disorders may result in vitamin D deficiency despite adequate synthesis of cholecalciferol in the skin.

The major target tissues for vitamin D action are the intestine, bone, and kidney. There is general agreement that 1,25-(OH)$_2$D is the main metabolite of vitamin D that regulates intestinal absorption and bone resorption. Whether other hydroxylated metabolites of vitamin D have physiologically important actions on any of the target organs remains unclear.

In the intestine, 1,25-(OH)$_2$D promotes calcium absorption by a mechanism that conforms generally to the steroid hormone model. Initially, the hormone is bound to a cytosolic receptor. The 1,25-(OH)$_2$D receptor complex becomes tightly associated with the cell nucleus and stimulates transcription of DNA to synthesize messenger RNA and translation of proteins to carry out the biological action of the hormone. Not all of the proteins that are synthesized in response to 1,25-(OH)$_2$D have been identified, but one of them is calcium-binding protein (CaBP), which cannot be detected in the intestines of vitamin D-deficient animals. The role of CaBP in intestinal calcium absorption is still controversial, but it probably facilitates calcium uptake by the intestinal cell. Vitamin D also stimulates phosphate and magnesium absorption by the intestine. A recently described human disorder known as vitamin D-dependent rickets type II points to the importance of the 1,25-(OH)$_2$D receptor in mediating the biological effects of vitamin D. This disorder is characterized by impaired intestinal calcium absorption resulting in rickets or osteomalacia, despite marked increases in circulating 1,25-(OH)$_2$D. It is caused by abnormalities in the receptor for 1,25-(OH)$_2$D.

Although 1,25-(OH)$_2$D produces bone resorption [100 to several thousand times more potent than 25-(OH)D], the mechanism by which this occurs is not known. Autoradiographic studies show uptake of labeled 1,25-(OH)$_2$D by osteoprogenitor cells and osteoblasts but not by osteoclasts. If bone resorption produced by 1,25-(OH)$_2$D is mediated by osteoclasts, it is unlikely to be a direct effect in view of the absence of a receptor for the hormone. There is evidence that osteoclasts may be derived from circulating peripheral monocytes that do have receptors for 1,25-(OH)$_2$D. Under the influence of 1,25-(OH)$_2$D, circulating monocytes may fuse into multinucleated macrophage-like cells that produce bone resorption and at least resemble osteoclasts. High-dose 1,25-(OH)$_2$D activated both osteoclasts and peripheral monocytes to produce bone resorption in a patient with malignant osteopetrosis, a disease characterized by impaired osteoclast function. Thus, 1,25-(OH)$_2$D may be important in regulating cellular differentiation in osseous tissue. PTH and 1,25-(OH)$_2$D clearly potentiate each other in bone resorption, and a maximum rate of resorption is observed only in response to the combined effects of both hormones. Present evidence indicates that the mechanism of action of 1,25-(OH)$_2$D in bone is identical to its mechanism of action in the intestinal mucosal cell, although the specific target cells in bone and the protein translation products induced by 1,25-(OH)$_2$D remain largely unknown.

Although the main defect in vitamin D deficiency is a failure of normal mineralization (e.g., rickets or osteomalacia), the exact role of vitamin D metabolites in bone formation remains uncertain. One view is that 1,25-(OH)$_2$D participates with PTH in regulating bone resorption but that bone mineralization is a passive process that proceeds normally given an adequate calcium $\times$ phosphate product in the extracellular fluid. Indeed, the infusion of calcium and phosphate into vitamin D-deficient humans sufficient to normalize the mineral ion product is associated with remineralization of bone. However, the mineralization that occurs is patchy and disordered, in contrast to the orderly mineralization process following vitamin D repletion. Perhaps a vitamin D metabolite controls the bone-forming cells responsible for mineralization (osteoid osteocytes). In addition to impaired mineralization, vitamin D-deficient states are characterized by retarded bone formation. The osteoblasts responsible for laying down the protein matrix of bone are flattened and inactive during vitamin D depletion. Following vitamin D treatment, the osteoblasts become cuboidal and active, and there is histologic evidence of increased bone formation. In human bone cells with many of the features of osteoblasts, 1,25-$(OH)_2$D stimulates production of alkaline phosphatase and osteocalcin, both of which may play a role in mineralization of newly formed bone matrix. In some human studies of vitamin D deficiency and end-stage renal disease, osteomalacic changes in bone could be completely cured only by administering 25-(OH)D, whereas 1,25-$(OH)_2$D produced only a partial reversal of the abnormalities. Thus, although most investigators regard 1,25-$(OH)_2$D as the vitamin D metabolite that mediates bone resorption, there is considerable evidence that one or more additional metabolites of vitamin D might directly or indirectly influence the processes of bone formation and mineralization.

In the kidney, both 25-(OH)D and 1,25-$(OH)_2$D have been reported to increase renal tubular reabsorption of both calcium and phosphate. The physiological relevance of these effects has been questioned. The potential effects of vitamin D on other tissues remain largely speculative and require additional investigation. Muscle weakness is a dominant symptom of vitamin D deficiency and responds rapidly to vitamin D repletion. This weakness is usually attributed to coincident hypophosphatemia, but may represent a more fundamental effect of vitamin D in muscle. The finding of receptors for 1,25-$(OH)_2$D in a number of cells and tissues other than those of established target organs indicates that the hormone may have a number of biological effects in addition to those related to mineral metabolism. A recently investigated role for this vitamin D metabolite may be in the regulation of cell growth and differentiation, particularly for those cells derived from bone marrow. 1,25-$(OH)_2$D receptors were demonstrated in promyelocytic leukemia cells (HL-60) and 1,25-$(OH)_2$D inhibited growth and induced differentiation of these cells into multinucleated macrophage-like cells that resorbed bone. 1,25-$(OH)_2$D and two fluorinated analogs induced macrophage differentiation of normal human and leukemic stem cells and prolonged survival of nude mice inoculated with leukemia cells. 1,25-$(OH)_2$D also has been shown to promote differentiation of human skin cells. This property has been utilized to reduce the abnormal turnover of skin cells in psoriasis. These observations have stirred excitement over the possible use of 1,25-$(OH)_2$D in the treatment of certain human leukemias and psoriasis.

Calcitonin

Calcitonin is a 32 amino acid peptide derived from a larger precursor called *procalcitonin*. Calcitonin's primary action is to inhibit bone resorption by inhibiting osteoclasts. It is synthesized by the parafollicular or "C cells" of the thyroid gland. C cells are actually neuroendocrine cells that originate in the neural crest. They are included in the amine precursor uptake and decarboxylation (APUD) cell family, which may explain the association of medullary carcinoma of the thyroid (C-cell adenocarcinoma) with another tumor of the APUD series, that is, pheochromocytoma in a polyglandular endocrine syndrome termed *multiple endocrine neoplasia type II* (Sipple's syndrome).

Several different products of the calcitonin gene have been recognized, each with different properties. The calcitonin gene generates multiple RNAs as a consequence of alternative processing events. One encodes the precursor of the 32 amino acid hormone, calcitonin, which also contains a flanking carboxy-terminal peptide, katacalcin, with hypocalcemic properties. The others encode precursors of two 37 amino acid peptides, α- and β-calcitonin gene-related peptide (CGRP). β-CGRP differs from α-CGRP at only three positions in the amino acid sequence, and both are equally effective in causing arterial vasodilation. CGRP is widely distributed in the central and peripheral nervous systems and in many organs, including the heart, lungs, thyroid, and gastrointestinal tract. It is found in association with the smooth muscles of blood vessels and perivascular nerves. CGRP is believed to function as a neurotransmitter and in the regulation of peripheral vascular tone.

Circulating calcitonin consists of heterogeneous forms owing to peripheral metabolism, as well as the secretion of multiple peptide fragments. However, detailed studies have revealed that the entire 32 amino acid sequence of calcitonin is required for biological activity and that even single amino acid deletions at either end of the molecule result in virtually complete loss of biological activity. According to most well-established radioimmunoassays, calcitonin normally fluctuates between 5 and 50 pg/ml in human serum. The half-life of human calcitonin is approximately 5 minutes. Salmon calcitonin, derived from the ultimobrachial body (an embryologic analogue of human C cells), is 10 times more potent than human calcitonin in its hypocalemic effect. The increased potency of salmon calcitonin appears to result both from a prolonged circulating half-time and from an increased affinity for and/ or an increased duration of binding to calcitonin receptors in the hormone's target tissues.

The main stimuli for the secretion of calcitonin are (1) an elevated serum calcium and (2) certain gastrointestinal hormones, particularly gastrin. The rate of calcitonin secretion is a direct function of the level of serum calcium above 9 mg/dl. Calcitonin secretion appears to proceed at a low basal rate within the normal range of serum calcium and is reduced to undetectable levels during hypocalcemia. Although the level of ionized calcium is the principal regulator of calcitonin secretion, there is evidence that gastrin also plays a role. For example, one study showed that calcitonin secretion could be elicited by oral calcium without a demonstrable change in circulating ionized calcium concentration, but with a rise in serum gastrin. An interpretation of this experiment is that the gastrin-secreting stomach cell senses calcium in the food contents and alerts the calcitonin-secreting parafollicular cell to its imminent absorption. The incoming calcium is met by a hormone that protects against alimentary hypercalcemia. The main organs responsible for the clearance and destruction of calcitonin are the kidney, liver, and possibly bone.

Acute administration of calcitonin causes hypocalcemia and hypophosphatemia. Both effects are due largely to the influence of the hormone on bone, although in pharmacological doses calcitonin also affects the renal handling of calcium and phosphate. The most important effect of calcitonin is the inhibition of bone resorption. There is little evidence for any direct effect of calcitonin on bone formation. Osteoclasts, the multinucleated bone-resorbing cells, contain membrane receptors for calcitonin. Within a few minutes after the administration of this hormone, the osteoclasts lose their ruffled borders (indicators of phagocytic activity), and there is a reduction in serum calcium and hydroxyproline excretion, indicating inhibition of both mineral and matrix resorption. In patients with Paget's disease, in whom both bone resorption and bone formation are excessive, calcitonin has a prolonged and marked inhibitory effect on osteoclasts. It is therefore very effective in the treatment of this disease. Because bone formation is "coupled" to bone resorption, chronic exposure to calcitonin leads ultimately to a decrease in all aspects of bone turnover, including osteoblast number and function. Calcitonin's ability to inhibit bone resorption acutely has also been utilized in the treatment of hypercalcemia associated with malignant disease. Within days, however, the inhibitory influence of calcitonin may be lost because of down-regulation of calcitonin receptors in bone ("escape phenomenon"). The reduction in serum phosphate produced by calcitonin appears to result from an inhibition of phosphorus mobilization from bone and from the renal phosphaturic effect.

There are calcitonin receptors in the kidney, and, infused in sufficient dosage, calcitonin leads to a transient increase in the excretion of calcium, phosphate, sodium, potassium, and magnesium. The calciuric and phosphaturic effects of calcitonin augment the hypocalcemia and hypophosphatemia that result from the actions of the hormone on bone. There is some evidence that calcitonin may increase the synthesis of $1,25\text{-}(OH)_2D$ by the kidney, even in parathyroidectomized animals. Cyclic AMP serves as the intracellular mediator of calcitonin's effects in both bone and kidney, although the exact mechanism by which the stimulation of cyclic AMP in these cells mediates calcitonin's effects on ion transport remains unknown. Calcitonin and PTH interact with different target cells in the renal tubule. Moreover, the two hormones differ in their effects on cyclic AMP excretion in the urine. Urinary cyclic AMP excretion provides a sensitive reflection of the effects of circulating, active PTH, whereas large doses of calcitonin have no influence on urinary cyclic AMP excretion (presumably because the target cells of calcitonin are less permeable to cyclic AMP).

The physiological significance of calcitonin in humans has been questioned because there are no known metabolic consequences of calcitonin deficiency or excess. Serum calcium is well regulated within normal limits after total thyroidectomy. When calcitonin is present in inappropriately high concentrations (as in medullary carcinoma of the thyroid), it appears to have little effect on mineral homeostasis or bone density. Clinically, calcitonin measurements have been useful in detecting the presence of medullary carcinoma of the thyroid, particularly after provocative tests involving infusion of calcium or pentagastrin. Measurements of basal and stimulated levels of serum calcitonin are also valuable as indices of the success or failure of treatment of these C-cell tumors. Salmon calcitonin injections have been utilized in the long-term treatment of Paget's disease, to prevent further bone loss in osteoporosis, and in the short-term treatment of hypercalcemia of malignancy. Salmon calcitonin is also available as a nasal spray.

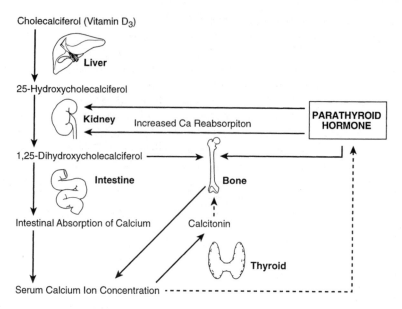

Fig. 15-4   Regulation of serum calcium homeostasis by the major calcium-regulating hormones. The solid arrows indicate activation; the dashed arrows indicate inhibition. When serum calcium concentration rises, for example, after a calcium-containing meal, parathyroid (PTH) secretion is inhibited. The normal actions of PTH to stimulate 1,25-$(OH)_2D$ production, increase renal calcium reabsorption, and promote bone resorption become suppressed. The reduction of 1,25-$(OH)_2D$ production lowers intestinal calcium absorption and bone resorption. Calcitonin release is stimulated by the rise in serum calcium concentration and inhibits bone resorption. These effects return the serum calcium level back to normal.

Overall Control of Calcium Ion Concentration

The maintenance of a steady serum calcium concentration by the major calcium-regulating hormones is diagrammed in Fig. 15-4.

OTHER HORMONES THAT AFFECT BONE AND CALCIUM METABOLISM

Although PTH, vitamin D, and, to a lesser extent, calcitonin are the major calcium-regulating hormones, a number of other hormones are known to have an important influence on bone and mineral metabolism. These include estrogens and androgens, glucocorticoids, thyroid hormones, and growth hormone.

Estrogens and Androgens

Gonadal steroids are involved in the pubertal growth spurt and in closure of the epiphyses. In childhood and puberty, the steroids of both sexes favor bone formation over resorption. In the adult female, estrogen inhibits PTH-mediated bone resorption by mechanisms that are incompletely understood. Estrogen receptors have been detected in osteoblasts, and conceivably these cells may release factors (e.g., transforming growth factor [TGF]-β) that inhibit osteoclastic resorption. Estrogens have also been demonstrated to reduce the amount of bone-resorbing cytokines such as interleukin (IL)-1 and IL-6 in bone and to reduce prostaglandin E_2 levels. Thus, estrogens protect the skeleton from the development of osteoporosis or decreased bone mass. Following oophorectomy or menopause, there is an increased rate of bone resorption that in many women causes osteoporosis, which can lead to vertebral compression fractures and hip fractures. Estrogen replacement therapy, although not without side effects, has been shown to slow or arrest the progress of postmenopausal osteoporosis. Interestingly, despite blocking the resorptive action of PTH on bone, estrogen appears to facilitate PTH action at the kidney. An increase in serum PTH concentration also occurs, owing to the hypocalcemic effect of the inhibition of bone resorption. Consequently, estrogen treatment results in greater urinary cyclic AMP and phosphate excretion and greater production of 1,25-$(OH)_2$D. The higher levels of this metabolite lead to greater intestinal calcium absorption and more positive calcium balance. These changes brought about by estrogen administration are beneficial for women at risk of developing osteoporosis and also for those with primary hyperparathyroidism who are unable to have surgical parathyroidectomy. Testosterone protects men from the development of osteoporosis, as evidenced by the occurrence of excessive bone loss in hypogonadal conditions such as Klinefelter's syndrome (see Chapter 8).

Glucocorticoids

At physiological levels, glucocorticoids are necessary for skeletal growth. When chronically present in excess, however, as in naturally occurring or iatrogenic Cushing's syndrome, these hormones have deleterious effects on nearly every target organ involved in calcium homeostasis. Glucocorticoids acutely decrease renal tubular calcium reabsorption, leading to hypercalciuria. More chronically, they interfere with intestinal calcium absorption, although it is not yet clear whether any alteration in vitamin D metabolism occurs. Because of the urinary calcium losses and reduced intestinal absorption of calcium, secondary hyperparathyroidism develops. Glucocorticoids also directly stimulate PTH secretion. Consequently, bone resorption is stimulated, but normal coupling does not occur because a major effect of glucocorticoids is to inhibit osteoblastic bone formation. Furthermore, glucocorticoids suppress gonadal estrogen and testosterone production. This combination of events explains the rapidly progressive and severe nature of glucocorticoid-induced osteoporosis.

Thyroid Hormones

The thyroid hormones are extremely important for the development and growth of the skeleton in infancy and childhood. A major feature of hypothyroidism is retarded bone age, as indicated by delayed ossification of the cartilaginous bone growth centers. In hyperthyroidism, both thyroxine and triiodothyronine in excess cause increased bone resorption. As excessive calcium enters the circulation from bone,

PTH is suppressed and hypercalciuria results. Since serum PTH is suppressed and serum phosphate may be increased, levels of 1,25-$(OH)_2D$ and intestinal calcium absorption are reduced. The enhanced bone resorption is of sufficient magnitude that 10–20% of patients with thyrotoxicosis have increased serum calcium levels, although usually not to a degree sufficient to cause symptoms. Negative calcium balance induced by excessive thyroid hormone levels, either due to thyrotoxicosis or overreplacement with thyroid hormone, may cause osteoporosis.

Growth Hormone

Growth hormone affects a number of target organs involved in mineral homeostasis. It has striking effects on bone growth by way of somatomedin C (or insulin-like growth factor [IGF]-I), a growth hormone-dependent peptide produced in the liver and bone, that is mitogenic for chondrocytes and osteoblasts. Somatomedin C generation is required for normal skeletal growth in children. Laron dwarfs are resistant to growth hormone (see Chapter 12) and are unable to produce sufficient somatomedin C levels. In adults, excess growth hormone produces striking bony deformities (acromegaly), which are associated with accelerated bone formation and resorption, the former process predominating. In growth hormone-deficient children, administration of growth hormone not only stimulates bone growth via somatomedin C but also increases intestinal calcium absorption by a vitamin D-independent mechanism, and increases renal tubular phosphate reabsorption. The latter effect appears to explain the high values for serum phosphorus noted in growing children and adults with active acromegaly.

LOCAL REGULATORS OF BONE REMODELING—EMERGING INSIGHTS INTO THE PATHOPHYSIOLOGY OF OSTEOPOROSIS

Bone is remodeled continuously during adulthood through the resorption of old bone by osteoclasts and the subsequent formation of new bone by osteoblasts. These two closely coupled events are responsible for renewing the skeleton while maintaining its anatomical and structural integrity. Under normal conditions, bone remodeling proceeds in cycles in which osteoclasts adhere to bone and subsequently remove it by acidification and proteolytic digestion. Shortly after the osteoclasts have left the resorption site, osteoblasts invade the area and begin the process of forming new bone by secreting osteoid (a matrix of collagen and other proteins), which is eventually mineralized. After bone formation has ceased, the surface of the bone is covered by lining cells, a distinct type of terminally differentiated osteoblasts. In adults, approximately 25% of trabecular bone is resorbed and replaced every year, as compared with only 3% of cortical bone, indicating that the rate of remodeling is controlled primarily by local factors.

In recent years, there have been important advances in our understanding of the origin of osteoclasts and osteoblasts, the interplay between them, and the systemic and local factors that regulate their development. The process of remodeling and its intimate relatiionship to the bone marrow (the source of the stem cells for osteoclasts and osteoblasts) is shown in Fig. 15-5. Osteoblasts originate from pluripotent mes-

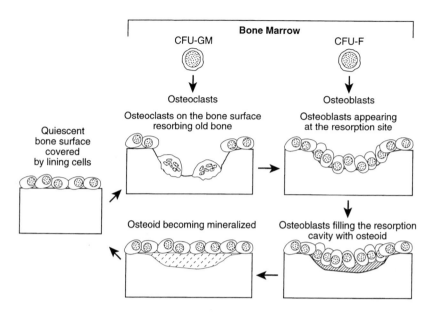

Fig. 15-5 The bone remodeling process and its relationship to bone marrow stem cells. Remodeling is accomplished by cycles involving the resorption of old bone by osteoclasts and the subsequent formation of new bone by osteoblasts. The osteoclasts and osteoblasts are replenished from their respective hematopoietic progenitors (granulocyte–macrophage colony-forming units [CFU-GM]) and mesenchymal progenitors (fibroblast colony-forming units [CFU-F]) in the bone marrow. (Adapted with permission from Manolagas SC and Jilka RL: Bone marrow, cytokines and bone remodeling. Emerging insights into the pathophysiology of osteoporosis. N Engl J Med 332:305–311, 1995. Copyright Massachusetts Medical Society. All rights reserved.)

enchymal stem cells of the bone marrow. Stromal cells with the potential to become osteoblasts, or to become fibroblasts, chondrocytes, adipocytes, or muscle cells as well, can be identified in marrow cell cultures as adherent colonies. The common progenitors that give rise to these colonies are termed *fibroblast colony-forming units* (CFU). Osteoclasts are derived from the hematopoietic granulocyte-macrophage CFU, which also give rise to monocytes and macrophages. Osteoclast progenitors move from bone marrow to bone either through the circulation or by direct migration from the marrow.

The effects of both systemic hormones and locally produced factors that stimulate the development of osteoclasts are mediated by cells of the stromal–osteoblastic lineage. Fig. 15-6 shows the pathways of differentiation of osteoclasts and osteoblasts, with the development of osteoclasts dependent on cells of the stromal–osteoblastic lineage. Interleukins are autocrine or paracrine mediators produced by hematopoietic and bone cells that permit communication between cells. IL-6 stimulates the early stages of osteoclastogenesis. IL-6 is produced by both stromal cells and osteoblastic cells in response to stimulation by systemic hormones such as PTH, PTH-related peptide, and 1,25-$(OH)_2$D, and also in response to bone-resorbing cytokines such as IL-1 and tumor necrosis factor (TNF). IL-6 has a pathogenetic role

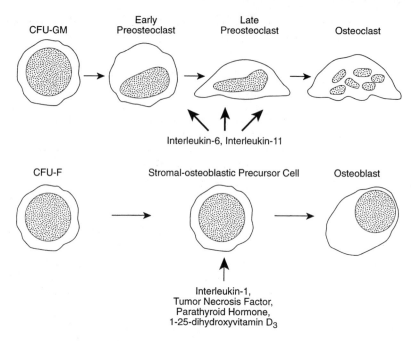

Fig. 15-6 Pathways of differentiation of osteoclasts and osteoblasts. Osteoclasts arise from granulocyte–macrophage CFU (CFU-GM), and osteoblasts from fibroblast CFU (CFU-F). The development of osteoclasts is controlled by stromal-osteoblastic cells that produce factors, including interleukin-6 and interleukin-11, that regulate the differentiation pathway of osteoclasts. Systemic hormones, such as parathyroid hormone and 1,25-dihydroxyvitamin D₃, and local factors, such as interleukin-1 and tumor necrosis factor, promote the development of osteoclasts through their ability to stimulate the production of cytokines such as interleukin-6 and interleukin-11. (Adapted with permission from Manolagas SC and Jilka RL: Bone marrow, cytokines and bone remodeling. Emerging insights into the pathophysiology of osteoporosis. N Engl J Med 332:305–311, 1995. Copyright Massachusetts Medical Society. All rights reserved).

in the abnormal bone resorption associated with multiple myeloma, Paget's disease, rheumatoid arthritis, and Gorham-Stout disease (also known as vanishing- or disappearing-bone disease)—four conditions characterized by excessive osteoclast development and focal osteolytic lesions.

Bone Remodeling and Osteoporosis

The hallmark of osteoporosis is a reduction in skeletal mass caused by an imbalance between bone resorption and bone formation. Loss of gonadal function and aging are the two most important factors contributing to the development of this condition. Starting around the fourth or fifth decade of life, both women and men lose bone at a rate of 0.13–0.5% per year. For about 5 years after menopause (or after castration in men), the rate of bone loss increases as much as 10-fold. In humans, the loss of bone mass that follows the loss of ovarian function is associated with an increase in the rates of bone resorption and bone formation, with the former exceeding the latter, and an increase in the number of osteoclasts in trabecular bone.

Role of IL-6 in Stimulating Osteoclastic Bone Resorption in Gonadal Deficiency

Evidence has accumulated that IL-6 plays a role in osteopenia associated with loss of gonadal function. The production of IL-6 by cultured bone marrow stromal and osteoblastic cell lines is inhibited by estrogens or androgens. These sex steroids act through a specific receptor in the promoter region of the IL-6 gene and inhibit transcription. The loss of gonadal function in mice stimulates osteoclastogenesis *in vivo* through increased production of IL-6 in the marrow. Administration of estrogen or an IL-6-neutralizing antibody prevents the osteoclastogenesis in ovariectomized mice. Similarly, the secretion of IL-6 is increased in human bone marrow cells after menopause or the discontinuation of estrogen replacement therapy.

Although IL-6 is an essential mediator of bone loss associated with loss of gonadal function, it does not seem to play a role in the development of osteoclast precursors in the marrow or to influence the number of osteoclasts in trabecular bone in animals with sufficient sex steroids. Under physiologic conditions (i.e., the estrogen-replete state), it appears that the production of IL-6 is below a critical threshold of sensitivity for osteoclastogenesis. Therefore, it appears that other cytokines are responsible for the development of osteoclasts in the estrogen-replete state. For example, the osteoclastogenic effects of IL-11 are independent of estrogen status.

IL-1 may also contribute to the bone loss associated with the loss of ovarian function. The production of IL-1 is increased in bone marrow mononuclear cells from ovariectomized rats. Administering IL-1 receptor antagonist to these animals prevents the late but not the early stages of bone loss induced by the loss of ovarian function. Estrogens do not seem to regulate the IL-1 gene directly.

Senescence and Osteoblast Development in Marrow

Bone loss associated with aging has characteristics that are different from those of bone loss associated with the early years of gonadal deficiency. The amount of bone formed during each remodeling cycle decreases with age in both sexes, most likely because the supply of osteoblasts is reduced in proportion to the demand for them. The cause of the impaired ability of the marrow to produce osteoblasts with aging remains uncertain, but one possibility is a deficiency of bone growth factors that normally stimulate osteoblast formation.

Several growth factors have now been identified in either extracts of bone matrix or conditioned mediums of cultured bone and bone cells. Among these bone-derived growth factors, insulin-like growth factor (IGF-I) seems particularly likely to be an important physiologic regulator. IGF-I present in the skeletal matrix may be derived from the systemic circulation (hepatic production under the influence of growth hormone) or it may be synthesized by a variety of cells present in bone, including osteoblasts (stimulated by growth hormone or parathyroid hormone). IGF-I has been shown to promote replication and differentiation of osteoblasts associated with increased production of osteocalcin and type I collagen (matrix proteins). It is now recognized that circulating concentrations of growth hormone decline with advancing age, and that this is accompanied by reduced levels of IGF-I in bone. Indeed, there is a 60% loss of human skeletal IGF-I between the ages of 20 and 60 years of age. Decreased IGF-I content in bone could have important relevance to the diminished osteoblast availability of aging and hence to the pathogenesis of senile osteoporosis. Small clinical trials involving treatment with recombinant human growth hormone or IGF-I have been disappointing, but treatment with intermittent PTH injections has shown some promise.

Another important local regulator of bone formation that declines with age is transforming growth factor-β (TGF-β). TFG-β is synthesized by many tissues, but bone and platelets are the major sources of this polypeptide. TGF-β induces cartilage formation but also stimulates replication and differentiation of osteoblast progenitors. TGF-β inhibited osteoclast formation in marrow cultures. When bone is resorbed, TGF-β is released from the bone matrix, and may act as the critical "coupling factor" in turning off further osteoclastic resorption and switching on osteoblastic bone formation. It is noteworthy that estrogen and fluoride, two agents that have proven highly effective in the treatment of osteoporosis, have been shown to stimulate TGF-β production or to increase TGF-β receptors in osteoblast-like cells.

EFFECT OF DIET AND ACTIVITY ON BONE AND MINERAL METABOLISM

Diet

In addition to all of the factors delineated above, bone and mineral metabolism is importantly influenced by nutrition and activity. Important dietary factors include calcium, sodium, animal protein, and simple carbohydrates. Because of reduced renal production of $1,25\text{-}(OH)_2D$ and a deficiency of intestinal receptors for this metabolite, elderly women suffer from low calcium absorption and are subject to negative calcium balance, especially at low calcium intakes. The amount of calcium intake required to achieve zero calcium balance increases by approximately 500 mg/day to a total of 1500 mg/day with the onset of menopause. Yet the estimated calcium intake of postmenopausal American women is only about 500 mg daily. Conflicting reports have appeared concerning the relationship between calcium intake and bone mass or occurrence of skeletal fractures, but in a widely quoted Yugoslavian study, the incidence of hip fracture was significantly higher in the region with low calcium intake than in a region with high calcium intake. Thus, prolonged adherence to a diet deficient in calcium may lead to secondary hyperparathyroidism and cortical bone loss.

An excessive intake of sodium may cause renal hypercalciuria by impairing renal calcium reabsorption, resulting in a compensatory increase of PTH secretion (Fig. 15-7). In young individuals and premenopausal women in whom the capacity for $1,25\text{-}(OH)_2D$ production is intact, stimulation of intestinal calcium absorption ensues from the sodium load consequent to PTH-dependent augmentation of $1,25\text{-}(OH)_2D$ synthesis. Thus, calcium balance is maintained. However, in postmenopausal women, the compensatory rise in intestinal calcium absorption does not de-

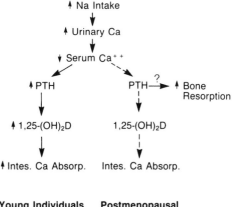

Fig. 15-7   Relationship of excessive dietary sodium intake to calcium homeostasis. In young individuals, there is an adaptive response to sodium-induced calciuresis. Stimulation of the parathyroid–vitamin D (PTH–1,25-(OH)$_2$D) axis from transient decrements in serum ionized calcium raises intestinal calcium absorption, thereby sustaining serum calcium concentration while permitting continued excessive urinary calcium excretion. Thus, in young individuals, the skeleton is protected but chronic hypercalciuria may pose a risk for nephrolithiasis. In elderly postmenopausal women, a sluggish response of the parathyroid–vitamin D axis prevents sufficient intestinal adaptation. Consequently, negative calcium balance and increased bone resorption would likely ensue in an effort to maintain normal serum ionized calcium concentration.

velop following sodium-induced calciuria because of impaired 1,25-(OH)$_2$D synthesis (related to aging kidney or estrogen deficiency) and reduced intestinal vitamin D receptors. Therefore, postmenopausal women are at least theoretically at risk for bone loss from habitual high sodium intake.

Administration of animal proteins increases urinary calcium and causes negative calcium balance, without altering intestinal calcium absorption. These effects are probably attributable to the acid load from the sulfur-containing amino acids that would stimulate bone dissolution. An excessive ingestion of readily metabolizable refined sugar has been shown to acutely increase urinary calcium excretion by altering renal tubular calcium reabsorption. It is not known whether the calciuric response to carbohydrate ingestion elicits the same hormonal and metabolic changes that have been described for sodium-induced hypercalciuria.

Activity

The importance of activity is indicated by the fact that normal people at total bed rest show a preponderance of bone resorption over bone formation and quickly go into negative calcium balance. In individuals with increased bone turnover such as growing children or patients with Paget's disease, sudden immobilization from injury may cause hypercalciuria and even hypercalcemia, often within weeks of the accident. Apparently, normal bone cell function requires the pressures generated in bone

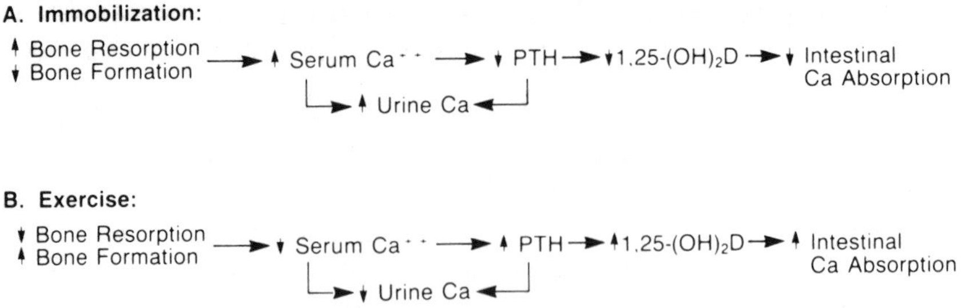

A. Immobilization:

↟ Bone Resorption ——→ ↟ Serum Ca ⁻ ⁻ ——→ ↡ PTH —→ ↡1.25-(OH)₂D —→ ↡ Intestinal
↡ Bone Formation   Ca Absorption
                   └—→ ↟ Urine Ca ◄—┘

B. Exercise:

↡ Bone Resorption ——→ ↡ Serum Ca ⁻ ⁻ ——→ ↟ PTH —→ ↟1.25-(OH)₂D —→ ↟ Intestinal
↟ Bone Formation   Ca Absorption
                   └—→ ↡ Urine Ca ◄—┘

Fig. 15-8   (A) Relationship of immobilization and exercise to calcium homeostasis. The scheme of "resorptive hypercalciuria" that occurs during immobilization has been well validated. Enhanced urinary calcium excretion following parathyroid (PTH) suppression, and the accompanying reduction in 1,25-(OH)₂D and intestinal calcium absorption are protective against development of hypercalcemia. At times, the amount of calcium entering the circulation is so great that these adaptive mechanisms are overwhelmed and hypercalcemia occurs. (B) The depiction of the sequence of events during exercise has also been largely confirmed. This scheme may account for the increased mineralization of bone that occurs during an exercise program.

by movement against gravity. During immobilization or the weightlessness of spaceflight, increased osteoclastic bone resorption is stimulated by an unknown mechanism. Moreover, there is reduced osteoblastic activity and less calcium enters bone from the circulation. Meticulous studies of patients paralyzed after spinal cord injury have shown that this abnormal bone turnover pattern leads to suppression of the PTH–vitamin D axis with an accompanying reduction in intestinal calcium absorption. Bone loss rapidly progresses and there is a further risk of kidney stone formation due to the ensuing hypercalciuria. In astronauts, calcium balance studies during Skylab flights demonstrated combined urinary and fecal calcium losses at a rate of 2.5% of total body calcium per month. Bone mineral loss and hypercalciuria are currently major impediments to long-term space travel.

On the other hand, several studies have shown that regular weight-bearing exercise may have beneficial effects on the skeleton leading to increased bone density. A regular exercise program has been recommended for women hoping to prevent postmenopausal osteoporosis. However, if female joggers or ballet dancers exercise excessively to the point at which gonadotropin suppression and amenorrhea occur, bone density diminishes because of the overriding effect of estrogen deficiency. The relationship of immobilization and exercise to bone and calcium homeostasis is summarized in Fig. 15-8.

INTEGRATED CONTROL OF CALCIUM HOMEOSTASIS

Under normal circumstances, the circulating concentration of ionized calcium is maintained within a narrow range by various homeostatic mechanisms. This integrated control is so precise that, in a normal individual, ionized serum calcium probably fluctuates by no more than 0.1 mg/dl in either direction from its normal set point throughout the day. Three types of homeostatic control mechanisms are involved in maintaining the serum calcium level: hormonal, physiological, and physicochemical.

Hormonal Regulation

The parathyroid glands are exquisitely sensitive to minor changes in serum ionized calcium. A decline in serum calcium stimulates secretion of PTH. The integrated actions of PTH on bone resorption, distal renal tubular calcium reabsorption, and 1,25-$(OH)_2$D-mediated intestinal calcium absorption are chiefly responsible for the precise regulation of serum ionized calcium in humans. The rapid mobilization of calcium from bone, requiring both PTH and 1,25-$(OH)_2$D, and the stimulation of distal tubular calcium reabsorption are the major control points in minute-to-minute serum calcium homeostasis. Adjustments in the rate of intestinal calcium absorption via the calcium–PTH–1,25-$(OH)_2$D axis require several days to be maximal. Another role of the increased PTH secretion in response to hypocalcemia is to decrease tubular phosphate reabsorption so that the increased amount of phosphate mobilized from bone and absorbed from the intestine along with calcium will be excreted into the urine. A decline in serum calcium also inhibits secretion of calcitonin and 24,25-$(OH)_2$D, which conceivably could help to restore serum calcium, although the physiological importance of the latter two metabolites in humans remains unclear. It is of interest that during a prolonged hypocalcemic challenge, the initial requirement for calcium mobilization from the skeleton is largely replaced by a more efficient absorption of calcium from the intestine. However, when severe intestinal malabsorption of calcium and vitamin D (e.g., steatorrhea) precludes intestinal adaptation, a marked secondary hyperparathyroidism develops to defend the serum calcium level at the expense of the skeleton.

A rise in serum calcium suppresses secretion of PTH and synthesis of 1,25-$(OH)_2$D, thereby returning serum calcium to normal as outlined in Fig. 15-4. This system functions quite well in meeting routine mild hypercalcemic challenges such as the absorption of calcium following a dairy meal. A small rise in the serum ionized calcium (<0.1 mg/dl) is capable of suppressing parathyroid function by 30% or more. The main result of this suppression is a decrease in distal tubular calcium reabsorption, so that the postprandial increase in the filtered load of calcium is "spilled" into the urine. In more severe hypercalcemic challenges such as enhanced bone resorption due to osteolytic metastases, despite suppression of the PTH–1,25-$(OH)_2$D axis, the quantity of mobilized calcium may overwhelm the renal capacity for calcium excretion (800 mg calcium/day) and result in hypercalcemia.

Physiological Regulation

As already noted, the kidney plays an important role in calcium homeostasis. This regulation is partly hormone independent since the renal excretion of calcium is a function of the filtered load of calcium, that is, increased serum calcium $\rightarrow$ increased filtered load $\rightarrow$ increased urinary calcium. Moreover, there is a calcium-sensing receptor in the renal tubule cells such that hypercalcemia leads to a reduced renal tubular reabsorption of calcium. It is this calcium sensor that is impaired in the inherited condition, familial hypocalciuric hypercalcemia (FHH), resulting in a sustained high renal calcium reabsorption despite the presence of hypercalcemia. (PTH levels are also inappropriately elevated in FHH because of a defective parathyroid calcium-sensing receptor).

Compensatory changes in intestinal calcium absorption occur in response to alterations in dietary calcium. Intestinal calcium absorption is most efficient at low calcium loads. As increasing amounts of calcium are ingested, the percentage of calcium absorbed declines. One may examine the amount of calcium appearing in the urine after an oral calcium load as an indirect measure of intestinal calcium absorption. There is a progressive increase in urinary calcium excretion as oral calcium loads are increased incrementally from 100 to 500 mg, but beyond 500 mg, there is little further change in urinary calcium excretion. It is thus more advantageous to provide calcium supplements as 500 mg four times daily than to provide 2000 mg of calcium all at once. In addition to the intrinsic ability of the intestine to absorb calcium less efficiently at higher calcium loads, a sustained higher calcium intake would also suppress the PTH-1,25-$(OH)_2$D axis and thereby reduce calcium absorption.

Physicochemical Regulation

It has been hypothesized that serum concentrations of ionized calcium and phosphate reflect the solubility of the bone mineral. This assumes that circulating concentrations of ionized calcium and phosphate are maintained at a constant product, representing saturation with respect to a particular phase of calcium phosphate in bone. This hypothesis could explain the decline in serum calcium that accompanies a rise in serum phosphate. Unfortunately, various physiocochemical studies have failed to identify the particular phase of calcium phosphate with which serum calcium and phosphate might be in steady state. Nevertheless, when the Ca × P product exceeds 60 (in mg/dl), soft tissue deposition of calcium salts is likely to occur.

LOSS OF CALCIUM HOMEOSTATIC CONTROL: HYPERCALCEMIA

Hypercalcemia may occur when there is a loss of calcium homeostatic control. For it to develop, there must first be a sufficient, unabated input of calcium into the circulation, as might occur from a continued excessive rate of bone resorption or of intestinal calcium absorption. Second, the extent of the influx of calcium into the circulation must exceed the capacity of the kidneys to remove the calcium load. These conditions are met in two fairly common disorders: primary hyperparathyroidism and hypercalcemia of malignancy. Together, these two disorders account for over 90% of the hypercalcemia that is clinically encountered.

PRIMARY HYPERPARATHYROIDISM

In primary hyperparathyroidism, calcium homeostatic control is lost as a result of excessive PTH secretion. Thus, PTH is secreted by adenomatous or hyperplastic parathyroid tissue in amounts that are inappropriately high for the level of circulating calcium. Hypercalcemia develops from the combined effects of PTH-induced stimulation of bone resorption, intestinal calcium absorption, and renal tubular reabsorption of calcium. Hypercalciuria also often occurs because even though PTH augments renal tubular reabsorption of calcium, this action of the hormone is usually insufficient to overcome the increased renal filtered load of calcium.

The major symptoms of primary hyperparathyroidism may be ascribed to the PTH-induced resorption of bone, hypercalciuria, and hypercalcemia. These abnormalities are in turn pathophysiologically related to the bihormonal defect, that is, an excessive production of both PTH and 1,25-$(OH)_2$D.

The classical bone disease of hyperparathyroidism is osteitis fibrosa cystica. This disorder is characterized by bone pain, pathological fractures of long bones, and compression fractures of the spine. Radiologically, the bones appear demineralized and may reveal a frayed erosion of the outer and inner cortical surfaces. These cortical erosions are referred to as *subperiosteal resorption* and are best demonstrated on the radial aspect of the middle phalanges and in the distal clavicles. Locally, destructive lesions known as *brown tumors* may appear as well-demarcated lucent areas. These brown tumors or bone cysts are composed largely of osteoclasts, intermixed with poorly mineralized woven bone. Histologically, extensive fibrosis replaces normal marrow and bone cell elements. The pathogenesis of osteitis fibrosa cystica is believed to be related to overstimulation of osteoclastic bone resorption by very high PTH levels. Increased 1,25-$(OH)_2$D levels may also activate bone remodeling. Osteitis fibrosa cystica appears to be a complication of severe and rapidly progressive hyperparathyroidism associated with large parathyroid adenomas or carcinomas. Following successful parathyroidectomy, the subperiosteal resorption, brown tumors, and marrow fibrosis all resolve. Hyperparathyroidism tends to affect the cortical bone more than trabecular bone. In mild, asymptomatic primary hyperparathyroidism, the spine is often spared.

Up to one-third of patients present with nephrolithiasis. They may form pure calcium phosphate stones, calcium oxalate stones, or mixed calcium stones. Hypercalciuria is an important factor in the pathogenesis of renal stone disease. Hypercalciuria probably contributes to stone formation by rendering urine supersaturated with respect to calcium phosphate and calcium oxalate. Patients with hyperparathyroidism who form kidney stones frequently have a pronounced hypercalciuria, but only a modestly elevated serum calcium level. Some investigators have found that the hyperparathyroid patients with marked elevations in serum 1,25-$(OH)_2$D, intestinal hyperabsorption of calcium, and resulting hypercalciuria are the most prone to stone disease. The greater intestinal absorption of calcium with associated parathyroid suppression could explain the tendency for smaller parathyroid glands and disproportionate hypercalciuria in these patients. It remains unclear why certain patients with hyperparathyroidism have near-normal circulating levels of 1,25-$(OH)_2$D, whereas in others the concentration of this active metabolite is markedly elevated. Moreover, hypercalciuria is not the only factor in stone formation, since some patients do not form stones despite hypercalciuria and supersaturation of the urine with respect to calcium salts. Perhaps the stone formers have an excess of as yet unidentified promotors or a deficiency of inhibitors of stone formation. In any case, nephrolithiasis generally ceases following successful parathyroidectomy.

Hypercalcemia may account for many of the other symptoms of primary hyperparathyroidism. Hypercalcemia is known to cause CNS depression, mental aberrations, fatiguability and weakness, as well as constipation and anorexia. Hypercalcemia may also play a role in the development of peptic ulceration in hyperparathyroidism, since it is associated with hypersecretion of gastrin and hydrochloric acid. Moreover, hypercalcemia may cause polyuria by impairing vasopressin-mediated renal concentrating ability. The fact that an increased calcium concentration inhibits the renal tubule's ability to conserve water may lead to a vicious spiral of dehydration, prerenal azotemia, and worsening hypercalcemia.

Bone disease, nephrolithiasis, and peptic ulcer disease constitute a clinical triad that should alert the physician to the possibility of primary hyperparathyroidism. When one or more of the triad is present in patients with hypercalcemia, a vigorous search for evidence supporting this diagnosis should be conducted. The hypercalcemia may be asymptomatic, or associated with mental aberrations, anorexia, constipation, or polyuria. There has been a dramatic change in the usual clinical presentation of hyperparathyroidism since the introduction of automated clinical chemistry techniques to measure serum calcium. Over half the patients now present with asymptomatic hypercalcemia, perhaps reflecting an earlier or milder form of the disease. Biochemically, the diagnosis may be made from determinations of serum calcium, phosphate and PTH, preferably by the IRMA assay.

Hypercalcemia of Malignancy

The underlying cause of hypercalcemia of malignancy is generally excessive bone resorption. Three major pathogenetic mechanisms have been recognized: (1) local osteolytic hypercalcemia, (2) humoral hypercalcemia of malignancy (HHM), and (3) $1,25\text{-}(OH)_2D$ synthesis by lymphomas. Each of these mechanisms will be briefly discussed.

Typically, patients with local osteolytic hypercalcemia (LOH) have extensive skeletal involvement with either breast cancer, multiple myeloma, or lymphoma. LOH accounts for about 20% of all hypercalcemia of malignancy. Hypercalcemia results largely from osteoclastic bone resorption, which is mediated by osteoclast-activating factors (IL-I, TNF) secreted by malignant cells. The mediator in breast cancer is not clear, but it may be one of the above cytokines, prostaglandin E_2, or other unknown factors. There is also evidence for direct resorption of the skeleton by mononuclear cells and by breast cancer cells. Biochemically, patients with LOH display reductions in circulating PTH and urinary cyclic AMP levels. Urinary calcium excretion is usually markedly increased, reflecting the increased filtered load of calcium together with parathyroid suppression. The serum $1,25\text{-}(OH)_2D$ concentration and, therefore, intestinal calcium absorption are also reduced as a consequence of parathyroid suppression.

The term *humoral hypercalcemia of malignancy* can be used generically to describe malignancy-associated hypercalcemia that occurs in the absence of skeletal metastases and results from the secretion by the neoplasm of a circulating, bone-resorbing, calcemic factor. Typical tumors where this occurs would include squamous cell carcinoma of the lung and renal carcinoma. A few cases have been reported of true native PTH production by ovarian carcinoma or small cell carcinoma of the lung. However, HHM generally refers to patients in whom hypercalcemia results from the overexpression of a novel class of peptide hormones, PTH-related protein (PTHrP). There has been explosive growth in the understanding of the pathogenesis of this syndrome in recent years.

The existence of PTHrP was first suspected clinically because cancer patients were evaluated and found to have hypercalcemia, hypophosphatemia, and elevated urinary cyclic AMP excretion, in the absence of immunologically detectable PTH. Moreover, the tumors causing the hypercalcemia did not contain mRNA encoding PTH. The patients with HHM also differed from patients with primary hyperparathyroidism in that they had reduced serum $1,25-(OH)_2D$ concentration and uncoupling between exaggerated osteoclastic bone resorption and reduced osteoblastic bone formation. Meanwhile, in the laboratory, the PTHrP obtained from tumor extracts was isolated and purified based on its ability to stimulate cyclic AMP production in canine renal cortical cells or rat osteosarcoma cells. The next step was the partial sequencing of the purified peptide from carcinomas of the lung, breast, and kidney, which provided uniform results. This permitted the synthesis of oligonucleotide probes for use in screening HHM tumor-derived complementary DNA libraries. Full length PTHrP complementary DNAs were isolated and found to encode products which were approximately twice the size of native PTH (Fig. 15-9). Actually, three forms of PTHrP were detected, one of 139, one of 141, and one of 173 amino acids in length (compared to PTH which has 84 amino acids). One of the most striking

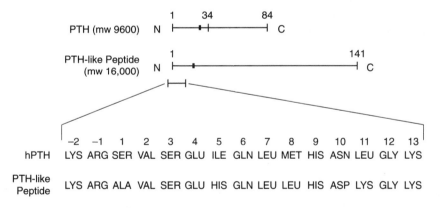

Fig. 15-9 Comparison of the structures of native human parathyroid hormone (PTH) with that of PTH-related protein (1 to 141), the first of the PTH-related protein complementary DNAs to be identified. The heavy portions of the two lines indicate the binding domain of parathyroid hormone (amino acids 25 to 27) and the putative receptor-binding domain of PTH-related protein at amino acids 19 to 21. Note the homology between PTH and PTH-related protein in the −2 to +13 region. Positions −1 and −2 are identical in both peptides and represent the proteolytic cleavage site between the mature N-terminal of the mature peptide and their "*pro*" sequences. Homology in the −2 to +13 region is equally striking at the nucleotide level. (Reproduced with permission from Stewart AF, and Broadus AE: In: *Advances in Endocrinology and Metabolism,* Vol. 1, EL Mazzaferri, RS Bar, and RA Kreisberg, eds., Mosby, St. Louis, pp. 1–21, 1990.)

features was that each of these tumor-derived peptides displayed marked homology with PTH through the first 13 amino acids (8 of the first 13 amino acid residues were identical). This structural homology to PTH allowed the PTHrP to attach to and activate the PTH receptor. However, beyond amino acid 13, the sequences of the two classes of peptides were completely different.

The availability of the complete PTHrP sequence from the complementary DNAs permitted the chemical synthesis and recombinant expression of a variety of PTHrPs for use in biologic and immunologic studies. Studies using these peptides have shown that they reproduce the full spectrum of PTH-like bioactivities, including the stimulation of adenylyl cyclase in renal membranes and bone cells, the stimulation of bone resorption *in vitro* and the production of hypercalcemia *in vivo,* the stimulation of renal 1α-hydroxylase activity, and the stimulation of renal tubular calcium reabsorption.

The evidence that PTHrPs are indeed the factors responsible for the clinical and biochemical manifestation of HHM is compelling. The peptides are present only in HHM-associated tumors and not in control tumors. In one elegant study, immunodeficient mice implanted subcutaneously with human HHM-associated tumors developed localized neoplasms associated with hypercalcemia, hypophosphatemia, and elevated urinary cyclic AMP excretion. Infusion of the mice with PTHrP antisera, but not control serum, reversed the hypercalcemia as well as the other manifestations of the HHM syndrome. Radioimmunoassays for PTHrP have been developed, and elevated circulating PTHrP concentrations have been found in the vast majority of patients with HHM. This mechanism is believed to account for at least 80% of the hypercalcemia of malignancy.

Another mechanism for hypercalcemia of malignancy that has been validated is $1,25\text{-}(OH)_2D$ synthesis by certain lymphomas. While some lymphomas produce hypercalcemia through the local osteolytic mechanism described above, hypercalcemia has been reported in at least a dozen patients with a variety of histologic types of lymphoma in whom skeletal involvement was either absent or negligible. In these patients, circulating $1,25\text{-}(OH)_2D$ concentrations were markedly elevated, but were normalized with either glucocorticoid therapy, surgical tumor resection, or chemotherapy, suggesting that the lymphomas were the source of the $1,25\text{-}(OH)_2D$. The overproduction of $1,25\text{-}(OH)_2D$ in these patients is analogous to the overproduction of $1,25\text{-}(OH)_2D$ that occurs in patients with granulomatous disorders (e.g., sarcoidosis and tuberculosis). It is believed to represent dysregulated production of $1,25\text{-}(OH)_2D$ by macrophages or lymphocytes within the lymphomatous tissue. Hypercalcemia in these patients results from $1,25\text{-}(OH)_2D$-induced intestinal hyperabsorption of calcium in addition to accelerated bone resorption. Biochemically, patients are characterized by hypercalcemia together with suppression of circulating PTH and urinary cyclic AMP levels. Circulating $1,25\text{-}(OH)_2D$ values are elevated.

Other Causes of Hypercalcemia

Although primary hyperparathyroidism and hypercalcemia of malignancy are the most commonly encountered forms of hypercalcemia, the clinician must keep in mind other potential causes of hypercalcemia. A detailed discussion of these disor-

Table 15-1 Classification Scheme for Hypercalcemias

Increased calcium input into circulation
High bone resorption
  Primary hyperparathyroidism
  Neoplasms
  Thyrotoxicosis
  Vitamin D toxicity
  Sarcoidosis and other granulomatous disorders
  Immobilization

High intestinal calcium absorption
  Primary hyperparathyroidism
  Vitamin D toxicity
  Sarcoidosis and other granulomatous disorders
  Milk-alkali syndrome
  Phosphorus depletion syndrome

Reduced calcium removal from circulation
Reduced renal calcium excretion
  Primary hyperparathyroidism
  Familial hypocalciuric hypercalcemia
  Extracellular volume depletion
  Thiazide therapy
  Milk-alkali syndrome

Reduced bone formation
  Immobilization
  Phosphorus depletion syndrome

ders is beyond the scope of this chapter, but a useful classification scheme is presented in Table 15-1. It is helpful to divide the various causes of hypercalcemia into those related primarily to increased calcium input into the circulation and those related to decreased calcium removal from the circulation.

LOSS OF CALCIUM HOMEOSTATIC CONTROL: HYPOCALCEMIA

Hypocalcemia occurs when there is an inadequate response of the parathyroid–vitamin D axis to a hypocalcemic stimulus, when there is target organ resistance to the major calcium-regulating hormones, or when excessive phosphate accumulates in the circulation. Various disorders may cause hypocalcemia (Table 15-2). In narrowing down the possibilities, a useful initial step is to consider whether the serum phosphate is increased or decreased.

    The cause of hypocalcemia is often multifactorial. In renal failure, for example, hypocalcemia occurs because of phosphate retention and inadequate production of $1,25\text{-}(OH)_2D$ by the damaged kidney. The low $1,25\text{-}(OH)_2D$ level reduces intestinal calcium absorption and impairs the mobilization of calcium from bone. In another condition, severe hypomagnesemia, there is reduced PTH secretion as well as resistance of bone and kidney to the actions of PTH. In endocrinology, the classic cause of hypocalcemia is hypoparathyroidism.

Table 15-2   Differential Diagnosis of Hypocalcemia Based on Serum Phosphate Level

Low calcium, high phosphate
Renal insufficiency
Hypomagnesemia
Excessive phosphate intake
Massive cell destruction (e.g., chemotherapy rhabdomyolysis)
Hypoparathyroidism

Low calcium, low phosphate
Vitamin D deficiency
Abnormal vitamin D metabolism
Vitamin D resistance
Acute pancreatitis

HYPOPARATHYROIDISM

The diagnosis of hypoparathyroidism must be considered when hypocalcemia and hyperphosphatemia occur in the presence of normal serum creatinine and magnesium concentrations, and in the absence of a source of massive phosphate leakage into the circulation. Hypoparathyroidism is a bihormonal disease in which the abnormalities in serum chemistry are the sequelae of reduced PTH action and $1,25\text{-}(OH)_2D$ deficiency. Hypocalcemia is invariably found in hypoparathyroidism; it is the result of (1) reduced PTH-dependent osteoclastic resorption, (2) decreased osteocytic calcium transfer (rapid bone calcium mobilization) consequent to $1,25\text{-}(OH)_2D$ deficiency and reduced PTH action, (3) low renal tubular reabsorption of calcium from reduced PTH action, and (4) impaired intestinal calcium absorption from $1,25\text{-}(OH)_2D$ deficiency. Hyperphosphatemia results primarily from the impaired renal phosphate clearance consequent to reduced PTH action. The retained build-up of phosphate in the serum lowers the serum calcium by physicochemical means and by further reducing synthesis of $1,25\text{-}(OH)_2D$.

Most of the symptoms of hypoparathyroidism may be ascribed to the abnormal levels of calcium and phosphate in the blood. Hypocalcemia is responsible for the increased neuromuscular irritability that may produce carpopedal spasm, prickling sensations in the lips, fingers, and toes, and occasionally muscle cramps. If the serum calcium decreases to very low levels, laryngospasm or seizures may occur. Latent tetany may be revealed by Chvostek's or Trousseau's sign (carpal spasm developing within 3 minutes of inflating the blood pressure cuff above the systolic level). The hyperphosphatemia may result in soft tissue calcifications. Basal ganglia calcification is not uncommon in hypoparathyroidism, and, if extensive, may lead to disturbances of the extrapyramidal motor system such as Parkinsonism or chorea. Other disorders that may accompany chronic hypocalcemia include cataracts, congestive heart failure, and mental impairment.

Within recent years, the development of assays for PTH, $1,25\text{-}(OH)_2D$, urinary cyclic AMP, and components of the adenylyl cyclase enzyme system has permitted the separation of hypoparathyroidism into several distinct types. Table 15-3 represents a current classification scheme for the various forms of hypoparathyroidism.

Table 15-3 Classification of Hypoparathyroidism

PTH-deficient hypoparathyroidism
Postoperative
Idiopathic
PTH ineffective hypoparathyroidism
PTH-resistant (pseudo)hypoparathyroidism
Type Ia
Type Ib
Type Ic
Type II

Each type has a unique pathogenetic basis, a discussion of which should nicely illustrate the variety of ways the calcium homeostatic system may be deranged.

PTH-Deficient Hypoparathyroidism

This condition is characterized by a reduced or absent synthesis of PTH; the target tissues are responsive to the action of PTH. Thus, the serum concentration of PTH is low or undetectable and the basal urinary cyclic AMP level is low. When challenged by exogenous PTH, PTH-dependent adenylyl cyclase in bone and kidney is stimulated, as shown by a marked increase in the renal excretion of cyclic AMP and phosphate. The circulating concentration of $1,25\text{-}(OH)_2D$ is low, accounting for the subnormal intestinal calcium absorption. During exogenous PTH challenge, there is a rapid rise in serum calcium (''calcemic response'') attributed to the combined action of PTH and newly synthesized $1,25\text{-}(OH)_2D$.

The most common cause of PTH-deficient hypoparathyroidism is the inadvertent removal of excessive parathyroid tissue during thyroid or parathyroid surgery. A rarer condition is idiopathic hypoparathyroidism, which results from absence, fatty replacement, or atrophy of the parathyroid glands. This condition may be inherited in families, or it may occur sporadically. In either case, it may occur in combination with pernicious anemia, Addison's disease (see Chapter 14), Hashimoto's thyroiditis (see Chapter 13), gonadal failure, and mucocutaneous candidiasis. The finding of circulating antibodies against parathyroid, thyroid, and other tissues has strengthened the suspicion that idiopathic hypoparathyroidism may be an autoimmune disease. Recently, a familial form of hypoparathyroidism has been attributed to an activating mutation in the calcium-sensing receptor gene. In affected patients, serum PTH levels are suppressed by lower than normal serum calcium levels.

PTH-Ineffective Hypoparathyroidism

This very rare condition is characterized by synthesis of PTH that is biologically inactive; the target tissues are responsive to the action of exogenous PTH. Typically, a patient presents with hypocalcemia, hyperphosphatemia, and normal renal function, but increased serum PTH. Infusion of exogenous PTH elicits normal cyclic AMP and phosphaturic responses.

PTH-Resistant Hypoparathyroidism (Pseudohypoparathyroidism)

Pseudohypoparathyroidism was first described by Albright as a disease characterized by the biochemical signs of hypoparathyroidism, hypocalcemia, and hyperphosphatemia, resulting from PTH resistance rather than PTH deficiency. Indeed, these patients have large parathyroid glands and very high circulating levels of PTH. In almost all patients, urinary cyclic AMP excretion after PTH infusion is significantly reduced, implicating a defect in the hormone receptor–adenylyl cyclase complex (pseudohypoparathyroidism, type I). As a consequence of the inability to generate cyclic AMP in the renal tubule, phosphate excretion and the production of 1,25-(OH)$_2$D are impaired. Therefore, intestinal calcium absorption is reduced and the rapid mobilization of calcium from bone in response to administered PTH is subnormal. Osteoclastic bone resorption responds to the high circulating PTH levels, however. Thus, in contrast to PTH-deficient hypoparathyroidism, bone density is usually diminished and hydroxyproline excretion is increased. Some patients with pseudohypoparathyroidism actually develop frank osteitis fibrosa cystica.

It is now known that peptide hormone receptors are coupled to adenylyl cyclase by guanine nucleotide-binding proteins (G proteins; see Fig. 15-10 and Chapter 3). Patients with pseudohypoparathyroidism type Ia have reduced activity of the stim-

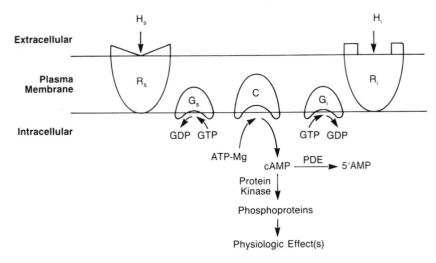

Fig. 15-10　The adenylyl cyclase–cyclic AMP system. H_s and H_i, stimulatory and inhibitory agents, respectively; R_s and R_i, stimulatory and inhibitory receptors; G_s and G_i, stimulatory and inhibitory guanine nucleotide-binding proteins; C, the catalytic unit of adenylyl cyclase; PDE, phosphodiesterase; cAMP, cyclic AMP. Hormones act by facilitating exchange of guanine nucleotides at a binding site on the regulatory G protein. Formation of a ternary complex consisting of hormone, receptor, and G protein allows GTP to exchange for GDP. The GTP-bound G protein in turn interacts with the catalytic unit of adenylyl cyclase. Most peptide hormones including PTH stimulate G_s resulting in cyclic AMP generation. The cyclic AMP acts as the second messenger to carry out the physiological effect of the hormone. The message is ended when cyclic AMP is broken down by phosphodiesterase. (Modified from Spiegel AM, Gierschik P, Levine MA, and Downs RW Jr: N Engl J Med 312:26–33, 1985.)

ulatory G protein (G_s) associated with adenylyl cyclase. As a consequence of this generalized G_s deficiency, patients are resistant not only to PTH, but also to many other peptide hormones including TSH, glucagon, and gonadotropins. Clinical hypothyroidism and gonadal dysfunction are quite common in these patients. Deficient G_s activity is presumed to limit cyclic AMP production and thereby cause resistance to multiple hormones in pseudohypoparathyroidism type Ia. Patients also tend to have a characteristic physical appearance that has come to be known as Albright's hereditary osteodystrophy. It consists of a short, stocky stature, round face, short neck, and short fingers and toes, most often affecting the fourth digits due to short fourth metacarpals and metatarsals. In contrast, subjects with pseudohypoparathyroidism type Ib have a normal appearance, normal G_s activity, and, in most instances, resistance to PTH alone. The pathogenesis of pseudohypoparathyroidism Ib is unknown, but since resistance is limited to PTH, it has been suggested that a defect in the PTH receptor could be involved. This concept has been supported by the observation that skin fibroblasts from many, but not all, subjects with pseudohypoparathyroidism Ib have selective resistance to PTH in terms of cyclic AMP response. Recently, pseudohypoparathyroidism type Ic has been recognized. These patients have Albright's hereditary osteodystrophy and multiple hormone resistance, but unlike pseudohypoparathyroidism type Ia, they have normal G_s levels. The defect appears to reside in an abnormal adenylyl cyclase catalytic subunit.

The type II form of pseudohypoparathyroidism is extremely rare. This type of patient presents with hypocalcemia, hyperphosphatemia, and increased serum PTH concentration. Exogenous PTH causes a marked rise in urinary cyclic AMP without provoking a phosphaturic response. In type II pseudohypoparathyroidism, the defect is thought to be distal to cyclic AMP generation, perhaps at the level of the protein kinase. It is also possible that a defect exists in a more distal site, for example, a substrate for the protein kinase, such as a phosphate transport protein in the renal tubular luminal brush border.

SUGGESTED READING

Bell NH: Vitamin D–endocrine system. J Clin Invest 76:1–6, 1985.

Breslau NA, McGuire J, Zerwekh JE, Frenkel E, and Pak CYC: Hypercalcemia associated with increased serum 1,25-dihydroxyvitamin D in three patients with lymphoma. Ann Intern Med 100:1–7, 1984.

Breslau NA, and Pak CYC: Clinical evaluation of parathyroid tumors. In: *Comprehensive Management of Head and Neck Tumors,* SE Thawley and WR Panje, eds., Saunders, Philadelphia, pp. 1635–1649, 1987.

Breslau NA: Pseudohypoparathyroidism: Current concepts. Am J Med Sci 298:130–140, 1989.

Breslau NA: Calcium, estrogen and progestin in the treatment of osteoporosis. In: *Rheumatic Disease Clinics of North America,* N Lane, ed., Saunders, Philadelphia, pp. 691–716, 1994.

Brown EM, Pollak M, and Hebert SC: Physiology and cell biology update: Sensing of extracellular CA^{2+} by parathyroid and kidney cells: Cloning and characterization of an extracellular Ca^{2+}-sensing receptor. Am J Kidney Dis 25:506–513, 1995.

Holick MF, MacLaughlin JA, Clark MB, Holick SA, Potts JT Jr, Anderson RR, Blank IH,

Parish JA, and Elias P: Photosynthesis of previtamin D_3 in human skin and the physiologic consequences. Science 210:203–205, 1980.

Liberman UA, Eil C, and Marx SJ: Resistance to 1,25-dihydroxyvitamin D. Association with heterogeneous defects in cultured skin fibroblasts. J Clin Invest 71:192–200, 1983.

Manolagas SC, and Jilka RL: Bone marrow, cytokines and bone remodeling. Emerging insights into the pathophysiology of osteoporosis. N Engl J Med 332:305–311, 1995.

Marcus R: Editorial: Bones of contention: The problem of mild hyperparathyroidism. J Clin Endocrinol Metab 80:720–722, 1995.

Naveh-Many T, and Silver J: Regulation of parathyroid hormone gene expression by hypocalcemia, hypercalcemia and vitamin D in the rat. J Clin Invest 86:1313–1319, 1990.

Nicolas V, Prewett A, Bettica P, Mohan S, Finkelman RD, Baylink DJ, and Farley JR: Age-related decreases in IGF-I and TGF-β in femoral cortical bone from both men and women: implications for bone loss with aging. J Clin Endocrinol Metab 78:1011–1016, 1994.

Pollak M, Brown EM, Chou Y-HC, Herbert SC, Marx SJ, Steinman B, Levi T, Seidman CE, and Seidman JG: Mutations in the human Ca^{2+}-sensing receptor gene cause familial hypocalciuric hypercalcemia and neonatal severe hyperparathyroidism. Cell 75:1297–1303, 1993.

Pollak MR, Brown EM, Estep HL, McLaine PN, Kilfor U, Park J, Hebert SC, Seidman CE, and Seidman JG: Autosomal dominant hypocalcemia caused by a Ca^{2+}-sensing receptor gene mutation. Nat Genet 8:303–307, 1994.

Silverberg SJ, Gartenberg F, Jacobs TP, Shane E, Siris E, Staron RB, and Bilezikian JP: Longitudinal measurements of bone density and biochemical indices in untreated primary hyperparathyroidism. J Clin Endocrinol Metab 80:723–728, 1995.

Silverberg SJ, Gartenberg F, Jacobs TP, Shane E, Siris E, Staron RB, McMahon DJ, and Bilezikian JP: Increased bone mineral density after parathyroidectomy in primary hyperparathyroidism. J Clin Endocrinol Metab 80:729–734, 1995.

Spiegel AM, Gierschik P, Levine MA, and Downs RW Jr: Clinical implications of guanine nucleotide-binding proteins as receptor-effector couplers. N Engl J Med 312:26–33, 1985.

Stewart AF, Adler M, Byers CM, Segre GV, and Broadus AE: Calcium homeostasis in immobilization: An example of resorptive hypercalcuria. N Engl J Med 306:1136–1140, 1982.

Stewart AF, and Broadus AE: Humoral hypercalcemia of malignancy. In: *Advances in Endocrinology and Metabolism,* Vol. 1, EL Mazzaferri, RS Bar, and RA Kreisberg, eds., Mosby, St. Louis, pp. 1–21, 1990.

16

Glucose, Lipid, and Protein Metabolism

DANIEL W. FOSTER
J. DENIS McGARRY

The maintenance of life requires a constant supply of fuels for body tissues. These fuels are used to generate energy and preserve organ structure. The process is in principle quite simple. Energy is derived when ingested food is oxidized to carbon dioxide and water in the presence of oxygen, generating adenosine triphosphate (ATP). A portion of the ingested foodstuff is also utilized, either directly or after transformation into other substrates, to replace cell membranes and organelles that turn over naturally during the process of living. The remainder is stored as potential energy in the form of glycogen or fat. Under normal circumstances, each individual remains in a near steady state with weight and appearance stable over prolonged periods. Despite the outward evidence of stability, however, fuel metabolism changes dramatically several times a day, particularly during the overnight fast. Two distinct phases, anabolic and catabolic, alternate one with the other. The *anabolic* phase begins with food ingestion and lasts for several hours. It is in this period, when caloric intake ordinarily exceeds caloric demands, that energy storage occurs. The *catabolic* phase usually begins 4–6 hours after the last food intake and lasts until the person eats once again. During this phase the body shifts from exogenous to endogenous fuels, a change heralded by the mobilization of substrate from storage sites in liver, muscle, and adipose tissue. This endogenous substrate sustains the organism between meals.

Both anabolic and catabolic phases exhibit characteristic biochemical processes (Fig. 16-1). In the anabolic phase glycogen synthesis occurs in liver and muscle. Fatty acid synthesis and triglyceride formation are stimulated in hepatocytes and adipose tissue, and protein synthesis is initiated. Flux of substrate is from intestine through the liver to storage and utilization sites. After a mixed meal glucose, triglyceride, and amino acid concentrations increase in plasma, whereas free fatty acids, ketones (acetoacetic and β-hydroxybutyric acids), and glycerol decrease. In the catabolic phase the biochemical activities are reversed. Glycogen breaks down, glycolysis is impaired, gluconeogenesis is activated (first in liver and then in kidney), lipogenesis and triglyceride formation cease, and lipolysis in the adipocyte increases manyfold. Free fatty acid oxidation is enhanced in most tissues and ketone body formation is initiated in the liver. Finally, proteolysis is activated in muscle to supply amino acids as substrate for gluconeogenesis. Flux of fuel is from storage depots to

349

State	Hormones	Fuel Source	Process
Anabolism	⬆ Insulin ⬇ Glucagon	Diet	Glycogen Synthesis Triglyceride Synthesis Protein Synthesis
Catabolism	⬇ Insulin ⬆ Glucagon	Storage Depots	Glycogenolysis Lipolysis Proteolysis Ketogenesis

Fig. 16-1   Fuel metabolism in anabolic and catabolic phases.

liver and other utilization sites. Glucose and triglyceride levels in plasma decrease, whereas certain amino acids, long-chain fatty acids, acetoacetate, and β-hydroxybutyrate increase.

When food is readily available (and the diet is not nutritionally unbalanced), the only risk of the anabolic phase in the normal individual is that it may be entered too frequently or maintained too long so that obesity results. Ingestion of food in certain inborn errors of metabolism such as fructose intolerance or intestinal lactase deficiency can cause symptoms, but these anabolic phase diseases are rare. In contrast, the absence of food is potentially dangerous since without major adaptations death or permanent brain injury may occur fairly rapidly because of a fall in plasma glucose concentrations to levels too low to sustain function in the central nervous system. Indeed, it can be stated that the primary aim of the acute metabolic changes accompanying the catabolic state is the prevention of hypoglycemia. Two changes are required to maintain plasma glucose within the normal range in the absence of food (Fig. 16-2). First, the liver must be converted to an organ of glucose production. This is accomplished initially by glycogen breakdown, but subsequently new glucose production occurs via gluconeogenesis, a process in which amino acids, lactate, pyruvate, and glycerol are converted to glucose by a linked series of enzymatic reactions. The second major change is the conversion of tissues other than the central nervous system to a lipid economy. Most organs utilize long-chain fatty acids (FFA) directly (adipose tissue → FFA → utilization sites), but a fourth to a third of the total fatty acids are taken up by the liver for conversion to ketone bodies (adipose tissue → FFA → liver → ketones → utilization sites). In contrast to long-chain fatty acids, the ketones can be oxidized by the central nervous system and, thus, can be considered a backup substrate to glucose; that is, in the presence of ketone bodies at concentrations usually found after several days of fasting, the brain can function normally despite levels of glucose low enough to cause loss of consciousness or convulsions in the absence of ketones. Since ketogenesis is not activated immediately by the absence of food, this protective mechanism is not available in the first few hours of a fast. Acetoacetic and β-hydoxybutyric acids are also readily utilizable by all other nonhepatic tissues. The liver, by contrast, has only a minimal capacity to oxidize ketones. The shift to lipid metabolism is critical to survival in a prolonged fast since the liver, even at maximal glucose production, cannot provide adequate carbohydrate to sustain all tissues in the body simultaneously. If fat oxidation is impaired for any reason, hypoglycemia supervenes.

When the adaptive mechanisms are functioning properly, human beings can

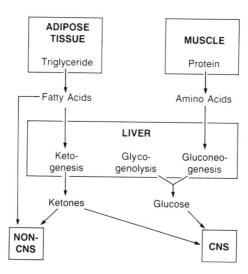

Fig. 16-2 The catabolic sequence. The liver becomes a net glucose producer first via glycogen breakdown and subsequently by gluconeogenesis. Glucose is preserved for use by the central nervous system (CNS) because other tissues (non-CNS) preferentially utilize long-chain fatty acids and their derivative products, the ketone bodies. Ketones can also be used by the brain. Glycerol (released by triglyceride breakdown) and pyruvate–lactate (from muscle glycogen) are additional substrates for gluconeogenesis (not shown in the figure). See text for details.

survive for long periods of time in the total absence of food, provided water is available. For example, most normal-sized (nonobese) adults have 100,000–150,000 kcal stored as fat that, at a utilization rate of 2000 kcal/day, would assure survival for 50–75 days.

REGULATION OF GLUCAGON AND INSULIN SECRETION

The primary regulatory organ for fuel metabolism is the endocrine pancreas and the two key hormones are insulin and glucagon. Insulin is the primary mediator of anabolism, whereas glucagon is the preeminent inducer of catabolism. There is no backup hormone for insulin, but four other hormones exhibit catabolic functions similar to those of glucagon. They are epinephrine, norepinephrine, cortisol, and growth hormone. Most norepinephrine reaches target sites via release from sympathetic nerve endings rather than through the bloodstream. Glucagon and the other four hormones are often designated *counterregulatory*, indicating that they oppose the actions of insulin and protect the body against hypoglycemia.

Insulin and glucagon release normally are smoothly coordinated, so that substrate flow in the alternating anabolic–catabolic phases maintains the plasma glucose levels in the normal range, avoiding both hypo- and hyperglycemia. Concentrations of the two hormones ordinarily vary reciprocally: when insulin concentrations in plasma rise, glucagon levels fall, and vice versa. Each hormone influences release of the other within the pancreatic islet. Insulin inhibits glucagon secretion directly. The venous drainage from the centrally located β (insulin-producing) cells appears to pass directly to the peripherally situated α (glucagon-producing) cells. There insulin binds to the insulin receptor and induces an inhibitory signal (nature unknown) that impairs glucagon release. Glucagon stimulates insulin secretion both directly and indirectly. The former mechanism involves binding of the α-cell hormone to its receptor on the β-cell with subsequent enhancement of insulin release, possibly mediated by cyclic AMP. The indirect effect stems from the ability of glucagon to cause a transient rise in the plasma glucose concentration (through its stimulatory action on hepatic glycogenolysis) which, in turn, promotes insulin release from the β-cell. Hypoglycemia inhibits insulin release and stimulates glucagon secretion. Hyperglycemia has an inhibitory effect on glucagon secretion that is independent of insulin, but the mechanism is poorly understood.

Hypoglycemia is such a powerful stimulus to glucagon secretion that it can override, at least acutely, the inhibitory effect of insulin on the α-cell. Thus, when plasma glucose is significantly lowered by the injection of exogenous insulin, glucagon levels rise to protect against that hypoglycemia despite high concentrations of insulin in plasma. Although it is presumed that the primary stimulus to the α-cell is hypoglycemia per se, acting through some form of a glucose sensor, hypoglycemia-induced catecholamine release doubtless also plays a role. The catecholamine initiation of glucagon secretion is independent of the glucose-sensing mechanism.

Overall it seems safe to say that the primary regulatory signal to both α- and β-cells is the plasma glucose concentration, although glucose-induced changes in glucagon and insulin modulate islet cell response. Stimulatory and inhibitory levels of glucose vary, depending on interactions with other nutrients and hormones, but in general, insulin release becomes negligible or absent when the plasma glucose concentration declines to the 80–85 mg/dl range. Above this level, insulin secretion increases in near linear fashion as the plasma glucose rises. Glucagon release begins to occur at a plasma glucose concentration of approximately 50 mg/dl. Maximal suppression of the α-cell hormone occurs when the plasma glucose reaches 150 mg/dl.

The mechanism of insulin release is much better understood than that of glucagon. Glucose enters the β-cell by a specific glucose transport molecule called GLUT-2 (which also transports glucose into the hepatocyte). Glucose is phosphorylated and subsequently metabolized with the generation of ATP. The ATP closes an ion channel, the (ATP)-sensitive K^+ channel, which depolarizes the β-cell membrane. This depolarization activates a voltage-dependent calcium channel that increases intracellular calcium and stimulates insulin release. Sulfonylureas, drugs that stimulate insulin release, interact with a receptor that appears to act through the (ATP)-sensitive K^+ channel.

While glucose appears to be the preeminent regulatory signal coordinating islet hormone traffic in the basal state, other fuels and hormones may play a role under certain conditions. For example, a high-protein, carbohydrate-free meal stimulates both glucagon and insulin release in contrast to carbohydrate-containing meals after which insulin increases and glucagon decreases as just described. The teleologic interpretation is that the protein-induced insulin release would cause hypoglycemia in the absence of carbohydrate were it not for the protective effect of glucagon that induces breakdown of hepatic glycogen to "cover" the insulin. This is the only physiological circumstance in which insulin and glucagon rise simultaneously. Long-chain fatty acids are now known to be important in enhancing insulin secretion after a fast.

Meals of all kinds induce hormonal and autonomic nervous system signals that enhance insulin secretion. These include the hormones cholecystokinin, secretin, gastrin, gastric inhibitory peptide, and acetylcholine. Acetylcholine is released in the pancreas from parasympathetic nerve endings. These regulatory molecules, controlled from the gastrointestinal tract, are thought to function in anticipatory fashion, poising the β-cell to release insulin promptly as the products of digestion enter the bloodstream. They also account for the fact that insulin response to oral glucose is greater than that seen when the same amount of glucose is given intravenously.

In stress and exercise, insulin–glucagon interrelationships are altered by adrenergic mechanisms. Epinephrine released from the adrenal medulla and norepinephrine reaching the islets from sympathetic nerve endings block insulin release and stimulate glucagon secretion so that hepatic glucose production rises as much as 4- to 5-fold over the rate seen after an overnight fast (approximately 10 g/hour overnight fast → 40–50 g/hour extreme exercise). Epinephrine and norepinephrine act via α_2-adrenergic receptors in the β-cell, decreasing cyclic AMP and inhibiting insulin release, whereas in the α-cell their action is mediated by β-adrenergic receptors that causes a rise in cyclic AMP with stimulation of glucagon secretion. Were it not for the marked increase in the [glucagon]:[insulin] ratio induced by the catecholamines, hepatic glucose output would not be sufficient to protect against hypoglycemia during strenuous exercise; that is, if muscle were using 40 g glucose/hour, whereas the liver was producing only 10 g/hour (the normal response to the hormonal signals operative in fasting), hypoglycemia would rapidly supervene.

The role of somatostatin in regulating insulin and glucagon release from the islets is problematic. Produced by the pancreatic δ-cell, somatostatin conceivably could act in a paracrine fashion (direct cell to cell interaction) within the islet, but in fact, this probably does not occur. Somatostatin is a powerful inhibitor of both insulin and glucagon secretion. Its concentration in pancreatic venous effluent is so high that essentially no glucagon or insulin would be released if the α- and β-cells were exposed to similar levels. For this reason it is thought that somatostatin acts as a true hormone, released into the systemic circulation for action at distal sites. One interpretation is that its central role is to slow intestinal absorption of fat. In animals, blockade of somatostatin action by the infusion of antisomatostatin antibodies results in the accelerated appearance of chylomicrons after a fat meal, suggesting enhanced absorption of triglycerides.

It seems probable that cortisol, growth hormone, and thyroid hormone modulate metabolism primarily through extrapancreatic effects, although all exert permissive regulatory actions in the islets. Cortisol, in particular, increases the gluconeogenic capacity of the liver, probably by stimulating synthesis of key enzymes.

MECHANISM OF ACTION OF INSULIN AND GLUCAGON

Despite intensive study over many years, the mechanism(s) by which insulin acts remains a mystery. As with other peptide hormones, the initial step involves binding to a specific receptor localized in the plasma membrane. The receptor itself is quite well understood, its gene having been localized to the nineteenth chromosome in humans. Cloning has made it possible to define its primary structure, which has two α-subunits (M_r 135,000) and two β-subunits (M_r 95,000). The hormone-binding site is on the α-subunit. The β-subunit expresses a specific enzyme, tyrosine kinase, that phosphorylates proteins on tyrosine molecules rather than the serines and threonines that are phosphorylated by the more common cyclic AMP-dependent protein kinase. Once insulin is bound, it stimulates autophosphorylation of the receptor and phosphorylation of other proteins by the tyrosine kinase. A candidate early substrate is insulin receptor substrate-1 (IRS-1), which acts through regulatory proteins Grb2 and SOS to activate the protooncogene *RAS* by exchanging GTP for GDP (GTP is the "on" signal). *RAS* in turn activates a related protein, RAF-1, which then activates a kinase sequence, mitogen-activated protein kinase kinase (MAPK-K) and mitogen-activated protein kinase (MAP-K) which presumably carry out the final phosphorylations leading to gene activation and other effects. Thus insulin acts through a complicated series of protein kinases to exert its actions. The kinase pathways are influenced by phosphatases; for example, a vanadate activated tyrosine phosphatase may reverse early tyrosine phosphorylations by the insulin receptor.

In hormone interactions, agonists often regulate other agonists, frequently by dominant negative action. By way of illustration, the stimulatory effects of cyclic AMP and glucocorticoids on the gluconeogenic enzyme phosphoenolpyruvate carboxykinase can be inhibited by insulin while insulin stimulation of the enzymes of glycogen synthesis and glycolysis can be overridden by glucagon.

It is of interest that *in vitro* insulin action on the liver is readily demonstrable only against a background of prior glucagon activation. Thus, in the hepatocyte, insulin can be considered primarily a counterregulatory agent for glucagon. In nonhepatic tissues, on the other hand, direct effects of insulin may predominate; that is, insulin is regulatory rather than counterregulatory.

Whatever the linkage between receptor binding and biological effect, perhaps the best known response to insulin administration is the transport of glucose from plasma across cell membranes. The mechanism is facilitated diffusion utilizing carrier molecules that are located in two sites: the plasma membrane and intracellular storage areas associated with microsomal membranes. Insulin-stimulated glucose transport appears to involve mobilization of intracellular ("cryptic") glucose transport units from microsomal storage sites to the plasma membrane, coupled with activation of the transporter once it has reached the cell surface. GLUT-4 is the transporter for muscle and fat, the predominant extrahepatic insulin-sensitive tissues.

The actions of glucagon are better understood than those of insulin. Its primary effect is to rapidly raise intracellular levels of cyclic AMP. The hormone binds to the glucagon receptor and stimulates adenylate cyclase through a coupling mechanism with a guanine nucleotide-binding protein usually called G_s (s = stimulatory). An inhibitory guanine nucleotide binding protein, G_i (i = inhibitory), also interacts with the system, and balance between stimulatory and inhibitory subunits determines net activity of the cyclase. Their interactions are very complex and need not be reviewed here (see Chapter 3). Cyclic AMP levels in liver rise very rapidly after exposure to glucagon. The cyclic nucleotide binds to inactive protein kinase, thereby dissociating the catalytic subunit that then initiates the phosphorylation of multiple proteins. Some of the effects of glucagon may be mediated by cyclic AMP-independent events involving changes in intrahepatic calcium stores. In either case, the net result is phosphorylation of key enzymes, an event that may either activate or inhibit them. For example, hepatic glycogen synthase is inactivated, whereas glycogen phosphorylase is activated, accounting for the fact that glucagon inhibits glycogen synthesis and stimulates glycogen breakdown.

ANABOLIC PATHWAYS

The absorptive process that follows food intake is initiated by intraluminal and brush border enzymes in the intestine which break down complex carbohydrates, lipids, and proteins to constituent sugars, fatty acids, and amino acids. In a normal mixed diet all three components are present (together with vitamins and minerals); their absorption occurs in parallel but at different rates.

Carbohydrate

Starch, sucrose, and lactose are the important carbohydrates in the normal diet. In Western societies about 60% of the daily carbohydrate intake is in the form of starch, 30% is sucrose, and 10% is lactose. About 20% of the starch is made up of amylose that consists of straight chains of glucose molecules in $\alpha 1 \rightarrow 4$ linkage. The remainder is amylopectin, which has branching chains that occur about every 25 molecules, the branch point representing an $\alpha 1 \rightarrow 6$ linkage. Digestion of starch begins with the action of pancreatic amylase. The initial products from the amylose chains are maltotriose and maltose, whereas branched oligosaccharides averaging 8 glucose units (called *α-limit dextrins*) are produced from amylopectin. The products of starch digestion together with dietary lactose and sucrose are hydrolyzed to single sugars by oligosaccharidases localized in the brush border of the intestinal mucosa. Glucose and galactose are transported across the intestinal membrane by an energy-requiring process after being bound to ''carrier'' protein molecules (distinct from the glucose carriers mentioned earlier that mediate insulin's action on glucose transport). The reaction depends on active pumping of the sodium ion out of the cell. Fructose is absorbed by the carrier as well, but this process, called *facilitated diffusion*, is energy and sodium independent.

As noted above, the ingestion of a meal is followed by anticipatory priming of the pancreas that makes insulin available to control the metabolism of glucose absorbed directly and of glucose produced by interconversion from fructose and galactose. Some dietary glucose is oxidized in the postprandial (after ingestion of a meal) period for energy production. The remainder is converted to glycogen and fat. In muscle the glucose molecule is phosphorylated and incorporated directly into glycogen via the sequence glucose $\rightarrow$ glucose 6-phosphate $\rightarrow$ glucose 1-phosphate $\rightarrow$ UDPglucose $\rightarrow$ glycogen. The same reaction sequence occurs in the liver, but in this organ glycogen is also formed by an ''indirect pathway'' (Fig. 16-3), in which glucose is first metabolized to lactate and the lactate is then converted into glucose 6-phosphate and subsequently into glycogen via gluconeogenesis. Most of the lactate is probably formed in nonhepatic tissues (e.g., muscle, intestine, red blood cells), but some could arise in the liver itself, particularly if the hepatic parenchyma is

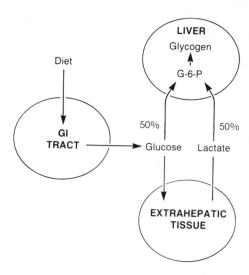

Fig. 16-3 The formation of glycogen in the liver. Glucose absorbed from the diet enters hepatic glycogen by two pathways. About 50% enters directly without degradation. The remainder enters indirectly after conversion to lactate in extrahepatic tissues. The precise site of this conversion is not known although muscle, intestine, and red blood cells have all been postulated. Lactate then returns to the liver and is converted to glucose 6-phosphate (G-6-P) via gluconeogenesis. Essentially all muscle glycogen is formed by the direct pathway.

architecturally divided into zones of predominant glycolysis and predominant gluconeogenesis as some have postulated. With refeeding after an overnight fast, the direct and indirect pathways contribute about equally to hepatic glycogen repletion.

Since release of glucose from the liver rapidly stops after a carbohydrate meal with concomitant insulin release, glucose-6-phosphatase is probably inhibited, thereby favoring conversion of glucose 6-phosphate to glycogen. A second factor assuring efficient glycogen synthesis is glucose-induced inactivation of glycogen phosphorylase, which slows or stops glycogen breakdown. The conversion of phosphorylase *a*, the active form, to phosphorylase *b*, the inactive form, has another effect. Phosphorylase *a* inhibits the specific phosphatase that dephosphorylates glycogen synthase and converts it from an inactive to active form. Thus a glucose-induced fall in glycogen phosphorylase *a* allows conversion of inactive glycogen synthase *b* to active glycogen synthase *a* with the result that glucose 6-phosphate is "pulled" into glycogen. The combined effect of inhibition of glucose-6-phosphatase and activation of glycogen synthase accounts for the rapid glycogen repletion that characterizes the postprandial state.

Substantial evidence now supports the view just articulated that at physiological concentrations of plasma glucose, such as those seen after a normal meal, a significant fraction of hepatic glycogen is formed by the indirect pathway. For many years it was believed that the gluconeogenic sequence (required by the indirect pathway) was shut down almost immediately after meal ingestion. This turns out not to be true, since it has been shown directly in experimental animals and indirectly in humans, using isotopic techniques, that gluconeogenesis continues unimpaired for several hours after a meal. Indeed, it is likely that some basal rate of gluconeogenesis operates at all times. In humans, gluconeogenic flux has been found to increase earlier in the postprandial state when was once believed. The control of gluconeogenesis is ordinarily thought to be vested in the regulatory glycolytic intermediate fructose-2,6-bisphosphate (F-2,6-P_2); high levels of F-2,6-P_2 normally inhibit gluconeogenesis (see Fig. 16-6 and discussion of gluconeogenesis below). In isolated hepatocytes, F-2,6-P_2 concentrations rapidly rise after exposure to glucose, but in intact animals its concentration in the liver does not increase until glycogen stores are repleted. This late rise readily explains continued gluconeogenesis. On the other hand, when fructose or sucrose is ingested, F-2,6-P_2 levels rise promptly to levels that should be inhibitory to gluconeogenesis, yet the process continues. This paradox remains to be explained.

As noted, dietary carbohydrate not utilized for energy production is converted to glycogen first and then to fat. Triglyceride synthesis from carbohydrate takes place primarily in the liver with lesser *in situ* synthesis occurring in adipose tissue. Thus, most glucose-derived fat is made in the liver and transported via lipoproteins to the adipocyte rather than being formed *de novo* in adipose tissue. The biochemical pathways are well understood. The primary substrate is lactate (pyruvate) derived from glucose. Pyruvate, which is formed in the cytosol, then enters the mitochondria and is decarboxylated to acetyl-coenzyme A (CoA) by the pyruvate dehydrogenase complex. Acetyl-CoA condenses with oxaloacetate to form citrate under the influence of citrate synthase. Citrate may proceed through the reactions of the tricarboxylic acid cycle, but a portion exits the mitochondria and is reconverted to acetyl-CoA and oxaloacetate in the cytosol by citrate cleavage enzyme. Acetyl-CoA is converted to malonyl-CoA under the influence of acetyl-CoA carboxylase and the sequential addition of malonyl-CoA to an anchoring acetyl-CoA, carried out by fatty acid synthase, leads to the formation of palmitic acid. The latter can then undergo changes such as elongation and desaturation. The sequence, in summary, is glucose → lactate/pyruvate → citrate → acetyl-CoA → malonyl-CoA → palmitate → other fatty acids. Once formed, the long-chain fatty acids are esterified to glycerol-3-phosphate and the newly synthesized triglyceride is incorporated into a very-low-density lipoprotein (VLDL) molecule. Fatty acids are transferred from VLDL into adipocytes after hydrolysis of VLDL-borne triglycerides by the endothelial-bound lipoprotein, lipase. Adipose tissue also has the capacity to synthesize long-chain fatty acids *de novo* by the mechanism just described.

Fat

Fat absorption is mediated by pancreatic lipase. Enzymatic activity is manifest only after the lipase is bound to triglyceride in complex formation with colipase, a small protein molecule that is formed from its precursor, procolipase, by the action of trypsin. Complexing of lipase with colipase prevents the former's inactivation by intraluminal bile salts whose concentration increases after a meal. Pancreatic lipase splits off long-chain fatty acids from triglyceride, leaving a monoglyceride with the fatty acid in position 2. In the presence of bile salts, fatty acids and the 2-monoglyceride form a soluble complex called a *micelle* that is then able to pass the diffusion barriers of the intestinal mucosa. Micelle formation, required because fatty acids and monoglycerides are poorly soluble in water, provides an efficient mechanism for delivery of these components to the absorptive surface of the intestinal epithelial cell. Inside the cell triglycerides are reformed and packaged into large, round lipoprotein particles called *chylomicrons* that are rapidly secreted into the lymph. Medium and short-chain fatty acids are not reesterified to triglycerides, nor are they transported out of the intestinal cell in chylomicrons; rather, they pass directly into the portal vein for transport to the liver as free fatty acids.

Chylomicrons are the largest lipoprotein particles found in the blood; about 90% of their bulk is made up of triglyceride. Clinically they can be separated from very-low-density lipoproteins, the endogenous fat transport molecule formed in the liver, by having plasma stand overnight in the cold. Chylomicrons then separate out as a cream layer at the top of the tube with clear plasma underneath, whereas VLDL remains distributed throughout the plasma volume, giving the appearance of dilute skim milk. Chylomicrons contain a variety of apolipoproteins including AI, AII, and AIV. They also have a characteristic apolipoprotein B that is designated *apo-B48*. Apo-B48 is clearly distinguishable from apolipoprotein B100 that is synthesized in liver and is the counterpart protein of VLDL. Once in the plasma, the chylomicron's apolipoprotein composition changes rapidly; most of the A proteins are lost and apolipoproteins CI, CII, CIII, and apolipoprotein E are added (Fig. 16-4). Apolipoprotein CII appears to play a critical role in activating lipoprotein lipase, thereby assuring rapid clearance of dietary fat from the plasma. Persons with a congenital absence of apolipoprotein CII characteristically have major hypertriglyceridemia when eating fat because of difficulty in clearing chylomicrons. Lipoprotein lipase, which is located on the luminal surface of capillary endothelial cells, acts in a fashion similar to pancreatic lipase described above, removing the long-chain fatty acids in positions 1 and 3 and leaving a 2-monoglyceride that is then hydrolyzed intracellularly. In adipose tissue, most of the long-chain fatty acids are reesterified to glycerol derived from glucose (the original glycerol released from the chylomicron passes via plasma to the liver), but in muscle, heart, and other tissues a significant portion of the fatty acid is oxidized for energy production. Most of the triglyceride in chylomicrons is removed in a single pass through tissues, only 10–15% of the original mass remaining in the "remnant" particle (Fig. 16-4). The small amounts of cholesterol and cholesteryl esters present are not deleted. The primary apolipoproteins left in the remnant are B48 and E. The remnant particle then passes to the hepatocyte where it is cleared from the plasma by a specialized receptor called the *apo-E receptor*. The apo-E receptor has now been renamed the *LDL receptor-related protein* (LRP). It is also sometimes called the *chylomicron remnant receptor*.

The pathway of fat metabolism just described is often designated the *exogenous* or dietary pathway. A similar series of events occurs in the *endogenous* or hepatic pathway in which free fatty acids derived from diet, adipose tissue triglyceride stores, or synthesis in the liver are reesterified to form triglycerides and packaged as VLDL, as mentioned previously. Lipoprotein lipase (activated by VLDL-bound apolipoprotein CII) acts on VLDL in a manner similar to that described for chylomicrons except that not all the triglyceride is removed in a single pass. Lipoprotein transformations in the endogenous pathway are complicated. The remnant particle of VLDL, called *intermediate-density lipoprotein* (IDL), exchanges apolipoproteins with high-density

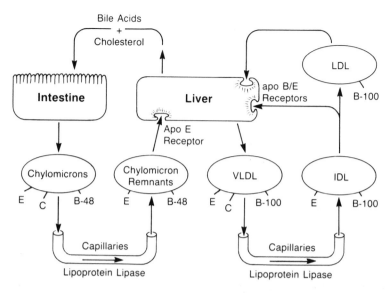

Fig. 16-4   The transport of dietary and endogenous fat. Capital letters stand for apoli-poproteins. B-48 and B-100 refer to apolipoprotein B arising in intestine and liver, re-spectively. Dietary fat is transported to peripheral tissues as chylomicrons that pass to the liver after being stripped of triglyceride through the action of lipoprotein lipase. Uptake in the liver is mediated by a specific receptor designated the "apo E" receptor. Very-low-density lipoprotein (VLDL) carrying endogenously synthesized fat follows a similar sequence except that the "remnant" particle is designated intermediate-density lipoprotein (IDL). IDL can be taken up directly by the liver but is also the precursor for low-density lipoprotein (LDL). Both IDL and LDL are taken up by the LDL receptor designated here as the "apo B/E" receptor (Redrawn with permission from Goldstein JL, Kita T, and Brown MS: Defective lipoprotein receptors and atherosclerosis. Lessons from an animal counterpart of familial hypercholesterolemia. N Engl J Med 309:288–296, 1983.)

lipoprotein (HDL) and is finally converted to low-density lipoprotein (LDL), the major cholesterol-carrying transport protein in the blood. IDL and LDL are then cleared from the plasma via a receptor on hepatocytes called the *apo-B/E* or *LDL receptor*. Identical receptors are present on other tissues of the body and binding of LDL to these receptors allows entry of cholesterol into cells for membrane synthesis and hormone formation. Partial or complete absence of the LDL receptor leads to the disease, familial hypercholesterolemia.

HDL molecules are not directly involved in fat transport but play a key role in reverse cholesterol transport out of tissues. LDL itself is innocent in atherogenesis but becomes atherogenic after oxidation to minimally modified LDL; the latter binds to a receptor distinct from the LDL receptor. The oxidized LDL receptor is known as the *acetyl LDL* or *scavenger receptor*. HDL functions as an antioxidant opposing the oxidation of LDL. Hence it protects against atherosclerosis by two mechanisms: by removing cholesterol from blood vessels and by preventing the formation of oxidized LDL.

Cholesterol Metabolism

Although cholesterol is not a fuel, a brief discussion of its metabolism is included here because it is intimately involved in triglyceride metabolism, and because elevation of plasma cholesterol is a major risk factor in the development of atherosclerosis. Cholesterol in the body arises from two sources: diet and endogenous synthesis. The latter takes place via a linked series of enzymatic reactions starting from acetyl-coenzyme A (CoA). An overview of the sequence is acetyl-CoA $\rightarrow$ acetoacetyl-CoA $\rightarrow$ hydroxymethylglutaryl-CoA $\rightarrow$ mevalonic acid $\rightarrow$ isopentenylpyrophosphate $\rightarrow$ squalene $\rightarrow$ lanosterol $\rightarrow$ desmosterol $\rightarrow$ cholesterol. The rate-limiting step is the conversion of hydroxymethylglutaryl-CoA to mevalonic acid, a reaction catalyzed by the enzyme hydroxymethylglutaryl-CoA reductase (HMG-CoA reductase). The general rule is that dietary and synthetic sources of cholesterol are reciprocally related: when dietary cholesterol intake is high, cholesterol synthesis is low and vice versa. Inhibitors of HMG-CoA reductase are the primary means of treating clinically significant hypercholesterolemia.

The average daily intake of cholesterol in Western diets is 500–1000 mg of cholesterol, mostly esterified, because of a high intake of eggs, dairy products, and meat. Esterified cholesterol is hydrolyzed before absorption by a pancreatic cholesteryl esterase that acts in the presence of bile acids. It is presumed that cholesterol is absorbed in the form of mixed micelles made up of bile salts, fatty acids, and phospholipid. The process has limited efficiency; absorption ranges from one-third to two-thirds of the ingested load. By means that are not well understood, cholesterol is transferred from the absorptive micelles to lipoproteins, the two major carriers being chylomicrons and HDL. The former predominates, particularly if the meal contains fat. As discussed earlier, absorbed cholesterol reaches the liver via the production of chylomicron remnants (Fig. 16-4). Cholesterol taken up by the liver has several functions and fates. It may be excreted into the bile directly or may be converted to bile acids. It also inhibits HMG-CoA reductase, thus slowing endogenous cholesterol synthesis in the hepatocyte. Simultaneously cholesteryl ester formation is stimulated through activation of acyl-CoA:cholesterol acyltransferase (ACAT). Finally, the lipoprotein-borne cholesterol suppresses synthesis of LDL receptors, thereby controlling access of LDL into the cell. Dietary cholesterol, like newly synthesized cholesterol, can be repackaged into VLDL molecules that lead to the production of LDL molecules, as described in the previous section. LDL is the principal means by which cholesterol reaches extrahepatic tissues for membrane synthesis and, in the endocrine glands, for steroid hormone formation.

Elucidation of the LDL receptor pathway has provided insight into a general mechanism whereby a variety of larger molecules pass from the plasma into cells. The system is called *receptor-mediated endocytosis*. Specificity is determined by the various receptors involved (i.e., insulin enters cells after binding to the insulin receptor, epidermal growth factor [EGF] by binding to the EGF receptor, etc.). LDL binds to its receptors in specialized areas of the plasma membrane called "coated pits." (The "coat" is a distinctive protein called clathrin.) The receptor, bearing its LDL, is then pinched off to form a coated vesicle. The vesicle next merges with lysosomes and lysosomal enzymes rapidly degrade LDL to constituent amino acids with release of cholesteryl esters. The latter are hydrolyzed to free cholesterol. Empty

LDL receptors cycle back to the cell surface. The speed of the process is quite remarkable, each receptor making the trip from coated pit to lysosome and back to the plasma membrane in 10 minutes. Normally about 60–70% of cholesterol traffic enters the cell via the LDL receptor, but about one-third traverses a nonspecific uptake pathway. Disturbances in the LDL receptor cycle ordinarily cause difficulty because cholesterol concentrations in plasma rise, leading to atherosclerosis via conversion of LDL to oxidized LDL. Interestingly, even the complete absence of the LDL receptor system is not accompanied by clinical hormonal deficiency states, presumably because the combination of endogenous cholesterol synthesis and entry of cholesterol by receptor-independent pathways satisfies cellular needs.

Protein

The absorption of protein is more complicated than the absorption of fat and carbohydrate. The process appears to start in the stomach through the action of pepsin in the presence of an acid environment. Pepsin is not necessary for protein absorption, however. Most of the absorptive process takes place in the small intestine with pancreatic proteases doing the bulk of the work. The major endopeptidases (trypsin, chymotrypsin, and elastase) act on peptide bonds to produce free amino acids and oligopeptides. Exopeptidases (aminopeptidase and carboxypeptidase) act after the initial formation of longer peptides by trypsin, chymotrypsin, and elastase to produce the final luminal products. Ordinarily the combined activities yield about 30% free amino acids and 70% oligopeptides with a chain length ranging from 2 to 6 amino acids. The fate of the oligopeptides is complicated, some being broken down by brush border enzymes and others being transported into the cell where cytoplasmic peptidases complete the degradative process. The high specific activity aminooligopeptidase of the brush border has its action blocked by the presence of a proline in the penultimate position of the oligopeptide chain. This regulatory proline thus appears to determine whether intraluminal/brush border digestion is complete or whether intracellular digestion is required. Surprisingly, the transport of free amino acids from the lumen into the cell is a slower process than the transport of oligopeptides. Absorbed amino acids pass to the liver via the portal vien.

The rise in amino acid concentrations in arterial blood that follows a protein meal does not reflect the makeup of the ingested protein. That is, all amino acids are not metabolized or transported in identical fashion. Although branched-chain amino acids make up no more than one-fifth of the total residues in meat, about 60% of the increase in amino acids following the protein meal is made up of leucine, valine, and isoleucine. Presumably they escape uptake by the liver and are transported to muscle where they serve as the major substrate for protein synthesis (Fig. 16-5). Insulin has the capacity to stimulate the process of protein synthesis in general and the utilization of branched-chain amino acids in particular. The mechanism is not completely understood, but at least one action of insulin is to stimulate the formation of "initiation complexes" that start the series of reactions leading to peptide formation. Some investigators believe that phosphorylation–dephosphorylation reactions may be important, as mentioned earlier in the discussion of insulin actions. Leucine may be a special stimulatory signal for protein synthesis, but again, the mechanism is not clear. Net buildup of muscle following a meal is also a con-

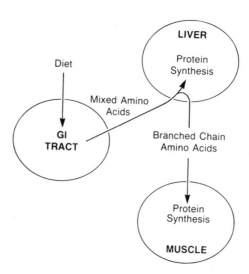

Fig. 16-5 Amino acid transport following a meal. Mixed amino acids reach the liver via the portal vein. Branched-chain amino acids preferentially pass through the hepatic circulation and account for 60% of the rise in amino acid concentrations in the general circulation. They appear to be utilized for protein synthesis under the influence of insulin.

sequence of insulin-induced diminution in proteolysis; that is, insulin has the capacity to simultaneously stimulate protein synthesis and block protein breakdown.

CATABOLIC PATHWAYS

To review briefly, the catabolic phase of metabolism is a series of metabolic adaptations designed to ensure adequate fuels for body tissues in the absence of exogenous substrate. The preeminent concern is to maintain the plasma glucose at safe levels for the central nervous system until ketone concentrations rise into the protective range. If acetoacetic and β-hydroxybutyric acids are not present in sufficient concentrations, the brain is vitally dependent on glucose and its absence is as devastating as the absence of oxygen. As noted earlier, the protective processes required are glycogen breakdown, gluconeogenesis, mobilization of amino acids from muscle, lipolysis, and ketogenesis.

Glycogenolysis and Gluconeogenesis

In the early postabsorptive state and the initial phases of fasting, maintenance of the plasma glucose depends almost entirely on glycogen stores. Although glycogenolysis can be induced by catecholamines acting as β-agonists (and, in humans, to a small extent via α-receptors), it is generally accepted that the bulk of glycogen breakdown occurs through a glucagon-mediated rise in cyclic AMP. Activation of cyclic AMP-dependent protein kinase phosphorylates phosphorylase *b,* converting it to the active *a* form. Simultaneously, phosphorylation of glycogen synthase *a* converts it to the inactive *b* form. In consequence, glycogen formation ceases, avoiding a futile cycle of coexistent glycogen breakdown and formation. It is likely that glucose-6-phosphatase activity also increases in the catabolic phase, assuring that glucose 6-phosphate (G-6-P) formed from glycogen will pass immediately into the plasma as free glucose. This would be a mirror image of the inhibition postulated to occur after feeding. The nature of regulation at the glucose-6-phosphatase reaction, if any exists, is not known. If the enzyme is not regulated, flux is probably substrate controlled; that is, a rise in G-6-P concentrations secondary to glycogen breakdown and impaired glycogen synthesis would drive the reaction.

The mechanisms by which carbon flow between G-6-P and pyruvate is controlled have only recently been clarified. As noted earlier, in the catabolic phase glycolysis (G-6-P → pyruvate) is blocked and gluconeogenesis (pyruvate → G-6-P) is enhanced. The key regulatory site appears to be the interconversion of fructose 6-phosphate (F-6-P) and fructose 1,6-bisphosphate (F-1,6-P_2) (see Fig.16-6). Although most of the enzymes of the glycolytic-gluconeogenic pathway are reversible, separate enzymes operate at this site. In the glycolytic pathway, F-6-P is converted to F-1,6-P_2 via the enzyme 6-phosphofructo-1-kinase, whereas the reverse reaction, the major regulatory step in gluconeogenesis, is mediated by fructose-1,6-bisphosphatase. Glucagon control of the opposing reactions is carried out by changing the concentrations of a regulatory intermediate, fructose 2,6-bisphosphate, which is itself not a direct participant in glycolysis. Fructose 2,6-bisphosphate is formed from fructose 6-phosphate by a bifunctional enzyme called 6-phosphofructo-2-kinase/fructose-2,6-bisphosphatase (PFK$_2$/FBPase$_2$). This remarkable enzyme shifts from a kinase to a phosphatase when phosphorylated. When insulin levels are high and glucagon levels are low, the enzyme is dephosphorylated and functions as a kinase that leads to the synthesis of fructose 2,6-bisphosphate. When insulin levels fall and

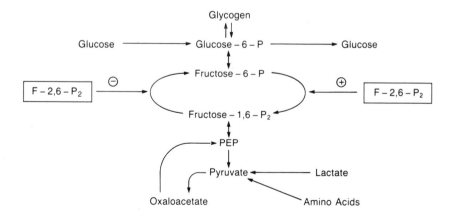

Fig. 16-6 The regulation of glycolysis and gluconeogenesis. Fructose 2,6-bisphosphate (F-2,6-P_2) stimulates glycolysis by activating phosphofructokinase, which converts fructose 6-phosphate to fructose 1,6-bisphosphate and inhibits gluconeogenesis by deactivating fructose-1,6-bisphosphatase, which converts fructose 1,6-bisphosphate to fructose 6-phosphate.

glucagon concentrations increase, the enzyme is phosphorylated, kinase activity is lost, and phosphatase activity appears. This causes, simultaneously, decreased synthesis and increased hydrolysis of fructose 2,6-bisphosphate with the result that its concentration in the hepatocyte rapidly falls. When fructose 2,6-bisphosphate levels are high, phosphofructokinase activity is stimulated and glycolysis is active, whereas fructose bisphosphatase is inhibited, blocking gluconeogenesis. The catabolic sequence can be summarized as follows: increased glucagon $\rightarrow$ increased cAMP $\rightarrow$ increased cAMP-dependent kinase $\rightarrow$ increased phosphorylation of $PFK_2/FBPase_2$ $\rightarrow$ decreased $F\text{-}2,6\text{-}P_2$ $\rightarrow$ decreased glycolysis + increased gluconeogenesis. Activation of gluconeogenesis assures sustained hepatic glucose production after preformed glycogen is depleted while the block in glycolysis indirectly allows the liver to become a ketogenic organ as described below.

In quantitative terms, hepatic glucose production averages about 2 mg/kg body weight/minute after an overnight fast. About 70–80% of the output is derived from glycogen and about 20% from gluconeogenesis under these conditions. If fasting is combined with exercise, glycogen stores are depleted much more rapidly and the percentage contribution from gluconeogenesis increases. The gluconeogenic substrates are lactate, glycerol, and amino acids. Lactate derives from two sources. The first is from glucose incompletely oxidized in peripheral tissues. That is, during the initiation of the catabolic state, glucose passes from the liver to peripheral tissues where it is metabolized to pyruvate–lactate. A portion of the pyruvate undergoes terminal oxidation to CO_2 and water with energy production, but the remainder of the pyruvate–lactate returns to the liver. Lactate concentrations in plasma are 10 times higher than pyruvate so that it is the principal member of the couple. The glucose $\rightarrow$ lactate $\rightarrow$ glucose sequence is known as the *Cori cycle*. Note that the Cori cycle does not result in net glucose synthesis in the body since the lactate was originally derived from preformed glucose. Net glucose synthesis from lactate requires its production from glucose derived from muscle glycogen. Glycerol, which is a relatively minor contributor to gluconeogenesis, is released from triglycerides hydrolyzed in adipose tissue (see lipolysis below). The major substrate for gluconeogenesis is amino acids that flow from muscle. As is the case in the anabolic phase, where the rise in amino acid concentrations in arterial blood does not reflect the amino acid makeup of ingested proteins, the output of amino acids from muscle is also asymmetrical. Thus, alanine and glutamine account for over 50% of the amino acids released, although alanine accounts for only about 7% of the total amino acid residues in muscle (Fig. 16-7). Glutamine is primarily taken up by extrahepatic splanchnic tissues and is also utilized for gluconeogenesis in the kidney after prolonged fasting. Studies in humans suggest that it may be a more significant substrate for hepatic gluconeogenesis than previously thought. Alanine is the primary amino acid utilized for glucose production by the liver. Its fractional extraction and total uptake are increased in the fasting state.

The explanation for the excess release of alanine from muscle is thought to be a transamination reaction wherein branched-chain amino acids (leucine, isoleucine, valine) donate their amino groups to pyruvate derived from muscle glycogen. A teleological explanation might be that alanine is the best precursor for hepatic gluconeogenesis, entering the gluconeogenic sequence after reversed transamination to pyruvate. The α-keto analogues of branched-chain amino acids remaining after

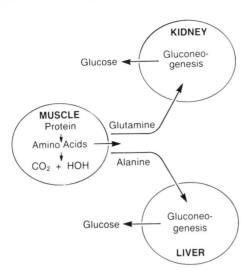

Fig. 16-7 Amino acids' release from muscle and their use in gluconeogenesis. Glutamine and alanine are the primary amino acids delivered to the liver and kidney for conversion to glucose. See text for details.

amino group donation can be oxidized in muscle for energy purposes, although some are released into the general circulation.

The signals for proteolysis in muscle are not completely understood. It is likely that both insulin deficiency and the presence of counterregulatory hormones play a role. For example, small physiological increments in plasma cortisol concentrations clearly stimulate proteolysis. Catecholamines (and perhaps glucagon) probably have similar effects. Protein breakdown occurs by both lysosomal and nonlysosomal pathways. At least a portion of the latter is calcium dependent.

The amino acids utilized in gluconeogenesis do not come from nonfunctional storage pools but from working muscle. It is assumed, therefore, that enzymes, transport molecules, and membrane proteins may be broken down. Their loss puts cellular integrity at risk. Indeed, in most cases death from starvation is due to protein wastage, often from cardiac arrhythmias. It thus becomes crucial that proteolysis be limited when fasting is prolonged. This in fact occurs. A great part of the physiological adaptation is the previously mentioned shift to a lipid economy in which most tissues of the body are sustained by the oxidation of fat, either as free fatty acids or ketone bodies. This markedly reduces the demand for glucose and, as a result, the requirements for gluconeogenic substrate diminish. This is reflected in a lessening of the negative nitrogen balance that is so characteristic of starvation. How the muscle is notified to diminish proteolysis is not clear. The thesis, once widely held, that ketone bodies represent the off signal is no longer tenable. One possibility is that rising free fatty acid and ketone body concentrations stimulate insulin release from the β-cell and that it is insulin that is the antiproteolytic signal. However, refeeding studies with fructose indicate that this sugar alone, in the absence of an increase in plasma insulin levels, also reverses all manifestations of starvation, including protein breakdown, suggesting that insulin is not an obligatory mediator for cessation of proteolysis. Conceivably the message is carried through amino acid concentrations in plasma. A diminished need for glucose either because of refeeding or because of fat oxidation might slow the uptake of alanine and other amino acids by the liver, leading to a transient rise in plasma and subsequently in muscle amino acid concentrations. This might somehow activate a negative feedback sequence that would blunt protein breakdown. Although it is clear that proteolysis slows with prolonged fasting, the mechanisms must be considered speculative at this time.

When the catabolic state induced by fasting is reversed by feeding there is a complete reversal of the metabolic changes just described. The triggering event is almost certainly a rise in the [insulin]:[glucagon] ratio. One interesting characteristic of the reversal in carbohydrate metabolism has been commented on earlier, namely, that gluconeogenesis continues for several hours after feeding, allowing hepatic glycogen repletion to take place in large measure via an "indirect" rather than a "direct" pathway. Continued gluconeogenesis is not difficult to explain after glucose feeding since fructose 2,6-bisphosphate levels do not rise for several hours as noted earlier. With sucrose refeeding, on the other hand, $F-2,6-P_2$ levels rise early on and yet gluconeogenesis is not impaired. Two possible explanations suggest themselves. First, there might be a regulatory intermediate substance (not yet identified) that can override normal $F-2,6-P_2$ inhibition of gluconeogenesis. Second, the measured rise in $F-2,6-P_2$ might not be uniform in all hepatocytes; that is, some zones of the liver would contain cells that do not manifest an early rise in $F-2,6-P_2$ and thus continue gluconeogenesis.

Lipolysis and Ketogenesis

The switch to a lipid economy in the catabolic phase of metabolism requires changes in both adipose tissue and liver. The hydrolysis of triglycerides in the adipocyte delivers long-chain fatty acids, together with glycerol, into the circulation. Two-thirds to three-fourths of the mobilized fatty acids are utilized directly, whereas the remainder are taken up by the liver for conversion to ketone bodies. Under normal anabolic conditions the liver produces only tiny amounts of acetoacetic and β-hydroxybutyric acids. Dietary fatty acids entering the hepatocyte are simply esterified to triglycerides and transported into the circulation as very-low-density lipoprotein. After a few hours of fasting, or under the influence of counterregulatory hormones induced by stress, fatty acid oxidative capacity is activated and significant ketone body production becomes possible, provided long-chain fatty acids are available. Conceptually, adipose tissue can be viewed as the substrate supply depot and the liver as the factory for ketone production. Full-blown ketosis requires both an adequate delivery of long-chain fatty acids to the liver and activation of the enzyme system that converts long-chain fatty acids to acetoacetic and β-hydroxybutyric acids. Failure of either substrate delivery or enzyme activation causes ketogenesis to fail. Once the liver is activated, rates of ketone body production are simply the consequence of the rate of delivery of long-chain fatty acids; that is, the higher the concentration of long-chain fatty acids in plasma, the greater the production rate of ketones by the liver until V_{max} rates are obtained. Qualitatively, there is no difference in the metabolic adaptations in adipose tissue and liver that accompany the very mild ketosis of a brief fast and the raging ketoacidosis of uncontrolled diabetes mellitus.

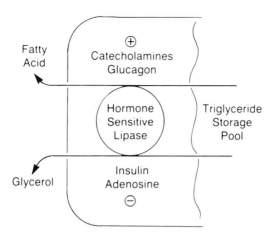

Fig. 16-8 Hormone-sensitive lipase in adipocytes and its regulation. +, activation; −, inhibition. Although not shown, ketone bodies may also inhibit the lipase.

The differentiating determinant is that free fatty acid concentrations are very much higher in the latter state.

The hydrolysis of triglycerides in adipocytes is due to activation of an intracellular lipase that is generally designated *hormone-sensitive lipase* to distinguish it from the extracellular lipoprotein lipase that is responsible for hydrolysis of triglyceride in plasma lipoproteins. The hormone-sensitive lipase is inhibited by insulin and adenosine and activated by counterregulatory hormones (Fig. 16-8). In humans, catecholamines appear to play the primary role in activation whereas glucagon, growth hormone, cortisol, and corticotropin (ACTH) have much lesser effects. Epinephrine and norepinephrine act through cyclic AMP-stimulated protein kinase in classic fashion. Removal of the insulin inhibitory effect, expected from a fall in plasma insulin accompanying the catabolic phase, also plays a role. The quantitative contribution of the removal of insulin inhibition in relation to the positive activating effects of catecholamines has not been completely worked out. Experimentally, the hormone-sensitive lipase can also be inhibited by adenosine and ketone bodies. Some evidence suggests that the effects of adenosine may be physiologically important.

The fatty acid–oxidizing system of the body is quite complicated and involves multiple enzymes. It can be divided into two components: a transport system required to get long-chain fatty acids into the mitochondria and the intramitochondrial oxidizing system (β-oxidation) that cleaves the long-chain fatty acid to acetyl-CoA. In most tissues, the fatty acid–derived acetyl-CoA undergoes terminal oxidation to CO_2 and water in the tricarboxylic acid cycle for the generation of energy. In the liver, however, the bulk of the acetyl-CoA in the catabolic phase (about 95%) is converted to acetoacetic and β-hydroxybutyric acids via the hydroxymethylglutaryl-CoA cycle. The transport step appears to be rate limiting under most circumstances so that any fatty acids traversing the inner mitochondrial membrane are rapidly oxidized. When long-chain fatty acids are taken up in liver or other tissues, they are converted to CoA derivatives with ATP as an energy source (Fig. 16-9). The long-chain fatty acyl-CoA cannot penetrate the inner mitochondrial membrane, however, requiring

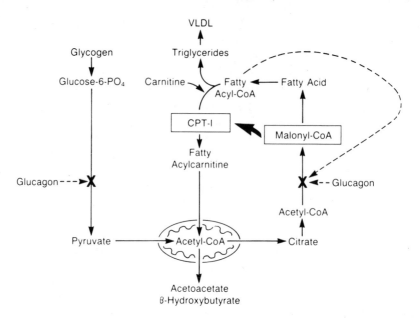

Fig. 16-9 The regulation of ketogenesis. In the transition from anabolic to catabolic phases glucagon blocks glycolysis, thereby interrupting flux over the sequence glucose 6-phosphate → pyruvate → acetyl-CoA → citrate → acetyl-CoA → malonyl-CoA. In addition, glucagon and long-chain fatty acyl-CoA cause inhibition of acetyl-CoA carboxylase. The fall in malonyl-CoA thus produced disinhibits carnitine palmitoyltransferase I (CPT I), the rate-limiting step in fatty acid oxidation, thereby activating the ketogenic pathway. Fatty acids delivered to the liver from adipose tissue are thus preferentially oxidized to acetoacetic and β-hydroxybutyric acids. The fall in malonyl-CoA also halts fatty acid synthesis.

a transesterification step in which CoA is exchanged for carnitine, a carrier molecule in lipid transport. The transesterification is carried out by an enzyme located on the outer mitochondrial membrane called *carnitine palmitoyltransferase I* (CPT I). Fatty acylcarnitine is then transported across the inner mitochondrial membrane by a translocase enzyme and reconverted to the CoA derivative under the influence of carnitine palmitoyltransferase II (CPT II). The newly reformed fatty acyl-CoA appears to be committed to oxidation in the β-oxidative sequence. CPT I and CPT II have been cloned and sequenced in both rodent and human tissues. CPT I in liver is distinct from CPT II in muscle, while CPT II appears to be the same in all tissues.

The activity of CPT I is regulated by malonyl-CoA. In the anabolic phase, malonyl-CoA concentrations in the liver, heart, and muscle are high and CPT I is inhibited. In tissues that synthesize fat, this assures that long-chain fatty acid synthesis and fatty acid oxidation do not occur simultaneously, thus avoiding a futile cycle. Under catabolic conditions malonyl-CoA concentrations rapidly fall, thereby activating fatty acid oxidation, whereas lipogenesis simultaneously ceases as a result of a deficiency of substrate. The fall in malonyl-CoA is glucagon mediated. One component of the mechanism is inhibition of glycolysis caused by the fall in fructose 2,6-bisphosphate previously described. The interruption of flux from glucose 6-phosphate to pyruvate reduces citrate production in the mitochondria, which in turn leads to failure of acetyl-CoA generation in the cytosol. Since malonyl-CoA is formed from the carboxylation of acetyl-CoA under the influence of the enzyme acetyl-CoA carboxylase, malonyl-CoA concentrations decrease, disinhibiting fatty acid oxidation. Glucagon also has the capacity to inhibit acetyl-CoA carboxylase (through cyclic AMP-mediated phosphorylation), as do long-chain fatty acyl-CoA concentrations, which also rise in liver during the catabolic adaptation. These events are summarized in Fig. 16-9. With refeeding, malonyl-CoA concentrations are restored to anabolic levels, thereby deactivating CPT I. The simultaneous rise in insulin, which inhibits lipolysis, normalizes the substrate wing of the process. Malonyl-CoA is also a regulatory molecule in the pancreatic β-cell. Its rising concentration after feeding enhances insulin secretion in response to glucose.

DERANGEMENTS IN GLUCOSE, LIPID, AND PROTEIN METABOLISM

In order to illustrate the consequences of dysfunction in the anabolic–catabolic regulatory system, which normally is tightly controlled, some selected clinical disorders will be described briefly.

Diabetes Mellitus

Diabetes mellitus is an extremely common metabolic disease affecting 1–2% of the population in Western societies. The prevalence is much higher in certain groups such as the Hispanic and Native American subpopulations in the United States. There are two main types of diabetes, usually categorized as insulin dependent and non-insulin dependent. The former makes up about one-quarter of the total and is the more severe.

Current understanding of insulin-dependent diabetes (also known as type 1 diabetes) suggests that one inherits a genetic susceptibility to disease that is permissive but not causal. That multiple genes are involved has been suspected for some time and is reinforced by a recent search of the entire human genome which revealed no less than 20 independent chromosomal regions with positive linkage to the disease. Of these, a particularly strong association is found with a gene located on the short arm of the sixth chromosome in the major histocompatibility complex. More than 90% of patients, when typed by a serologic test, exhibit human leukocyte antigen (HLA) DR3, DR4, or the combination DR3/DR4. It is currently thought that genes on the DQ β-chain may be more closely linked to diabetes susceptibility than those in the DR region. Some alleles appear to be susceptibility genes and others resistance genes while still others are neutral. HLA associations appear less informative in persons developing diabetes after the age of 20.

The pathogenesis of autoimmune insulin-dependent diabetes is not precisely known. Leakage of cryptic antigens following a viral infection is one possibility. Alternatively, the mechanism of *molecular mimicry* may be involved. In molecular mimicry, homology between viral or food antigens and β-cell antigens fool the body's immune system into attacking itself. For example, a Coxsackie virus infection elicits an immune response. The cytotoxic T lymphocytes, armed against the viral antigens, cannot distinguish between the virus and the homologous amino acid segment in the β-cell, and thus attacks and destroys the latter.

Symptoms appear when 90% or more of the insulin-producing capacity of the islets of Langerhans is destroyed. Immune destruction is probably carried out both by humoral and cellular mechanisms, but the latter is considered dominant. Islet cell antibodies can be demonstrated in the plasma long before symptomatic diabetes appears. The most important predictive antibody recognizes glutamic acid decarboxylase (GAD) in the β-cell. The consequences of the immune attack are an absolute deficiency of insulin and a major overproduction of glucagon that is secondary to removal of the normal insulin restraint. The predictable metabolic consequences of a high [glucagon]:[insulin] ratio are overproduction of glucose by the liver, underutilization of glucose in nonhepatic tissues, and initiation of ketone body production in the liver, exactly as occurs in a normal fast. The only difference is that in normal persons who do not eat, intact β-cells remain. Whereas a rising glucose concentration in plasma is the usual signal for insulin release, both ketone bodies and free fatty acids can also act in stimulatory fashion. Thus, in prolonged fasting, even though glucose concentrations are low, free fatty acids and ketone bodies reach a threshold that stimulates insulin release, limiting lipolysis and thereby maintaining ketone production within the safe range. In uncontrolled diabetes, because insulin production is impossible, hyperglycemia steadily increases and a life-threatening ketoacidosis results.

The more common noninsulin-dependent diabetes mellitus (NIDDM, also called type 2 diabetes) is generally associated with obesity. Insulin resistance is a major problem and insulin deficiency is relative rather than absolute. Obesity is a major cause of this resistance, but it can also be demonstrated in offspring of two parents with NIDDM, even when they are not overweight or diabetic. The mechanism of the intrinsic insulin resistance has not been worked out. Early in NIDDM, insulin levels are high in absolute terms but are insufficient to overcome insulin resistance. Later, insulin levels fall as hyperglycemia worsens. The nature of the β-cell defect leading to diminished insulin reserve is not known. Because of the insulin resistance, glucagon levels are high and the [glucagon]:[insulin] ratio is elevated in terms of biological activity, even though the actual ratio may be near normal when assessed by radioimmunoassay. That is because the insulin present does not work well, leaving glucagon's actions unopposed. Once again the hormonal changes are those characteristic of the catabolic state and major hyperglycemia supervenes. For reasons that are not yet clear, ketoacidosis does not develop in noninsulin-dependent diabetic subjects under most circumstances. Acetoacetic and β-hydroxybutyric acid levels may rise after prolonged periods of poor medical control, but in general, they never exceed concentrations seen during the fasting state in normal people. Hyperglycemia may become extreme, reaching levels as high as 5000 mg/dl. It is not surprising that under these circumstances the blood has been described as "syrupy." Such profound elevations of the blood sugar result in severe body fluid losses through the kidney and ultimately lead to the syndrome called *hyperosmolar, nonketotic diabetic coma.* The former term refers to the fact that massive water deficits caused by glucose-driven urine production lead to increases in the plasma osmotic pressure from a normal of 280–300 mOsmol/liter to values as high as 400–500 mOsmol/liter. The term *nonketotic* does not mean that ketone bodies are not measurable in the urine or plasma but simply that their production is not high enough to give ketoacidosis. Patients with hyperosmolar, nonketotic coma usually manifest altered consciousness or coma, convulsions, impaired kidney function, and sometimes pancreatitis and infections with gram-negative organisms. Mortality rates are very high.

The profound metabolic derangements of uncontrolled diabetes underline the importance of the control systems that normally govern fuel metabolism in the body. Although the metabolic disturbances of diabetes can usually be controlled well enough to prevent symptoms by dietary restriction and insulin therapy, it is almost impossible to maintain the plasma glucose within the normal range. Consequently, after many years of diabetes patients become vulnerable to late degenerative complications involving the eyes, the nerves, the kidneys, and the blood vessels. Thus, diabetes is a leading cause of blindness, kidney failure, amputations, strokes, and heart attacks. It is these complications that make it such a devastating illness.

Hyperglycemia of Severe Injury

Nondiabetic subjects who sustain extreme injuries such as massive burns or major trauma may develop hyperglycemia even in the face of normal pancreatic function. The stress of such injuries leads to an outpouring of epinephrine and norepinephrine that block insulin release from the β-cell, stimulate glucagon production, activate glycogenolysis, stimulate lipolysis, and induce proteolysis. In short, severe injury may quickly convert a normal state, even in a person who has just eaten, into the catabolic response. Stress overrides normal metabolism and does so by provoking sustained release of counterregulatory hormones.

Systemic Carnitine Deficiency

Systemic carnitine deficiency is a relatively rare syndrome in which carnitine concentrations are below the lower limits of normal in plasma and multiple tissues including muscle and liver. As noted above, protection against hypoglycemia requires the conversion of most tissues in the body from a carbohydrate to a lipid-utilizing state. This in turn requires that long-chain fatty acids be transported into the mitochondria for oxidation by the carnitine palmitoyltransferase system of enzymes. When carnitine is deficient, fatty acid oxidation is prevented, with the consequence that long-chain fatty acids cannot undergo terminal oxidation for energy generation or, in the liver, for the production of acetoacetic and β-hydroxybutyric acids. This means that all tissues remain dependent on glucose. Since the liver, even at maximal levels of gluconeogenesis, cannot deliver sufficient glucose in the plasma to meet body needs, profound hypoglycemia supervenes. Deficiencies of carnitine palmitoyltransferase or any of the enzymes of β-oxidation may lead to a similar syndrome. Children with systemic carnitine deficiency exhibit, in addition to hypoglycemia, massive storage of fat in all tissues of the body and also have the clinical picture of Reye's syndrome (hepatic encephalopathy, nausea, vomiting, increased arterial ammonia levels, and elevated liver enzymes in plasma).

Carnitine deficiency and related diseases of fatty acid oxidation illustrate the critical role of fat oxidation in protecting against hypoglycemia. Similar hypoglycemia can be seen in profound cachexia in the starvation belts of the world or in fat depletion consequent to cancer or other wasting illnesses.

Insulinoma

Insulin-producing tumors of the pancreas are not an uncommon medical problem and enter into the differential diagnosis of hypoglycemia. These tumors secrete insulin into the plasma in unregulated fashion. One might predict that counterregulatory hormones would protect against hypoglycemia under these circumstances, but persistent insulin secretion keeps glucagon suppressed to levels insufficient to overcome the actions of the β-cell hormone with the result that blood sugar levels fall to concentrations that cause loss of consciousness, convulsions, or death. Just as severe injury can cause sufficient release of counterregulatory hormones to overcome the capacity of the β-cell to respond, so can abnormal production of insulin abrogate the ability of counterregulatory hormones to respond. This emphasizes again how normal metabolism requires finely tuned and balanced interaction between the various components of the endocrine system.

Familial Hypercholesterolemia

Familial hypercholesterolemia (FH) is an autosomal dominant illness of humans that produces marked elevation of the plasma cholesterol. The latter, in turn, causes premature atherosclerosis with heart attacks occurring at an early age. In the more severe homozygous form of the disease (two abnormal genes for the LDL receptor), plasma cholesterol levels are in the range of 600–1200 mg/dl (normal, <200 mg/dl). In heterozygous patients (one mutant gene), levels are in the 300–550 mg/dl range. It is the LDL cholesterol fraction that is elevated. In homozygous familial hypercholesterolemia heart attacks commonly occur between the ages of 5 and 30 and have been reported as early as 18 months. In men with the heterozygous form of the illness there is a 5% chance of having a heart attack by the age of 30, 51% by the age of 50, and 85% by the age of 60. An interesting feature of the hypercholesterolemic states is the presence of cholesterol deposits in skin, tendons, and over bony prominences such as the elbows. These deposits, called xanthomas, are strong evidence of hypercholesterolemia. Careful palpation of the Achilles tendon may reveal beaded xanthomas in patients who do not show typical deposits in the skin or over bony protuberances.

The elevation in plasma cholesterol is due in all cases to a genetic defect that produces an abnormal LDL receptor. Several types of defects are now known. Class 1 mutants synthesize no receptors. Class 2 mutants do not transfer LDL receptors efficiently from the endoplasmic reticulum to the Golgi apparatus so that the receptors are degraded in the cell before they can reach the cell surface. Class 3 mutants have LDL receptors that reach the cell surface but are unable to bind LDL. Class 4 mutants produce receptors that do not cluster in coated pits, thereby interrupting the normal metabolism of cholesterol.

It is possible to increase LDL receptors in heterozygous patients by giving bile-salt-binding resins. If endogenous cholesterol synthesis is then blocked by administering nicotinic acid or drugs that inhibit HMG-CoA reductase (e.g., lovastatin), substantial drops in plasma cholesterol can be obtained. Unfortunately, the drugs do not work in the homozygous form since no normal gene is available to stimulate.

Hypertriglyceridemia

Elevated plasma triglycerides may appear as primary genetic disease or secondary to other illnesses. The most common defect is a deficiency of lipoprotein lipase (LPL). Acquired deficiencies of LPL are seen in diabetes mellitus, hypothyroidism, alcoholism, lupus erythematosus (caused by antibodies against LPL or its cofactor, heparin), and uremia. Elevated triglycerides have two major detrimental effects. Acutely, they predispose to acute pancreatitis. Chronically, they lower HDL levels, predisposing to atherosclerosis.

SUGGESTED READING

Atkinson MA, and Maclaren NK: The pathogenesis of insulin dependent diabetes mellitus. N Engl J Med 331:1428–1436, 1994.

Bonnefont JP, Demaugre F, Tein I, Cepanec C, Brivet M, Rabier D, and Saudubray JM: Clinical deficiencies of carnitine palmitoyltransferase. In: *Current Concepts of Carnitine Research.* AL Carter, ed., CRC Press, New York, pp. 179–187, 1992.

Breslow JL: Genetic basis of lipoprotein disorders. J Clin Invest 84:373–380, 1989.

Brown MS, and Goldstein JL: A receptor-mediated pathway for cholesterol homeostasis. Science 232:34–47, 1986.

De Feo P, Perriello G, De Cosmo S, Ventura MM, Campbell PJ, Brunetti P, Gerich JE, and Bolli GB: Comparison of glucose counterregulation during short-term and prolonged hypoglycemia in normal humans. Diabetes 35:563–569, 1986.

Foster DW: From glycogen to ketones—and back. Diabetes 33:1188–1199, 1984.

Gelfand RA, and Sherwin RS: Nitrogen conservation in starvation revisited: Protein sparing with intravenous fructose. Metabolism 35:37–44, 1986.

Havel RJ, and Rapaport E. Drug therapy: management of primary hyperlipidemia. N Engl J Med 332:1491–1498, 1995.

Kahn CR: Banting lecture. Insulin action, diabetogenes, and the cause of type II diabetes. Diabetes 43:1066–1086, 1994.

Kather H, Bieger W, Michel G, Aktories K, and Jakobs KH: Human fat cell lipolysis is primarily regulated by inhibitory modulators acting through distinct mechanisms. J Clin Invest 76:1559–1565, 1985.

McGarry JD: What if Minkowski had been ageusic? An alternative angle on diabetes. Science 258:766–770, 1992.

McGarry JD, Kuwajima M, Newgard CB, Foster DW, and Katz J: From dietary glucose to liver glycogen: The full circle round. Annu Rev Nutr 7:51–73, 1987.

McGarry JD, Woeltje KF, Kuwajima M, and Foster DW: Regulation of ketogenesis and the renaissance of carnitine palmitoyltransferase. Diabetes/Metab Rev 5:271–284, 1989.

Philipson LH, and Steiner DF: Perspectives. Pas de Deux or more: The sulfonylurea receptor and K^+ channels. Science 268:372–373, 1995.

Pilkis SJ, and Granner DK: Molecular physiology of the regulation of hepatic gluconeogenesis and glycolysis. Annu Rev Physiol 54:885–909, 1992.

Schuit FC, and Pipeleers DG: Differences in adrenergic recognition by pancreatic A and B cells. Science 232:875–877, 1986.

Taylor SI, Accili D, and Imai Y: Insulin resistance or insulin deficiency—which is the primary cause of NIDDM? Diabetes 43:735–740, 1994.

Unger RH: Diabetic hyperglycemia: link to impaired glucose transport in pancreatic β-cell. Science 251:1200–1205, 1991.

Index

Male reproductive tract, structural
organization of, 201–203
Malabsorption, 254
Malnutrition, 97, 246, 254
Mammary gland, 191
development and hormonal control, 8
in female reproductive function, 172
Mammary lines, 153
Mammilary body, 103
Maternal compartment, in pregnancy, 239
McCune-Albright syndrome, 63
MCPs. *See* Monocyte chemotactic
proteins
MCR. *See* Metabolic clearance rate, 151
Median eminence, 102–103, 105
Meiosis, 225
Melanocyte-stimulating hormone
functions, 101, 106, 113–114
structure, 112
Melatonin, 5
Menarche, 186, 188
critical weight at, 187–188
Menopause, 190–191
osteoporosis, 332
Menstruation, 173–183
and hypothalamic-anterior pituitary
unit, 120, 121
and LHRH, 119–120
mechanism of, 82–83
and thyroid hormone deficiency, 274
Mesonephric duct, 150
Mesonephros, 151
Messenger RNA (mRNA)
cytokine, 75
posttranscriptional regulation, 18–19
Mestranol, 198
Metabolic clearance rate, 13
Metabolism
general principles, 13–14
integrated control, 145
Metanephrine, 310
6α Methylprednisone, 292
Metyrapone, 98
Mifepristone, 199
Milk
ejection, 193
"let down," 193
major components, 194
synthesis, 192
Mineralocorticoids, 302–306. *See also*
Aldosterone
domain structure, 54

free, 56
functions, 304–305
in genetic disease, 27
molecular mechanisms of action, 294
receptor regulation of gene expression,
58
regulation of secretion, 302–304
Mitogen-activated protein kinases, 49
Mitogens, and inflammatory response, 70
Mitotic spindle, 225
Mixed function oxidases, 167
Molecular mimicry, 369
Monocyte chemotactic proteins, 68, 70
Monocytes
glucocorticoid effect on, 4
and inflammatory response, 70, 73
Monokines, definitions and categorization,
67
Morula, 226
mRNA. *See* Messenger RNA
MSH-inhibiting factor, 1–6
MSH-releasing factor, 106
Müllerian duct structures, 151
absent, in androgen resistance, 63
Müllerian-inhibiting hormone, 149–150,
155, 158, 161
Multiple endocrine neoplasia type II, 326
Muscarinic receptors, 34
Myasthenia gravis, 62
Myometrium, 153
Myxedema, 274

Naloxone, 122, 139
Natriuresis, 141
Natriuretic peptide receptors, 142–143
Natural rhythm, 198
Negative feedback
androgen resistance, 64
in female puberty, 188
hormone secretion, 14
ovarian gonadotropin, 179
Neonates, hypothalamic-pituitary unit,
183–185
Nerve growth factor, 252
neu oncogene, 47
Neural lobe. *See* Posterior pituitary gland
Neuroendocrine-immune system
relationship, 4, 69–70
Neuroendocrinology, definition, 4
Neurohumoral hypothesis, 104–105
Neurohypophysis, 102. *See also* Posterior
pituitary gland